OECD Review of Agricultural Policies

China

ORGANISATION FOR ECONOMIC CO-OPERATION AND DEVELOPMENT

ORGANISATION FOR ECONOMIC CO-OPERATION AND DEVELOPMENT

The OECD is a unique forum where the governments of 30 democracies work together to address the economic, social and environmental challenges of globalisation. The OECD is also at the forefront of efforts to understand and to help governments respond to new developments and concerns, such as corporate governance, the information economy and the challenges of an ageing population. The Organisation provides a setting where governments can compare policy experiences, seek answers to common problems, identify good practice and work to co-ordinate domestic and international policies.

The OECD member countries are: Australia, Austria, Belgium, Canada, the Czech Republic, Denmark, Finland, France, Germany, Greece, Hungary, Iceland, Ireland, Italy, Japan, Korea, Luxembourg, Mexico, the Netherlands, New Zealand, Norway, Poland, Portugal, the Slovak Republic, Spain, Sweden, Switzerland, Turkey, the United Kingdom and the United States. The Commission of the European Communities takes part in the work of the OECD.

OECD Publishing disseminates widely the results of the Organisation's statistics gathering and research on economic, social and environmental issues, as well as the conventions, guidelines and standards agreed by its members.

> *This work is published on the responsibility of the Secretary-General of the OECD. The opinions expressed and arguments employed herein do not necessarily reflect the official views of the Organisation or of the governments of its member countries.*

> *This document has been produced with the financial assistance of the European Union. The views expressed herein can in no way be taken to reflect the official opinion of the European Union.*

© OECD 2005

No reproduction, copy, transmission or translation of this publication may be made without written permission. Applications should be sent to OECD Publishing: *rights@oecd.org* or by fax (33 1) 45 24 13 91. Permission to photocopy a portion of this work should be addressed to the Centre français d'exploitation du droit de copie, 20, rue des Grands-Augustins, 75006 Paris, France (*contact@cfcopies.com*).

Foreword

The Review of Agricultural Policies: China was undertaken as part of an initiative to provide analyses of agricultural policies for four major agricultural economies outside the OECD area, the others being Brazil, India and South Africa. The study examines the policy context and trends while measuring the extent of support provided to agriculture using the same method the OECD employs to monitor agricultural policies in OECD countries. In addition, it focuses on key interactions between China and OECD countries, including the economic impact of trade and agricultural policy reforms. The aim of the country study is to strengthen the policy dialogue with OECD members on the basis of consistent and internationally comparable analysis, and to provide an objective assessment of the opportunities, constraints and trade-offs that confront China's policy makers.

The study was carried out by the OECD Directorate for Food, Agriculture and Fisheries. The principal author was Andrzej Kwiecinski, who received valuable contributions from Stephen Apted (ABARE), Dirk Bezemer (University of Groningen), Fabrizio Bresciani (FAO), Brian Morrissey (Canadian consultant) and John Nash (World Bank) as well as from OECD staff members including Joe Dewbre, Hsin Huang, Wayne Jones, Wilfrid Legg, Pete Liapis, Claude Nenert, Stefan Tangermann and Martin von Lampe. Research and statistical support were provided by Florence Mauclert, editorial support by Michèle Patterson, and technical and secretarial assistance by Anita Lari.

The study benefited from the substantive input from Chinese experts. Data and policy information for the evaluation of the level of support was provided by Guoqiang Cheng from the Development Research Center of the State Council (DRC), and who also acted as the main contact and liaison person on all aspects of the study. Information on policy context and on the restructuring of the agro-food sector was provided by Xiande Li from the Institute of Agricultural Economics of the Chinese Academy of Agricultural Sciences (CAAS). Information on domestic and trade policies was provided by Tian Weiming from the College of Economics and Management of the China Agricultural University.

This study benefited greatly from the support provided by the DRC, in particular the Department of Rural Economic Research led by Han Jun. The DRC experts as well as experts from the Ministry of Agriculture, Ministry of Commerce, Ministry of Finance, National Development and Reform Commission, State Administration of Grain and Chinese Academy of Sciences provided valuable comments on the draft report.

This study was made possible through voluntary financial contributions by several OECD countries, in particular Australia and Canada as well as funding from the European Union. It was reviewed in a roundtable with Chinese officials and experts in Beijing in April 2005. Chinese agricultural policies were examined by the OECD's Committee for Agriculture at its 141st session in June 2005, bringing together policy makers from China, OECD member countries, and some non-OECD countries. This report is published under the authority of the Secretary-General of the OECD.

Stefan Tangermann

Director

Directorate for Food, Agriculture and Fisheries

Table of Contents

Highlights and Policy Recommendations 9
    1. Reforms and their impacts 11
    2. Agricultural policy trends 17
    3. The benefits of future policy reforms 22
    4. Policy challenges 24

Chapter 1. **The Policy Context** 27
    1.1. General aspects 28
    1.2. Agriculture's importance to China's economy 31
    1.3. Structural change in the agro-food sector 36
    1.4. The effects of economic reforms on China's agriculture 47

Chapter 2. **Policy Trends** 73
    2.1. Agricultural policy framework 74
    2.2. Domestic policies 85
    2.3. Trade policies 110
    2.4. Evaluation of support to Chinese agriculture 134

Chapter 3. **Policy Impacts** 149
    3.1. Welfare impacts of trade and agricultural policy reforms 150
    3.2. The impact of liberalisation on Chinese agricultural commodity markets 159
    3.3. Domestic and world market implications of alternative grain stock
         estimates and trade policies in China 166

Annex A. Labour Mobility and Rural Poverty in China 177
Annex B. Agricultural Policies and Support for Individual Commodities 189
Annex C. China's Approach to Food Safety 216

Acronyms and Abbreviations 226

List of Boxes

1.1. China's political system 29
1.2. Trade reform and factor mobility in China: the long and the short of it 35
1.3. The distribution of land rights across levels of authority 40
1.4. Problems with Chinese agricultural statistics 51
1.5. Social security in rural China 61
2.1. The Agricultural Law of China 82
2.2. A brief history of the "peasant burden" 99
2.3. Public debt funds in agricultural development 103
2.4. VAT assessment on imported agricultural products 118
2.5. Major WTO accession commitments by China – agricultural trade 123
2.6. OECD indicators of support to agriculture: definitions 134
2.7. China's PSEs: what and how? 136
2.8. Transfer efficiency in agricultural support policies 142
3.1. The baseline projections for Chinese and world agricultural markets 160
A.1. The role of rural industries 180

List of Tables

1.1.	World top-10 countries by GDP (current USD), 2002	30
1.2.	China: selected macroeconomic indicators, 1990-2003	30
1.3.	Arable land and yields in selected countries, 2000-2002 average	36
1.4.	Food industry in China between 1999 and 2003	44
1.5.	The composition of food industry and tobacco enterprises in China by type of ownership, 2002	44
1.6.	Agricultural machinery in China per 100 rural households, 1990-2003	49
1.7.	Changes in the composition of the primary sector production, current prices, 1990-2003, %	53
1.8.	China's agricultural trade, 1992-2003	57
1.9.	Relative labour productivity by sector, 1978-2001	59
1.10.	Rural poverty in China, 1978-2003	60
1.11.	Rural household incomes by source, 1985-2003	61
1.12.	Food consumption in China, 1990-2002 (kg/person/year)	64
1.13.	Shifts in energy sources in the Chinese diet, ages 20-45	65
2.1.	Major laws and regulations in the agro-food sector	83
2.2.	Prices of electricity by different users in selected provinces in 2002	92
2.3.	Rates of railway shipment for selected goods (since July 2000)	92
2.4.	Comparison of water prices among different usages	93
2.5.	Annual interest rates of selected types of loans (%)	95
2.6.	The rates of tax on special agricultural products – % of value	97
2.7.	Agriculture-related taxes	98
2.8.	Total national aggregate budgetary support to agriculture	110
2.9.	Changes in MFN tariffs for basic commodities	115
2.10.	China's TRQ performance	124
2.11.	Evolution of producer support (% PSE) and consumer support (% CSE) in China and selected countries, 1993-2004	138
2.12.	Total support to Chinese agriculture	143
3.1.	Tariffs levied and faced (%)	150
3.2.	Welfare effects of multilateral policy reform, USD millions	151
3.3.	The 2000 rural household survey frame	154
3.4.	Structure of household income, provincial averages (% of total income)	155
3.5.	Consumption and poverty, provincial averages	156
3.6.	Proportional welfare effects of price changes	157
3.7.	Money metric welfare effects of price changes	158
3.8.	Principal assumptions of the liberalisation scenarios	162
3.9.	Development of Chinese grain tariff rate quotas after WTO accession	169
A.1.	Rural poverty rates by region, China, 2002	177
A.2.	Comparison of income structure in rural areas, 2000	178
A.3.	TVEs in China's economy, 1990-2002	180
A.4.	Migrants and remittances	185
A.5.	Duration of migration and remittances	185
B.1.	Changes in grain prices (CNY/kg in current price)	190
B.2.	China's WTO grain trade commitments	195
B.3.	The state-set guidance prices for sugar beet and cane (CNY/tonne)	200
B.4.	China's WTO sugar trade commitments	201
B.5.	The state-set prices of cotton	204
B.6.	China's WTO cotton trade commitments	205

List of Figures

0.1.	China's agricultural trade, 1992-2004	15
0.2.	Net trade in land and labour intensive agricultural commodities	15
0.3.	Producer Support Estimate in China and selected countries, 2000-2003 average	20
0.4.	Composition of Producer Support Estimate, 1993-2003	21
0.5.	China's Producer Support Estimate by commodity, 2000-2003 average	21
0.6.	Welfare gains (losses) by source of liberalisation	23
1.1.	The share of agriculture in GDP, employment, total exports and imports, 1990-2003	32

1.2.	Agriculture's share in GDP versus GDP per capita (2000-2002)	33
1.3.	Agriculture's share in employment versus GDP per capita (2000-2002)	33
1.4.	Regional distribution of agricultural labour and cultivated area, 2003	37
1.5.	Nominal price indices, 1990-2003, 1990 = 100	48
1.6.	Chemical fertiliser use in China (active substance kg/ha of sown area), 1985-2003	49
1.7.	Chemical fertiliser use in selected countries (active substance kg/ha of sown area), 2002	50
1.8.	Growth in Gross Agricultural Output, 2003 (1989-1991=100)	50
1.9.	GAO yearly growth rates in China, %, 1990-2003	52
1.10.	Output indices for main crops, 1990-2003, 1990 = 100	53
1.11.	Total cereal production and nominal farm gate prices, 1993-2003	54
1.12.	Composition of the sown area, 1991 and 2003, %	55
1.13.	Indices of livestock production, 1990 = 100	55
1.14.	Crop yields for selected crops, 1990-2003	56
1.15.	Evolution of employment in Chinese agriculture, 1990-2003	58
1.16.	Rural household income per person, 1981-2004	59
1.17.	Wages and net incomes per person in peasant families across provinces, CNY, 2003	62
1.18.	Urban to rural per capita income and living expenditures ratios, 1978-2004	63
2.1.	Central institutions with oversight over China's agro-food sector	80
2.2.	Comparison of different types of grain and soybean prices in China	86
2.3.	Simple average MFN tariffs on agricultural products	114
2.4.	Dispersion of China's agricultural tariffs in 2002 and 2004	116
2.5.	China's agricultural trade, 1992-2004	128
2.6.	Net trade in land and labour intensive agricultural commodities	129
2.7.	China's main agro-food imports, 2003	130
2.8.	China's main agro-food exports, 2003	130
2.9.	China's agro-food exports (including fish and fish products) by region	131
2.10.	Main export markets for Chinese agro-food products (including fish and fish products), 2003	132
2.11.	China's agro-food imports (including fish and fish products) by region	133
2.12.	Main suppliers of agro-food products (including fish and fish products) to China, 2003	133
2.13.	Percentage PSEs for China and selected countries, average 2000-2003	139
2.14.	Percentage PSE for crops and livestock products in China, 1993-2003	140
2.15.	Composition of producer support estimate, 1993-2003	141
2.16.	Total support estimate in China and selected countries, average 2000-2003 – as per cent of GDP	143
2.17.	Chinese % PSE by commodity, average 2000-2003	144
2.18.	Distribution of producer support by commodity, 2000-2003 average	145
3.1.	Welfare gains (losses) by source of liberalisation	152
3.2.	Changes in factor returns to agriculture and non-agriculture resulting from multi-sectoral reduction in trade protection	153
3.3.	Chinese grain market developments: past and projections	160
3.4.	Development of Chinese meat production	161
3.5.	Impact of 50% liberalisation on world crop markets, average 2005-2013	163
3.6.	Impact of 50% liberalisation on Chinese crop markets, average 2005-2013	163
3.7.	Recent developments in the Chinese grains balance, 1990/91-2003/04	167
3.8.	Chinese total grain stocks: recent FAO revisions	168
3.9.	Impact of Chinese grain stock revisions: import projections of wheat, coarse grains and rice	169
3.10.	Impact of restricted and extended import quota access on Chinese and world grain prices, average 2004-2013	170
3.11.	Impact of restricted and extended import quota access on Chinese grain consumption, average 2004-2013	171
3.12.	Impact of unlimited import quota extension on Chinese grain imports, 2002-2013	172

3.13.	Self-sufficiency rates of Chinese grain markets, 1991-2013, in different scenarios	173
B.1.	Percentage PSEs, producer and reference prices for wheat, 1993-2003	196
B.2.	Percentage PSEs, producer and reference prices for maize, 1993-2003	196
B.3.	Percentage PSEs, producer and reference prices for rice, 1993-2003	197
B.4.	Percentage PSEs, producer and reference prices for soybean, 1993-2003	199
B.5.	Percentage PSEs, producer and reference prices for rapeseed, 1993-2003	199
B.6.	Percentage PSEs, producer and reference prices for peanuts, 1993-2003	200
B.7.	Percentage PSEs, producer and reference prices for sugar, 1993-2003	202
B.8.	Percentage PSEs, producer and reference prices for cotton, 1993-2003	206
B.9.	Percentage PSEs, producer and reference prices for apples, 1993-2003	209
B.10.	Percentage PSEs, producer and reference prices for beef and veal, 1993-2003	212
B.11.	Percentage PSEs, producer and reference prices for pigmeat, 1993-2003	212
B.12.	Percentage PSEs, producer and reference prices for sheepmeat, 1993-2003	213
B.13.	Percentage PSEs, producer and reference prices for poultry, 1993-2003	213
B.14.	Percentage PSEs, producer and reference prices for milk, 1993-2003	214
B.15.	Percentage PSEs, producer and reference prices for eggs, 1993-2003	214

Highlights and Policy Recommendations

During the reform period, which started in 1978, China made huge progress in meeting its objectives: agricultural production rose sharply, rural industries absorbed a large part of farm labour, poverty fell dramatically, and the level and quality of food consumption improved significantly. The commune system was replaced by one where individual families lease land from the collectives, ensuring that almost all rural households have access to land and are, at minimum, food self-sufficient.

Currently, China has about 200 million farm households with an average land allocation of just 0.65 ha. Limited arable land and a large rural labour force mean that, in general, China tends to have a comparative advantage in the production of labour intensive crops, such as fruits and vegetables, and a disadvantage in the production of land intensive crops, such as grains and oilseeds.

One of the most striking features of China's development in the reform period has been a large and growing income disparity between the rural and urban populations. This is largely due to limited factor mobility, especially of labour and capital. Surplus farm labour and low labour productivity have resulted in low agricultural incomes and hidden unemployment in rural areas. The rural-urban gap in living standards is further accentuated by differences in access to education, health care and other social services.

The level of support to agriculture from policies fluctuated at low levels through the 1990s, rising to 8% in 2003, still well below the OECD average of 30%. Support levels are highest for import-competing commodities such as sugar and milk, but also for exportable maize. Grain markets remain distorted, mostly due to state trading which drives a wedge between domestic and world prices. The Total Support Estimate (TSE) is relatively high at 3.7% of GDP, reflecting though large expenditure on general services, in particular investments in agricultural infrastructure to improve productivity.

In line with the improving economic situation and sectoral performance, government priorities have shifted from increasing production, especially of food grains, to rural income support and recently to environmental concerns. In the medium term, the main challenges for China's policy makers include closing the large income gap between rural and urban populations; integrating small-scale farmers, who are dominant, into markets; stimulating internal reallocation of resources to create more efficient farm structures; reducing the negative impacts of increasing agricultural production on the environment; improving the competitiveness of agricultural and food products on domestic and international markets; and improving the governance of institutions in designing and implementing agricultural policies.

Analysis of the impact of further trade liberalisation suggests China would benefit significantly from liberalisation of non-agricultural trade in OECD countries. Multilateral agricultural liberalisation would affect China's overall welfare to a limited degree only, but would improve the lot of China's farmers. The vast majority of rural households would gain.

Policy performance would benefit from a focus on economy-wide measures such as further relaxation of administrative barriers to rural/urban migration, improved access to education in rural areas, better health care, pension and other social security services, enhanced land property rights and rural tax reforms. The provision of modern research and extension services, food safety agencies and agricultural price information, services which provide widespread benefits to producers and consumers throughout the economy will also be of crucial importance.

1. Reforms and their impacts

China is a key player in the global economy

With a GDP of USD 1 270 billion China is the world's sixth largest economy at current exchange rates, but second in purchasing power parity (PPP) terms. It is the most populous country with 1.3 billion inhabitants, 21% of the world's total. China is relatively scarce in agricultural land and water, having only 10% of the world's arable land and its water resources per capita are around one quarter the world average.

China has carried out fundamental reforms of its economic policies since 1978, resulting in a gradual transition from a centrally planned economy towards a socialist market economy. Its economic performance during the reform period has been remarkable. Growth in real GDP averaged above 9% between 1990 and 2004 and has been accompanied by even more rapid growth in trade and investment. China's accession to the World Trade Organisation (WTO) in 2001 confirmed a further strengthening of the reform course it has been following for the last 25 years.

Rural areas have provided major boosts to China's development...

Even though the importance of agriculture in China's economy has declined, it is still an important sector, accounting for almost 15% of GDP and above 40% of employment. The share of the rural population in the total population is falling, but still remained high at 60% in 2003. China's agro-food sector is less integrated with international markets than are other sectors of its economy. While the ratio of China's agro-food trade to the Agricultural Gross Value Added (AGVA) remained relatively stable, the share of agro-food products in the total value of exports and imports more than halved between 1992 and 2003, reflecting an expansion of trade in other products. As a result, while the sum of agro-food exports and imports to the AGVA is about 19%, the ratio of total exports and imports to China's GDP is much higher at 55%.

Rural areas provided two major boosts to China's development during the reform period. The first came from a major transformation in the policy environment in agriculture in the early 1980s and the second began in the late 1980s with the expansion of rural industries.

In the early 1980s, the tightly controlled commune system was replaced by the Household Production Responsibility System (HPRS) in which individual farmers lease their land from the collectives, are largely autonomous in their decisions, and bear the profits or losses from their operations. In order to employ workers leaving agriculture and to avoid large-scale migration to the cities, sub-national governments were encouraged to foster the growth of rural non-agricultural industries, commonly known as township and village enterprises (TVEs). These enterprises have been the main vehicle for absorbing workers leaving agriculture and provided a second major source of China's growth and development. A uniqueness of China's experience is that the bulk of the shift in employment has taken place within the rural economy rather than through migration from rural to urban areas.

... despite a divide in China's economy between rural and urban areas

China's exceptionally good economic performance has been favourable for the development of agriculture through stimulating demand for agro-food products and creating non-agricultural employment opportunities. Yet, there has been a wide range of economy-wide policies that impede rural areas, including agriculture, from reaping the benefits from economic growth:

- China's population registry (*hukou*) includes provisions that create disincentives for migration of rural-born workers into urban areas.
- Education and health care resources per capita in rural areas are significantly lower than in urban areas.
- Pensions and other social benefits provided in urban areas are largely unavailable to the rural population.
- The tax system in rural areas differs from that applied in urban areas and includes charges, fees and levies on rural households which are not imposed on urban families.

Agricultural production structures are dominated by small-scale farming

China's agriculture is characterised by scarce land in relation to labour and small-scale production using little mechanisation. In 2005, there were about 200 million farm households with an average land allocation of just 0.65 ha. Moreover, each farm typically consists of several tiny and separated plots. As in some other East Asian countries, agricultural output per unit of land is high by international standards but output per worker is low. Limited arable land and a large rural labour force mean that in general China tends to have a comparative advantage in the production of labour intensive crops such as fruits and vegetables and a disadvantage in the production of land intensive crops such as grains and oilseeds. However, the situation is quite strongly diversified regionally and depends, among other things, on availability of land and water, climatic conditions, transportation costs and access to markets.

The land tenure system is based on land lease contracts. Farmland is *de facto* owned by village collectives which extend land lease contracts to individual farm households, currently for 30 years. Households have most of the property rights: they can use, sub-lease and transfer land, but they cannot sell it. To a growing extent the new lease rights are being backed by written contracts.

In the early stages of reform, land tenure policies were based on the HPRS and the principle of egalitarian access to land brought a number of positive effects. Compared to the communal system, the HPRS provided farmers with stronger rights to land and to production, which stimulated growth in agriculture and rural incomes. Through equal distribution of land use rights, China avoided having large numbers of rural landless workers vulnerable to famine or other economic shocks. It has also ensured that the vast majority of rural households are, at minimum, food self-sufficient.

However, as China develops, the land tenure system increasingly reveals its limitations. Land market transactions are limited and land rental arrangements between farmers tend to be informal, short-term and most often between relatives. There is also scope for arbitrary decisions by local leaders, and for conflictual situations to arise between village leaders and farmers, including when local regulations limit land-transfer rights. There are also a large number of cases, when local leaders, assuming the role of landowners, decide to lease or sell land to external investors without consensus from local farmers and without farmers receiving proper compensation for lost access to land.

Upstream and downstream sectors are becoming increasingly competitive

Until the mid-1980s, China's upstream and downstream sectors were determined by the centrally planned economy system. By the beginning of the 2000s, both sectors had started to operate in line with market principles.

On the upstream side, the Chinese Supply and Marketing Co-operatives (SMC) have preserved an important role in the market. However, there are three other main channels through which farmers may purchase inputs: "three agricultural stations" (*nongye san zhan*), direct sales by input producers and private traders. Evidence suggests that prices have become competitive and farmers do not have difficulties in accessing a wide variety of inputs.

Farmers continue to have difficulties in accessing credit, however. The two main formal channels of credit for farmers include the state-owned Agricultural Bank of China (ABC) and Rural Credit Co-operatives (RCCs). Farmers lack collateral and obtaining formal credit involves high transaction costs. Another factor is the closing of many local branches of financial institutions and the failure of new ones to emerge. As a result, more than 70% of loans are obtained through informal channels, while less than 30% from formal financial institutions.

On the downstream side, while government enterprises and agencies remain the main purchasers of grains, farmers have the option of selling their products through diversified private channels. The most widespread are "big marketing households" (*yunxiao dahu*), which are family based traders specialising in the intermediation between farmers and wholesale markets. Other main channels include direct sales by farmers to retail or wholesale purchasers, including supermarket chains, direct sales to local markets, direct contracting with food processors/exporters and sales through farmers' professional associations or co-operatives. Changes in the downstream sector have been accelerated by the development of supermarket chains. While their share in overall food sales is below 20% (compared to 60-80% in the United States and the European Union), this share will increase rapidly as retail chains are spreading to all cities, including China's interior provinces.

Up to the mid-1990s, the state trading enterprises (STEs) were the only channel linking producers with foreign markets. Since then the government has changed the foreign trade regulatory framework, streamlined the existing STEs, allowed non-state entities to undertake foreign trade activities, and have simplified registering procedures for entities wishing to become foreign trade operators. Nevertheless, STEs remain important players in China's agro-food trade, especially for grains.

Agricultural production expanded and...

During the reform period China's agriculture experienced phenomenal rates of growth. The HPRS boosted production incentives, encouraged farmers to reduce costs, take risks, and enter new lines of production. At the beginning of the 1990s, when the economy grew very rapidly, consumers shifted their preferences from quantity to quality. A new phase of adjustments started in the late 1990s and in the early 2000s when oversupply appeared on most agricultural markets, causing grain prices to fall, and increasing exposure to international competition stimulated further structural adjustments. The main policy objective shifted to raising farmers' incomes.

Between 1990 and 2003, the Gross Agricultural Output (GAO) increased by almost 90%, with crop production up 60% and livestock production up 145%. However, in line with a slowdown in demand, the rates of growth slowed from 6.2% per year in the first half of the 1990s to 3.6% per year between 2000 and 2003.

... its composition shifted

Reflecting changes in consumer demand, the composition of production continues to shift from crops to livestock and fish production. While crops accounted for 65% of the total value of primary production in 1990, the share fell to 50% in 2003. During the same period, the share of livestock production increased from 26% to 32%. The performance of the fishery sector was even more spectacular with the share rising from 5% to 14%. The share of forestry remained stable at 4%.

While cereals remain the key crop, their share in total crop production and in the area sown declined quite substantially between 1990 and 2003 as other crops became more profitable and the government relaxed most of the policy measures which had previously forced farmers to produce grains. Impressive increases in vegetable and fruit production compared to performance of other crops reflects China's comparative advantage and adjustments in land use in response to changes in domestic demand as well as to emerging export opportunities for selected products such as garlic, onions, apples, and pears. Within the cereal sector there was also an important restructuring as land sown to rice and wheat tended to decline, while land sown to maize increased. This reflects a shift in focus from food to feed in line with changes in food consumption patterns.

Traditionally, the overwhelming majority of meat was produced by small, part-time "backyard" operations. But full-time "specialised" household operations and commercial operations have grown rapidly and in 2003 accounted for around 30% of pork and almost 70% of poultry production. Although specialised and commercial operations are far more efficient at converting feed to meat, they are very intensive users of feed grains. Accordingly, the net demand for feed grains to support the livestock sector is expected to grow as the structure of the livestock industry continues to adjust.

Changes in trade flows tend to reflect China's comparative advantage

The agro-food sector (including fishery) remained a net export earner at about USD 5 billion a year until 2002. In 2003 and 2004, following several consecutive falls in grain production and having overcome some early problems with the implementation of the tariff rate quota (TRQ) system agreed within the WTO, the agro-food imports increased at high rates leading to net imports of agro-food products at about USD 11 billion in 2004 (Figure 0.1).

During the period 1992-2003, agricultural trade has shown notable structural change (Figure 0.2). Clearly, the trade balance in land intensive commodities has been trending down in the period 1992-2003. This is mainly due to a large increase in cereal imports in 1995 and 1996, and an increasingly high level of oilseed imports since 1997. Labour intensive commodities have posted a trade surplus throughout the period 1992-2003, with a virtually flat trend. The shift in the broad composition of agro-food trade is in-line with China's advantage in producing labour intensive agricultural products.

Figure 0.1. **China's agricultural trade, 1992-2004**

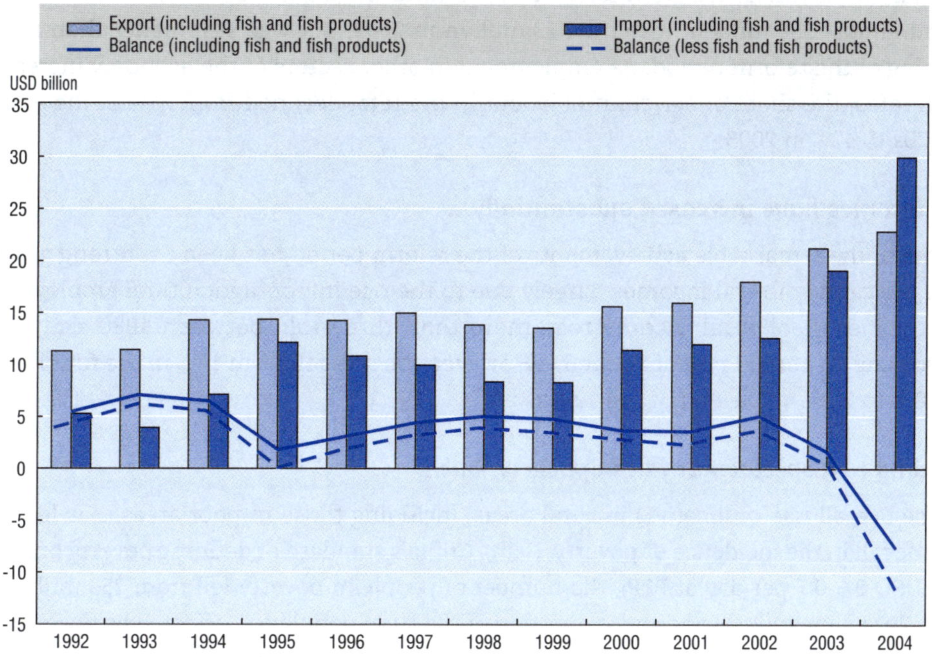

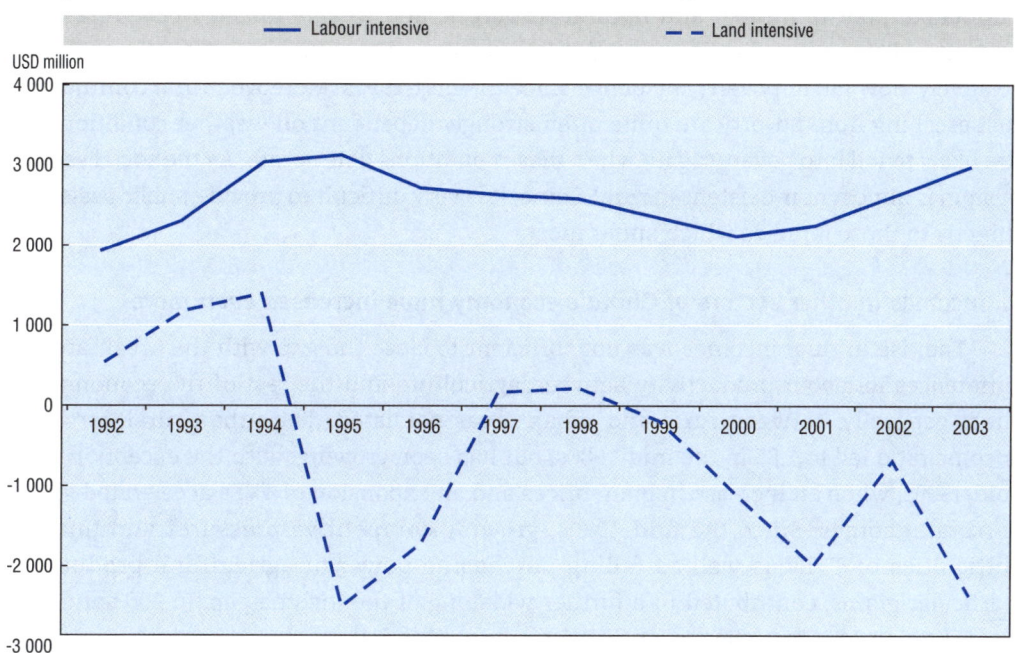

Source: Comtrade database; China Customs Statistics for 2004.

Figure 0.2. **Net trade in land and labour intensive agricultural commodities**

Notes: Land intensive defined as HS10 – cereals and HS12 – oilseeds. Labour intensive defined as HS07 – vegetables, HS08 – fruits and nuts, HS09 – tea, and HS24 – tobacco.
Source: Adapted from Huang and Chen (1999); derived from Comtrade database.

Agricultural employment has started to fall

Total employment in agriculture increased until the beginning of the 1990s when it reached almost 350 million. It fell to 313 million in 2003, but was still higher than at the beginning of the reform period. As employment in other sectors of the economy increased at high rates, the share of agricultural labour in the total declined from 71% at the end of the 1970s to 42% in 2003.

Rural incomes have increased substantially...

One of the remarkable achievements of the reform period has been the strong growth of real per capita of rural incomes, largely due to the rise in non-agricultural employment opportunities. Real rural income rose more than three-fold between 1980 and 2000, representing an annual rate of about 6%. However, the rate fell from 14% in the first half of the 1980s to 2-3% at the end of the 1990s.

... leading to a spectacular fall in poverty, but...

The overall rise of incomes in rural areas, including those in poor areas, has led to a dramatic fall in the incidence of poverty. Using China's standard of defining poverty (income below USD 0.6-0.7 per day at PPP), the number of people in poverty fell from 250 million in 1978 to 29 million in 2003 and the proportion of the rural population affected by poverty fell from 31% to 3% over the same period. At the World Bank definition of poverty line (income of USD 1 per day at PPP), the number of people living in poverty in rural China was still high at 88 million in 2002, but the progress in eliminating poverty is still remarkable, as about 400 million people rose above the poverty line between 1979 and 2002.

Of those defined as poor, some 99% live in rural areas. They are mostly concentrated in the western provinces and in localities that lag in terms of growth, are remote, sparsely populated, poor in human and natural resources, and weakly linked to the rest of the economy. This makes further reduction of poverty more difficult and partly explains a relatively slow fall in poverty incidence since the mid-1990s. Moreover, rural communities just escaping from poverty are quite often strongly dependent on weather conditions and are likely to fall into poverty again when these conditions deteriorate. As the social security system is almost non-existent in rural China, it is very difficult to provide public assistance directly to those families which suffer most.

... incomes in other sectors of China's economy have increased even more

The rise in rural incomes was not sufficient to close the gap with the urban areas as differences in labour productivity between agriculture and the rest of the economy and, more generally, between rural and urban areas persisted. The urban-rural per capita income ratio fell to 1.85 in the mid-1980s, but has been growing, with the exception of the mid-1990s, when an increase in grain prices and an expansion of TVEs accelerated growth in rural incomes. Since the mid-1990s, growing competitive pressures and financial difficulties of TVEs as well as falling purchasing power of agricultural products, in particular grains, contributed to a further widening of the income gap. In 2003 and 2004, the ratio was 3.2, the highest over the whole reform period.

As it is unlikely that the high rates of agricultural production growth obtained over the first two and a half decades of the reform period are sustainable, the prospects for substantial gains in agricultural labour productivity reside in the outflow of labour from agriculture to other sectors of the economy. This will secure higher incomes for those remaining in the agricultural sector and those leaving should benefit from higher incomes due to more efficient use of labour in non-agricultural sectors. In fact, the income distribution across rural areas in different parts of China depends increasingly on the availability of non-agricultural job opportunities in rural areas and out-migration. As the availability of jobs is highly uneven, the variation in per capita income by province is much greater for rural than for urban areas.

Food consumption has increased

Measured in energy terms, average food consumption in China is high at 2 951 calories per capita per day in 2002 compared to the world average of 2 804 (2 761 in Japan and 3 058 in South Korea). Protein consumption is higher than the world average and just below the levels in Japan and South Korea. Less than 10% of the population is undernourished and this is the result of poverty rather than any physical lack of food. On average, households spent 40% of their total expenditures on food in 2002, a significant fall compared to 55% in 1990.

Environmental pressures may reduce long-term agricultural productivity

Historically, increased food production was achieved by expanding agricultural land. In recent years, however, the area of cultivated land has been decreasing and output increasing due to higher productivity thanks to greater use of fertiliser, pesticide and mechanical inputs. This reflects the move away from traditional farm practices towards more intensive chemical-use farming systems. China's use of fertilisers at 280 kg per hectare is one of the highest in the world. Pressure on utilisable agricultural land and water resources has also arisen from urban and industrial growth, while agriculture is itself affected by pollution from other human activities, especially industrial production.

Some of the ecological problems are location-specific, such as desertification in northern China, whereas others are common across the country, such as land degradation, soil erosion and water pollution. These problems risk contributing to a reduction in agricultural productivity in the long term. This is a crucial element in future projections of agricultural production and incomes. The protection of the environment and the ability to improve agricultural productivity are integrally linked, and where markets are not able to take environmental externalities into account, there is a need for effective policies.

2. Agricultural policy trends

Agricultural policy objectives…

Generally, the 1990-2005 period can be divided into two stages. Up to 1997, the principle objective was to increase agricultural production, especially production of food grains, while since 1998 more attention has been given to supporting rural incomes. In recent years environmental concerns have also started to gain more attention.

Ensuring an adequate supply of affordable food, in particular of grains, remains a top priority of China's policy makers. Although not clearly defined in government policy, it is widely interpreted to mean that China should produce 95% of its own grain requirements. This is a key factor in understanding how agricultural policies have evolved.

During the late 1990s and early 2000s, the growing income gap between urban and rural populations, and between developed and underdeveloped rural areas, became an important policy issue. Policy makers began to address the integration of urban and rural development and devised regional development programmes to accelerate economic growth in less developed regions. Policies aimed at raising agricultural incomes nationwide were also adopted, with a fundamental shift from taxing agriculture to supporting it. These policies were strengthened in 2004 through the adoption of "The Suggestions of the Central Committee of the Communist Party of China and the State Council on Policies for Boosting Growth in Farmers' Income", the highest priority document of central authorities for 2004.

... and policy instruments have evolved

The main agricultural policy measures employed by the government cover producer support measures, general services and consumer support measures. In turn, producer support measures cover both domestic and trade policy measures.

Domestic policy measures include:

- *State pricing*: in place for major agricultural commodities for much of the period 1990-2004. From 2004, centrally set state pricing only applies to tobacco (under a state monopoly). For most of the period 1990-2004, state pricing was accompanied by state procurement.

- *Input subsidies*: charges for water, electricity and transport tend to be lower for farmers, but the level of subsidy is difficult to assess as the cost of provision is different across various users. To lower prices of fertilisers, producers have been given access to lower priced inputs, such as electricity. Since 2002, farmers have been subsidised for the cost of purchasing improved quality soy seed. In 2004, this scheme was extended to include subsidies for purchasing improved seed for production of wheat, corn and rice, as well as soybeans.

- *Credit subsidies*: until the end of the 1990s, preferential loans were provided mostly to state marketing organisations to fund purchase and storage of key agricultural products. In the 2000s, most of these programmes were discontinued, but are still applied for grains.

- *Direct payments*: initiated as a trial in 2002, and implemented nationally in 2004. Farmers engaged in growing grains have received a direct budgetary financed subsidy based on the area of land they sow to rice, wheat or corn.

- *Payments for returning farmland to forests*: also known as the "grain for green" programme, commenced in 1999. Farmers cultivating ecologically vulnerable land received a cash subsidy and a grain allocation for each *mu* (1/15 hectare) they retired from agricultural production. Subsidised seedlings were also available for afforestation. In 2004, the grain allocation was converted to a cash equivalent.

- *Agricultural taxes*: between 1990–2004, farmers were required to pay agricultural taxes either in cash or in kind. In addition, they also paid various fees to local governments and collectives and provided "labour accumulation" for the construction of communal facilities. Agricultural tax reform was initiated as a trial in 2000, and is being phased in across rural China from 2004.

Trade policy measures include:

- *Tariffs*: the simple average import tariff for agro-food products fell from 45% in 1992 to 15% in 2005, remaining at that level under the agreed terms of China's accession to the WTO.

- *Tariff rate quotas* (TRQs): first introduced for major grain and oilseed commodities in 1996. Under the terms of China's WTO accession, China can apply TRQs to wheat, rice, corn, sugar, cotton, wool and some vegetable oils, but oilseeds themselves are subject only to a tariff. China's TRQ system includes criteria for allocating the import quotas to state trading enterprises (STEs) and non-STEs.

- *State trading*: dominating until the mid-1990s. Its role has been diminishing since then, but it is still important for key commodities, in particular those covered by TRQs.

- *Export subsidies*: prior to becoming a member of the WTO, China provided export subsidies for corn (maize) and rice. In line with its WTO accession commitments, China is not allowed to apply export subsidies.

General services, provided to the agricultural sector as a whole, include:

- *Agricultural infrastructure*: investment in agriculture-related projects is a major tool for the government to achieve development targets and is by far the largest component in government's budgetary support for agriculture. The government has continued to accept primary responsibility for pollution control, land rehabilitation, transport and irrigation infrastructure maintenance and development.

- *Research and development*: government funding for this element of agricultural support is relatively small, and tending to decrease.

- *Agricultural schools*: government funding for agricultural schools is also a small expenditure item, but unlike research funding, agricultural school funding has been increasing.

- *Inspection services*: while China has funded food inspection services throughout the period 1990-2005, in the latter part of that period, food safety has become a higher priority concern of policy makers. Not only have expenditures on inspection services increased, China has also undertaken significant work to upgrade food safety standards.

- *Public stockholding*: China mainly engages in public stockholding of food grains. In line with food security policies, governments at the national and sub-national levels are active in maintaining buffer stocks of food grains.

Consumer support measures include:

- *Food price subsidies*: since 1992, China has paid subsidies to urban consumers to offset the price increases of staple food products. Although some of the subsidies are still paid, there has been a significant decline in the level of budgetary expenditure on them.

The level of producer support is low, but has increased in recent years

The Total Support Estimate (TSE) is the broadest indicator of support representing the sum of transfers to agricultural producers (the PSE), expenditure for general services (the GSSE), and direct budgetary transfers to consumers. The TSE reached USD 41 billion per year in 2000-03 which is equivalent to 3.3% of China's GDP in this period. This is much higher than the OECD average and suggests a relatively high burden of agricultural support on the Chinese economy. However, it reflects the economic importance of agriculture in a relatively poor economy, and is partly due to large budgetary expenditures on general services, some of which may be directed more to rural areas in a broader sense than just to agriculture. While the share of general services in the TSE tends to fall in the 2000s and was 51% in 2003, it is still high, mostly due to large investments in agricultural infrastructure. As consumer subsidies are of marginal importance, about half of total support is provided through transfers to producers (the PSE).

As measured by the percentage PSE, producer support fluctuated within a range of minus 14% to plus 6% between 1993 and 1998, and then, after falling to minus 3% in 1999, it increased each year and was positive 8% in 2003 (Figure 0.4). A comparison of producer support for China and selected OECD and non-OECD countries, including principal world players, indicates that China is a country with a low level of producer support. The percentage PSE in China, at 6% on average in 2000-03, is above that in countries with the lowest support (New Zealand, Brazil, Australia), but much lower than the OECD average (31%) and far below that in Japan and Korea (58% and 64%, respectively), the closest OECD neighbours and China's main export markets for agro-food products (Figure 0.3).

As seen from Figure 0.4, the level of producer support is determined predominantly by Market Price Support (MPS). However, the contributions of MPS to the PSE varied, in particular in the 1990s, reflecting fluctuations in the levels of domestic prices relative to world prices. While budgetary support has almost constantly been growing in absolute terms, its share in the aggregate fell notably between 2000 and 2003.

While producer support is low, the level of support varies significantly across commodities. The highest levels of support are for import-competing commodities, such as sugar, milk, sheep meat, cotton and soybeans, as well as some export commodities such as maize and rice (Figure 0.5). The distortions on grain markets are still relatively high, mostly due to state trading, which continues to drive a wedge between domestic and world prices. In contrast, for the majority of exportables, such as pig meat, beef and veal, eggs, peanuts and apple, the level of support is low or even negative.

Figure 0.3. **Producer Support Estimate in China and selected countries, 2000-2003 average**

As per cent of gross farm receipts

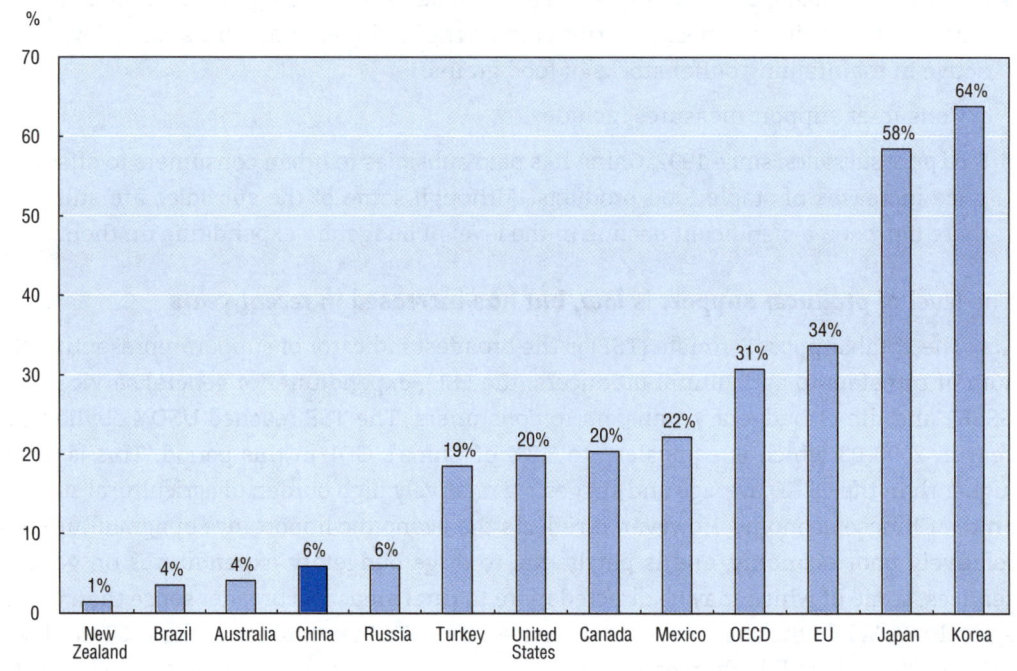

Note: EU15.
Source: OECD PSE/CSE databases 2005.

HIGHLIGHTS AND POLICY RECOMMENDATIONS

Figure 0.4. **Composition of Producer Support Estimate, 1993-2003**

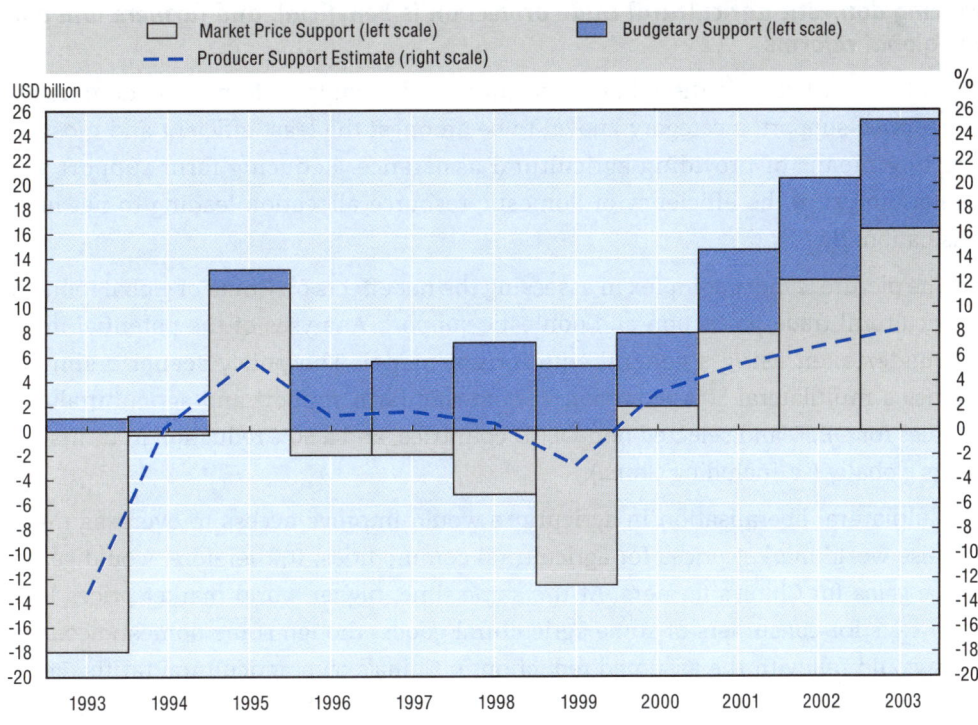

Source: OECD PSE/CSE databases 2005.

Figure 0.5. **China's Producer Support Estimate by commodity, 2000-2003 average**
As per cent of gross farm receipts

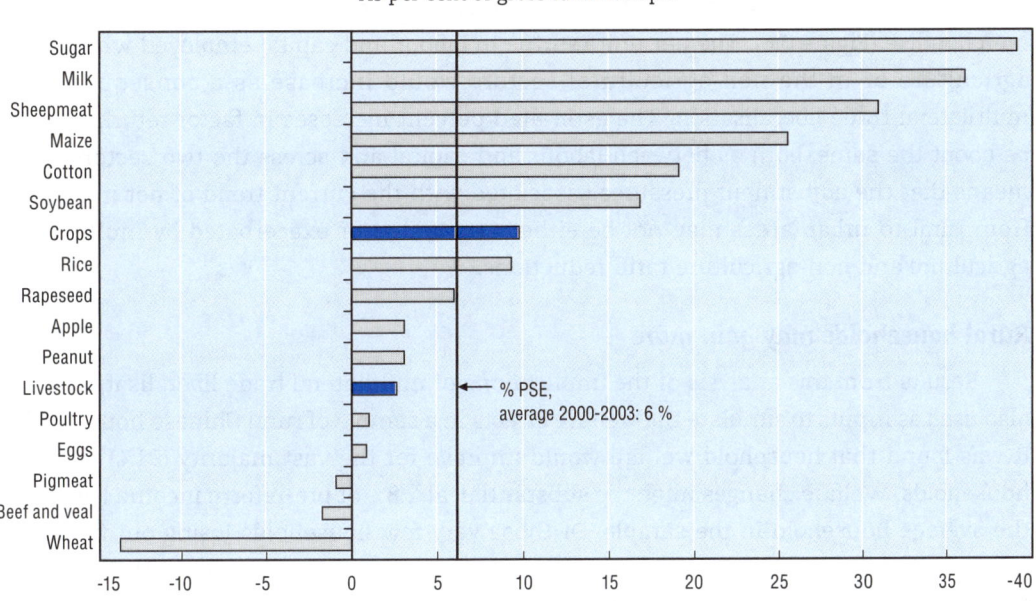

Source: OECD PSE/CSE databases 2005.

OECD REVIEW OF AGRICULTURAL POLICIES – ISBN 92-64-01260-5 – © OECD 2005　　21

3. The benefits of future policy reforms

Reducing domestic agricultural trade protection is beneficial, and farmers will gain from global reforms

As Figure 0.4 reveals the mix of measures used to support farmers is dominated by market price support, a category known to be amongst the least efficient and most trade distorting means of providing agricultural assistance. Reducing farm support would therefore improve the efficiency of domestic resource allocation leading to net income gains nationally.

The picture is more complex in assessing the net effects on China of global reductions in agricultural trade protection and domestic support. Analyses of the potential impacts were undertaken using a general equilibrium model. The policy scenario simulated assumes a multilateral 50% reduction in domestic farm support and agricultural export subsidies for OECD and selected non-OECD countries, and a 50% reduction in tariffs for all sectors globally (i.e. including China).

Multilateral liberalisation in agriculture would improve access to overseas markets and raise world market prices for agricultural commodities. It, therefore, would generate income *gains* for China's farmers. At the same time, higher world market prices lead to higher *costs* for consumers of some agricultural goods, though some domestic consumer prices would fall with the assumed reduction in China's own agricultural tariffs. Because net trade constitutes only a small fraction of China's agricultural output, the gain in farm incomes would be just about the same as the consumer loss due to higher prices, and China's overall welfare would not be much affected.

China benefits from non-agricultural liberalisation in OECD countries

Results of the same analysis suggest quite strongly that China would derive significant welfare gains from reductions in the tariffs OECD countries charge on non-agricultural merchandise (Figure 0.6). The per unit returns to labour and capital employed whether in agriculture or in the non-agricultural sectors would increase as a consequence of multilateral trade liberalisation. The estimated percent increases in factor returns would be about the same, both as between labour and capital and across the two sectors. This means that the adjustment pressures associated with the current trend of net migration from rural to urban areas may not be either attenuated or exacerbated by multilateral agriculture and non-agriculture tariff reductions.

Rural households may gain more

Results from this analysis of the implications of multilateral trade liberalisation were also used as inputs to simulate the welfare effects in a sample of rural Chinese households. It was found that household welfare would improve for the vast majority (91%) of these households. Welfare changes might be substantial, at 2.8% of pre-reform income levels for the average household in the sample. Of those very few households losing out from the reform (9% of the sample), the percentages of those who lose are equal among the poor and the non-poor. Those who lose out typically live in communities poorly endowed with agricultural potential, physical infrastructure and human capital. In absolute value terms, poor households receive welfare gains lower than the non-poor. But in percentage terms, the welfare gains of the poor (4.6%) are much higher than those of the non-poor (2.6%).

Figure 0.6. **Welfare gains (losses) by source of liberalisation**

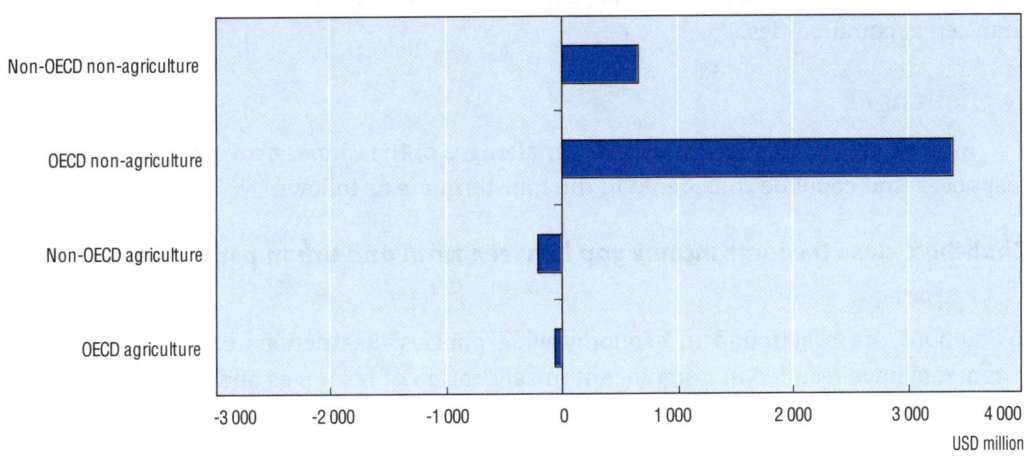

Source: OECD Secretariat.

Liberalisation is likely to bring offsetting effects for various agricultural commodities...

Changes in China's production, consumption and net trade of major agricultural commodities resulting from OECD, Chinese and multilateral liberalisation have also been analysed with the use of a partial equilibrium model. A key finding is that the impact on China's grain markets strongly depends on the actual implementation of Chinese TRQ policies. Within WTO, large shares of the agreed grain TRQs remain under control of STEs, and hence of the government. As in the past these entities have been found to be rather restrictive with respect to grain imports, it remains unclear whether, in case of grain shortages on the Chinese markets, TRQs would become filled. Therefore, actual grain imports could be more limited than suggested by the TRQ quantities even if domestic markets would ask for larger supplies from exporting countries. On the other hand, Chinese authorities may opt to allow for larger than agreed imports at in-quota tariffs if the domestic market shows significant grain deficits.

The scenario results show that a restrictive import policy would have severe implications on domestic markets, with livestock producers as well as consumers suffering from significantly higher grain prices. Clearly, such implications cannot be in the interest of China, which repeatedly has made clear the high priority it accords to food security issues. In contrast, allowing for larger imports at low tariffs would benefit grain users through lower prices. While this would have only small consequences for food consumption, lower grain prices would be particularly beneficial for livestock producers: with higher feed consumption due to larger meat production, both the livestock industry and domestic meat consumers would gain, and net meat imports could be reduced.

... but self-sufficiency in grain production is likely to remain high

Self-sufficiency rates for grain markets while likely to be below the 95% used as a benchmark for food security are still expected to remain above 90%. Only fairly restrictive assumptions on the China's import policies would result in self-sufficiency rates above 95%. Much of the imports would be of coarse grains. As international markets for coarse grains are large, imports of such grains should remain readily available. At the same time, more liberalised markets would benefit from a more efficient allocation of resources: in particular farmers close to urban areas and export ports could play off their comparative advantage in more labour intensive products, such as livestock, but also vegetables, rather than cereal commodities.

4. Policy challenges

As identified in this study, the main challenges of the Chinese government and policy responses that could be considered in the mid-term are as follows:

Challenge: close the large income gap between rural and urban populations

Responses

- Solutions are best found in economy-wide policies. Restrictions on labour and land markets have resulted in a significant misallocation of resources and contributed to the large disparity in urban and rural incomes and across regions. Policies need to facilitate more efficient allocation of resources and to strengthen the functioning of markets. The administrative barriers to rural/urban migration, permanent residence, land markets, and tenure security should be removed or at least relaxed.

- The transfer of huge numbers of workers from low productivity agriculture to higher productivity manufacturing is one of the basic ingredients of China's economic growth. The key issue is how to manage the rural migration process properly so as to avoid undesirable consequences. The continuing transfer of rural labour to non-agricultural employment would create favourable conditions for the structural adjustment of agriculture, such as farm land consolidation and intensification of farm production, but attention needs to be paid to the social and environmental pressures on expanding urban areas.

- The rural population requires better access to education and health care services as well as to the old age pension and other social security measures. Huge differences in public per capita expenditures on education, health care, pension and social security measures between rural and urban populations are one of the major sources of welfare disparities.

- Rural tax reform should be made sustainable. The success of reform depends on continued and growing tax revenue transfers from central government to provinces and counties to compensate them for lower tax revenues. This necessitates the implementation of an effective vertical and horizontal allocation system of tax revenues across various levels of administration and regions. Moreover, the "peasant burden" seems to be underestimated. Therefore, even if some taxes are phased out and some illegal practices discontinued, the reform is unlikely to eliminate miscellaneous fees, levies and fines paid to various government institutions at different hierarchical levels.

- The role of agricultural policies in increasing rural incomes is rather limited, but agricultural policy measures should be consistent with the overall goal of integrating the rural economy into the national development process. Measures which are intended to protect producer incomes should be selected on the basis of maximising efficiency in transferring payments from taxpayers and consumers to farmers and minimising both trade distorting effects and negative effects on environment.

Challenge: integrate small-scale farmers into markets

Responses

- Provision of public services should be improved. The high share of general services can be viewed as a positive feature of China's policy as such support is provided through measures characterised by relatively low production-distortions. However, further progress in the provision of a wide range of public service institutions – such as a rural education system; an agricultural research and extension system for farmers; an agricultural price information system; a sound, scientific, consistent and transparent system to determine the risks of pest and disease; agricultural and food product safety standards and institutions enforcing their implementation – is needed to improve the integration of farmers into domestic and international markets.

- While the government has made significant efforts to improve rural physical infrastructure, such as roads, more should be done in this respect, in particular in central and western provinces, to facilitate integration of farmers into the rest of the economy.

- Chinese farmers should be able to organise themselves on the basis of autonomous large scale peasant organisations for marketing their products to meet their own interests. Although farmers need to take the initiative to meet and act as a group, the government can create a legal and regulatory framework to promote the creation of autonomous peasants' marketing organisations.

Challenge: create more efficient farm structures

Responses

- While the extension of land-lease contracts to 30 years is a step in the right direction, further growth in land rental transactions may be constrained by the remaining ambiguity over land tenure rights, thus slowing the natural process of movement of land resources to the most efficient farmers. However, "economies of scale" and "economies of specialisation" in farming cannot be imposed from above. The circumstances in which they exist involve a process of discovery to be exploited by farmers themselves, but it is the government's role to allow land markets to function.

- Access to formal rural finance institutions should be improved. However, the persistence and success of informal rural credit institutions is an indication of restrictive finance policies and of the failures and lack of flexibility and responsiveness within the formal financial sector. In this context, the incidence and magnitude of informal financial transactions suggest that these activities play a positive role in improving the lives and prospects for China's farmers and rural households.

Challenge: enhance the environmental sustainability of agriculture

Responses

- Finding a balance between the objectives of expanding food production, raising rural welfare, opening up the domestic market to international trade, and protecting the environment (agricultural and environmental policy coherence) is needed. The evidence suggests that the environmental and resource depletion problems caused by agriculture are serious, and are likely to become more severe unless more effective efforts are made by policy makers in the future, which involve a combination of taxes and payments, regulations, and voluntary and co-operative approaches.

- The complex system of administration of environmental legislation at the different levels of government where it applies (administrative level co-ordination) should be simplified.

- Agri-environmental issues should be addressed as far as possible through the creation of markets and shifted from administrative towards market-based policy instruments to ensure that the polluter pays and the provider of environmental benefits is paid.

Challenge: improve the competitiveness of China's agricultural sector

Responses

- Continued efforts to promote choice for producers and consumers and encourage competition across geographic boundaries and institutional jurisdictions will foster growth and improve competitiveness in some commodity sectors. The interventions that still exist, weak enforcement of contracts, lack of transparent information and open bargaining among several buyers and sellers indicate that more is required to foster a climate for competitiveness.

- To allow for further reallocation of resources in line with China's comparative advantages, the current interpretation that China should produce 95% of its own grain requirements needs to be reconsidered. Such interpretation limits the scope of adjustments, in particular in switching more resources to labour intensive products such as fruits and vegetables, and thereby also constraints the productivity and income generation capacity of agriculture.

- Continued efforts should be made to diminish the role of STEs. Despite reforms undertaken so far, STEs remain important players in China's agro-food trade, especially for commodities covered by TRQs. As a result, available price information suggests that prices of commodities exposed to state trading, are largely de-linked from the level and evolution of world market prices – as seen by the level of producer support – thus not providing producers with adequate incentives.

Challenge: improve the design and implementation of agricultural policies

Responses

- The relationship between central government and sub-national governments needs to be redefined. The current trend is toward downward devolution of some aspects of government authority and increasing local-level responsibility without the commensurate transfer of funds and other resources. This increases off-budget and extra-budgetary activities, undermines transparency and accountability, and increases opportunities for corrupt or opportunistic behaviour. In this respect, the delivery of public goods and services at the local level is in need of serious review, taking care to ensure consistency across jurisdictions and that the responsibilities do not exceed the resources available.

- Government institutions should concentrate on the creation of the enabling framework for the agro-food sector and rural economy, rather than on planning, which has dominated in the past.

- Policy decisions should be based on adequate and accurate information. Reliable statistics are therefore necessary to assess the results of reforms undertaken so far and to formulate proper policy responses. To date, user orientation of statistics is only beginning and data collection methods are not sufficiently transparent. Moreover, the accuracy of data, particularly on budgetary support for agriculture, agricultural commodities' prices, livestock production and consumption, grain stocks, overall farmland area, farm structures is far from adequate. Until a more comprehensive and coherent system of monitoring, analysing and reporting of China's agricultural sector emerges, it will continue to be difficult to analyse, assess and improve policy performance.

Chapter 1

The Policy Context

> As China becomes a key player in global agro-food markets it is essential to understand the framework that has shaped developments in Chinese agriculture since the introduction of reforms in 1978. This chapter will focus on the processes undertaken since 1990 and identifies the constraints, opportunities and challenges to China's agricultural development.
>
> Section 1.1 provides a brief overview of basic information. Section 1.2 assesses those aspects of China's macroeconomic performance that have had a crucial impact on developments in the agro-food sector, and then examines the importance of the agro-food sector and its contribution to the development of the economy as a whole. Section 1.3 contains a review of structural issues in the agro-food sector, including agricultural production structures, land tenure systems and policies, as well as reforms and responses in agricultural upstream and downstream sectors. Finally, Section 1.4 provides an assessment of agriculture's performance in terms of output, trade, employment, productivity and incomes followed by a review of food consumption and the agro-environmental situation.

1.1. General aspects

Background data

China is the world's sixth largest economy at current exchange rates, but second in purchasing power parity (PPP) terms, behind the United States (Table 1.1). With 1.3 billion inhabitants, it is the most populous country accounting for 21% of the world's population. Its territory is 9.6 million km^2, making it the third largest country by area in the world, after Russia and Canada. With a GDP per capita of USD 1 000 (4 580 in PPP terms), China qualifies as a "lower middle income country". Yet, life expectancy at birth was 71.4 years in 2000 (69.6 for men and 73.3 for women), which was above the average level achieved by countries of the same group.

Although China's land area is vast, farmland (cultivated land) accounts for only 130 million ha, or 13.5% of China's land surface. Forests account for 16.6%, grassland for 41.7%, and other uses for 28.2%. Cultivated land is mainly located in the east, grasslands in northern and western China, and forests in the remote north-eastern and south-western areas. It is a widely recognised fact that with just 10% of the world's farmland, Chinese agriculture is able to feed 21% of world's population.

China's population density is 134 people per km^2, about one-third of the density in Japan and less than that of many other countries in Asia and Europe. But the population is unevenly distributed. Along the densely populated eastern region, there are more than 400 people per km^2; in the central areas, over 200; and in the sparsely populated plateaus in the west there are less than 10 people per km^2. Over 90% of inhabitants live on less than 40% of the land area. The share of urban population is 41% (2003) which is low compared with countries with a similar level of development. China has 171 cities with population greater than 1 million, of which 11 cities each have more than 4 million inhabitants. Beijing, China's capital is the third largest city with 11 million, behind Chongqing (at least 14 million) and Shanghai (13 million). China has 56 ethnic groups, the great majority belonging to the Han (92%).

For administrative purposes, China is divided into provinces with an average population of 45 million, which corresponds to the population of South Korea or Spain. Below the provinces are prefectures, counties and then townships. Villages are the basic units in the countryside, and number about 0.7 million. A measure of direct electoral democracy is in place at the village level with village officials elected by local inhabitants. However, the elected village head (*cun zhang*) is officially subordinate to the village Communist Party Secretary (*cun zhi shu*), who is still a nominated official.

The policy making process in China is structured by long (usually five years) and short (one year) plans. The five-year plans for national economic and social development, the five-year legislative plan, the national economic and social annual plans, the budgets and yearly legislative programmes are defined within the framework provided by the Communist Party of China (CPC) (Box 1.1).

> Box 1.1. **China's political system**
>
> The Communist Party of China (CPC) occupies the dominant role in the political life of China. The CPC, by its own definition, represents the population at large and interprets and expresses the will of the people. As they represent the will and voice of the people, Communist Party officials and institutions outrank and can overrule officials and institutions that are part of the formal state administrative apparatus. The CPC governs all central and sub-national level State organs. Currently, the head of the CPC is also the President of the State.
>
> According to the Constitution, the National People's Congress (NPC), and its permanent office the Standing Committee, is the "highest organ of state power". The NPC exercises the power of legislation, decision, supervision, election, appointment and dismissal. The President is head of State and who, amongst other functions, promulgates laws, appoints the premier, vice-premiers, state councillors, ministers of various ministries and state commissions according to decisions of the NPC and its standing committee. He also ratifies or abrogates treaties and important agreements with foreign states. The State Council is the highest body of state administration, supervising ministries, commissions and bureaus.
>
> *Source:* OECD (2005a).

Changes in the macroeconomic environment[1]

The development of China's agriculture and rural economy is dependent on conditions and policies affecting the economy as a whole. China's macroeconomic performance during the economic reform era that started in 1978 has been remarkable in many respects.

Economic growth has averaged 9.5% over the last two decades and seems likely to continue at that rate for some time (OECD, 2005c). This has been accompanied by even more rapid growth in trade and investment. As a result, China has become one of the world's largest economies and one of the largest recipients of foreign direct investment (FDI). In the second half of the twentieth century only Japan and Korea achieved a similar rate of economic progress. These achievements are even more remarkable taking into account China's gradual approach to reforms and the fact that development has occurred despite extensive, even if declining, state ownership and intervention in the economy (OECD, 2002b).

China's accession to the World Trade Organisation (WTO) in December 2001 was not a new direction of reform but rather an important landmark along the reform course China has been following for the last twenty five years. Since the mid-1980s China has been liberalising its international trade and investment policies and is now more open and more dependent on trade than many present WTO members. For example, China's trade (merchandise exports plus imports) was 55% of GDP in 2002, just above the level in France, Italy and the United Kingdom and more than twice that in Japan or the United States (Table 1.1).

The depth and scope of China's WTO commitments to liberalise access to its domestic economy are recognised to be more extensive than those agreed to by previous adherents to WTO (OECD, 2002b). This willingness to liberalise is in line with China's domestic policy goals such as promotion of market disciplines and access to technology, but also with its expectations to gain significantly from easier access to export markets.

Table 1.1. **World top-10 countries by GDP (current USD), 2002**

	GDP, current USD	GDP, PPP	Foreign direct investment, inflows	GDP per capita, PPP	(Exp. + imp.)/GDP
	USD billion	USD billion	USD billion	USD	%
1 United States	10 383	10 308	40	35 750	24
2 Japan	3 993	3 425	9	26 940	21
3 Germany	1 984	2 236	37	27 100	67
4 United Kingdom	1 566	1 549	28	26 150	53
5 France	1 431	1 601	52	26 920	52
6 China	**1 266**	**5 861**	**49**	**4 580**	**55**
7 Italy	1 184	1 525	15	26 430	53
8 Canada	714	925	21	29 480	82[1]
9 Spain	653	878	21	21 460	58
10 Mexico	637	905	15	8 970	56

1. 2001 data.
Source: World Development Indicators CD ROM, World Bank (2004c).

China's external balances have remained healthy (Table 1.2). It has undergone several short periods of overheating and inflation during the reform period (e.g. between 1993 and 1995 and most recently in 2004), but has avoided the prolonged phases of very high inflation suffered by many other developing countries.

Table 1.2. **China: selected macroeconomic indicators, 1990-2003**

		1990	1991	1992	1993	1994	1995	1996	1997	1998	1999	2000	2001	2002	2003
GDP growth	Annual %	3.8	9.2	14.2	13.5	12.6	10.5	9.6	8.8	7.8	7.1	8.0	7.5	8.0	9.1
GDP, current USD	USD billion	388	406	483	601	543	700	817	898	946	991	1 081	1 175	1 270	1 412
Exchange rate, period average	CNY/USD	4.78	5.32	5.51	5.76	8.62	8.35	8.31	8.29	8.28	8.28	8.28	8.28	8.28	8.28
Inflation, consumer prices	Annual %	3.1	3.4	6.4	14.7	24.1	17.1	8.3	2.8	–0.9	–1.4	0.4	0.7	–0.8	1.2
Unemployment rate (urban)	%	2.5	2.3	2.3	2.6	2.8	2.9	3.0	3.1	3.1	3.1	3.1	3.6	4.0	4.3
Budget balance	% of GDP	–0.8	–1.1	–1.0	–0.8	–1.2	–1.0	–0.8	–0.8	–1.2	–2.1	–2.8	–2.6	–3.0	–2.5
Export	USD billion	62	72	85	92	121	149	151	183	184	195	249	266	326	438
Import	USD billion	53	64	81	104	116	132	139	142	140	166	225	244	295	413
Trade balance	% of GDP	2.3	2.0	0.9	–2.0	1.0	2.4	1.5	4.5	4.6	2.9	2.2	1.9	2.4	1.8
External debt	% of GDP	13.6	14.9	14.4	13.9	17.1	15.2	14.2	14.5	15.2	15.3	13.5	14.7	13.6	13.7p

p: preliminary.
Sources: World Bank (2004b); China Statistical Yearbook, NBSC, various editions; China Statistical Abstract, NBSC (2004).

Up to 1993, the foreign exchange system was a two-tier pricing system, with an official overvalued exchange rate usually kept constant for extended periods and a secondary market rate determined by supply and demand in legal markets called Foreign Exchange Adjustment Centres (FCACs). To offset the adverse effects of overvaluation of the official exchange rate, exporting enterprises were required to surrender only 20% of their foreign earnings at the official exchange rate and the remaining 80% could be sold at the secondary market rate. The official rate was devalued and currency unified in 1994. The Yuan (CNY) was pegged initially at CNY 8.7 per USD,[2] but the foreign exchange rate has since been managed. In 1996, the remaining exchange rate restrictions on current accounts were eliminated and on 1 December 1996, China formally accepted the obligations of Article VIII of the IMF's Articles of Agreement. The currency appreciated to CNY 8.28 per USD in 1998 and, despite the lack of any official announcement, continues to be pegged at the same level against the US dollar despite the strong pressure to devalue during the Asian financial crisis. In recent years, the Yuan has gradually become undervalued in real terms and there is growing pressure on China to revalue its currency. A relatively fixed exchange rate has also exposed the economy to inflationary or deflationary pressures stemming from fluctuations in the effective rate of the dollar. Therefore, a more flexible exchange rate would support a stable macroeconomic environment in China (OECD, 2005c).

Despite these successes, a wide range of structural issues needs to be resolved. The banking sector remains largely state owned. The non-performing loans ratio is high (25-30%). The reform of state owned enterprises (SOEs) has made limited progress. Labour market and social security system reforms have been slow. Employment growth is slower than the rate needed to absorb new entrants into the labour force, rural-to-urban migrants and those laid off from SOEs. The result has been a rise in urban unemployment, even if the official rate of registered unemployed workers in urban areas remains low at 4.3% in 2003.

There is growing concern over economic inequality, which in turn raises questions about the sustainability of current policies. While China's level of inequality appears to be lower than in countries such as Brazil, Mexico, Russia, South Africa and Turkey, it is well above the average of most OECD countries. However, a distinctive feature of the Chinese case is the speed at which the level of inequality has increased. Similar increases in inequality over such a short period of time have been observed only in some of the central Asian republics and in Russia (OECD, 2004b).

China's exceptionally good macroeconomic performance has been favourable for the development of agriculture through stimulating demand for agro-food products and the creation of non-agricultural employment opportunities. Yet, there has been a wide range of economy-wide policies breaking the Chinese economy into rural and urban sectors, and impeding rural areas, including agriculture, from reaping the benefits from the economic growth achieved.

- The property rights regime in rural areas, in which land, the main asset of agricultural production, is owned collectively (Section 1.3), is distinct from that in non-agricultural activities in which the main asset, capital, has increasingly been owned privately.
- China's population registry (*hukou*) includes provisions that create disincentives for migration of rural-born workers (Annex A).
- Education and health care resources per-capita in rural areas are significantly lower than in urban areas.
- Pensions and other social benefits provided in urban areas are largely unavailable to the rural population.
- Tax system in rural areas is different from that applied in urban areas and includes charges, fees and levies on rural households which are not imposed on urban families (see Chapter 2).

The fragmentation of factor markets, among business segments, and among regions has lead to large income gaps across regions and between urban and rural areas. Therefore, rather than focusing on particular sectors, reforms now need to emphasise more economy-wide policies to promote more efficient allocation of resources and to strengthen the effectiveness of markets (OECD, 2002b).

1.2. Agriculture's importance to China's economy

Even though the importance of agriculture in China's economy has fallen, it is still an important sector accounting for almost 15% of GDP and providing above 40% of

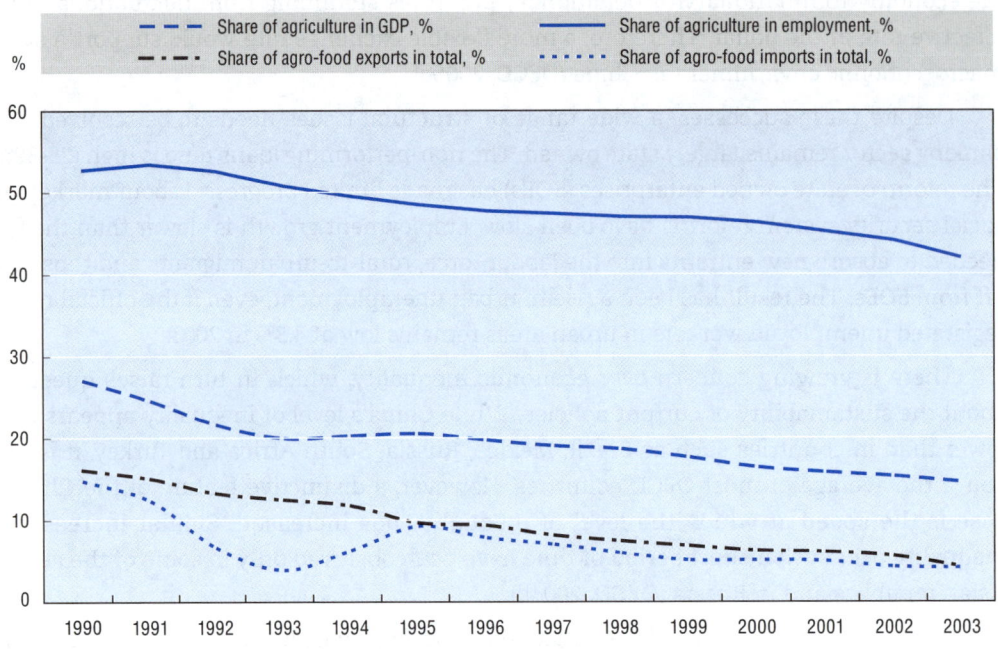

Figure 1.1. **The share of agriculture in GDP, employment, total exports and imports, 1990-2003**

Sources: China Statistical Yearbook, NBSC (2004); China Agricultural Development Report, MOA; China Customs Administration (2003); World Development Indicators CD Rom, World Bank (2004c).

employment (Figure 1.1).[3] The share of rural population in the total is falling, but remains very high at 59% in 2003. The declining role of agriculture in total trade is particularly striking. Even if the real values of agro-food imports and exports have been increasing, their shares fell from around 15% in 1990 to just 4% in 2003, reflecting an expansion of trade in other products.

While the sector's share in GDP is just at the level one would expect given China's GDP per capita (Figure 1.2), the share in employment is significantly higher than that (Figure 1.3). This reflects a relatively low level of labour productivity in agriculture and indicates the buffer role of agriculture in labour markets (Section 1.4).

China is the world's biggest producer and consumer of such products as cereals (in total), potatoes, garlic, onions, apples, pears, duck meat, eggs, goose meat, honey, pigmeat, sheep and goat meat. While China's share of world population is about 21%, its share of world production for the above mentioned products varied from 18% for cereals to 47% for pigmeat in 2003 (FAOSTAT).

Limits on integration with the rest of economy did not prevent the rural sector from providing two major boosts to China's development during the reform era. The first came from a major transformation in the policy environment in agriculture in the early 1980s and the second in the late 1980s/first half of the 1990s with the expansion of rural industries.

In the early 1980s, the tightly controlled commune system of pre-reform times was replaced by a household based system in which individual farmers lease their land from the collectives, are largely autonomous in their decisions, and bear the profits or losses from their operations. Market forces have largely replaced government plans and targets.

Figure 1.2. **Agriculture's share in GDP versus GDP per capita (2000-2002)**

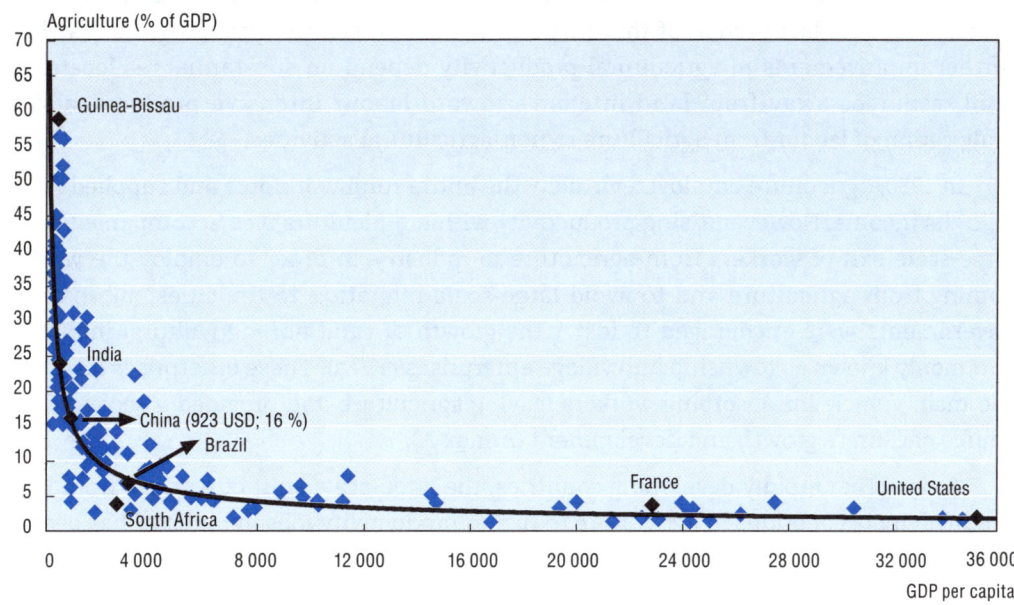

Source: World Development Indicators CD Rom, World Bank (2004c).

Figure 1.3. **Agriculture's share in employment versus GDP per capita (2000-2002)**

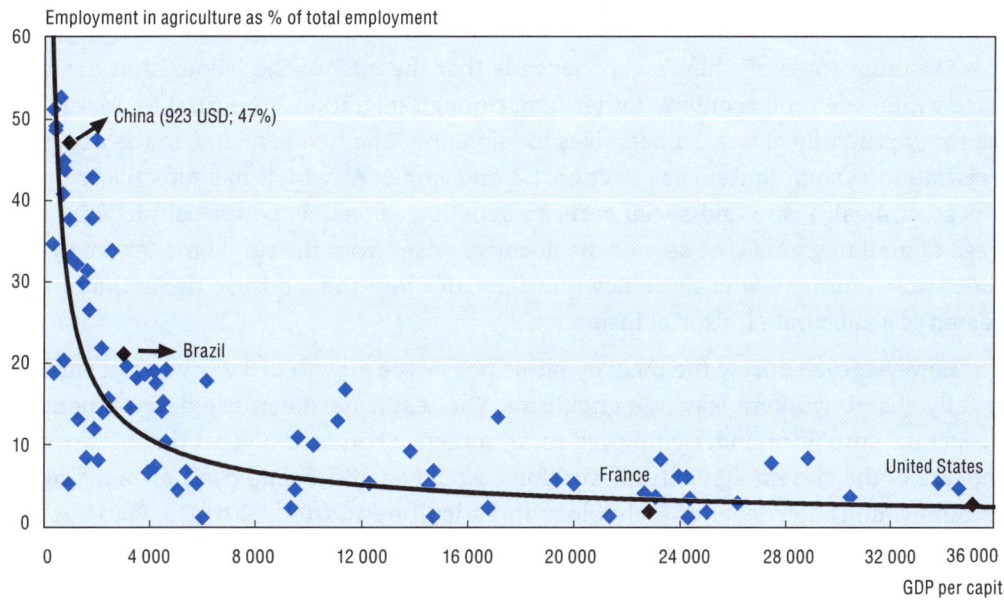

Source: World Development Indicators CD Rom, World Bank (2004c).

With the exception of grains, direct government intervention in the production, pricing, and marketing of agricultural products is now limited (Chapter 2 and Annex B).[4]

These policies have been instrumental in raising agricultural productivity and rural living standards during the reform era. The increase in agricultural productivity provided the first major boost to China's growth during the first half of the 1980s. However, the physical constraints on China's land and environmental resources make it difficult to increase land productivity further. Fertiliser use is already exceptionally high and the scope for increasing the use of pesticides is limited by their adverse environmental impact. Water shortages and other environmental problems pose increasing barriers to higher land productivity (see last section of this chapter). While bio-technological progress may help, further improvements in agricultural productivity depend on substantial reallocation of land resources away from land intensive toward labour intensive products, and on reallocation of labour from agriculture to non-agricultural activities.

In 1980, agriculture employed virtually the entire rural workforce and supplied nearly all of its income. However, rising productivity within agriculture was accompanied by the large-scale exit of workers from agriculture to industry. In order to employ the workers coming from agriculture and to avoid large-scale migration to the cities, sub-national governments were encouraged to foster the growth of rural non-agricultural industries, commonly known as township and village enterprises (TVEs). These enterprises have been the main vehicle for absorbing workers leaving agriculture and provided a second major source of China's growth and development (Annex A).

As for other rapidly developing countries, the large-scale shift of workers from lower productivity occupations in agriculture to higher productivity jobs in industry has been an important engine of China's growth. Moreover, the rapid expansion of rural industry puts competitive pressure on SOEs (Naughton, 1995) and helped finance the provision of local public goods (*e.g.* roads, bridges, schools) thus further improving the performance of the Chinese economy. The overall effect has been to increase the interdependence between the rural and urban economies, even though the traditional administrative impediments have largely remained.

The uniqueness of China's experience is that the bulk of the labour shift has taken place within the rural economy rather than through migration from rural to urban areas, the result primarily of two disincentives to migration. The first is related to the household registration system (*hukou*; see Section 1.3 and Annex A) which has effectively denied education, health care and social security benefits, as well as better paid jobs in urban areas to rural migrants. The second disincentive arises from the rural land tenure system, under which farmers who settle down in the cities lose the land-use rights that are still treated as a substitute for social insurance.

However, even during the most dynamic phase, the growth of TVEs was not sufficient to fully absorb workers leaving agriculture. The result has been the development of a substantial surplus of under-employed rural workers. Estimates suggest that between 30% and 40% of the current agricultural workforce would be under-employed, even taking into account China's low level of technology in agriculture (Carter, 2001; Du, 2001). A large portion of these workers – as many as 100 million – have become "floating" migrants who have taken up informal sector employment in urban areas.

Estimates suggest that even with no further opening of markets, nearly 70 million additional workers will leave agriculture between 2000 and 2010 (Zhai and Li, 2000). China's WTO accession is expected to add 2 to 3 million to this exodus from agriculture. The analysis indicates that even under optimistic assumptions about how much their performance can be improved, TVEs are unlikely to be able to take up more than a fraction of the rural workers who will need to find jobs outside the agricultural sector (OECD, 2002b). This further emphasises the fact that the development of the rural economy is increasingly dependent on conditions and policies affecting the economy as a whole. The importance of factor mobility for enhancing positive welfare effects of policy reform is discussed in Box 1.2.

Box 1.2. **Trade reform and factor mobility in China: the long and the short of it**

John Maynard Keynes is famously quoted as saying that "In the long run, we are all dead." But when it comes to modelling the welfare effects of reforms, it generally turns out that in the long run everyone is better off than in the short run. This is true because the longer the time frame, the more effectively can individuals adapt their behaviour in response to the new incentives – provided they are allowed – and by doing so take advantage of new opportunities to maximise their well-being. In micro-economic terms, this is why elasticities of supply and demand are higher in the long run.

A case in point is the impact of trade reform on the Chinese economy. China's accession to the WTO has been estimated to have a hugely positive effect on the global economy – USD 74 billion per year – with about half of that going to China itself (Ianchovichina and Martin, 2004). Estimates of the impact on poverty in China, however, have been ambiguous, and to a large extent depend on the timeframe being modelled. In a study of the short-run impact (i.e. using a model in which factors are not mobile) by Chen and Ravallion (2004), WTO accession and associated reforms were shown to leave the overall national poverty rate virtually unchanged, but to cause a loss of income for some of the most vulnerable households, especially in certain rural areas, in some cases 3-5%. The magnitude of this loss is small compared to other shocks suffered by rural populations, but is of special concern because of the lack of social safety nets for this group and their extreme poverty. This focuses attention on the questions of whether this impact is likely to persist, and what kinds of policies could mitigate such negative effects.

A study by Hertel, Zhai and Wang (2004) used a long-run model in which – unlike the Chen and Ravallion study – households adjust their labour and capital allocations in response to relative price changes. With high labour mobility, the impact of accession was shown to be much more salutary than in the short-run model, with virtually all groups gaining, although the urban still benefited more than the rural populations. The key to the gains by the rural poor is their ability to move from these areas where labour is in surplus to areas which benefit from accession-related reforms – urban areas, other provinces, more prosperous farming regions, etc. – rather than remaining mired in their current low-return activities. The movement of labour out of surplus areas, in turn, helps to raise wages in those regions as well.

The importance of this kind of mobility is underscored by studies that explicitly model reforms of policies that impede factor movement. Shi (2002), for example, found that the *hukou* residence permit system has imposed an implicit tax equivalent to 34% of rural wages on transfers of labour out of agriculture. A study that investigated the effect of abolishing the hukou system and improving the land tenure system so that it no longer acts as a disincentive to leave the land found both reforms to be hugely beneficial (Hertel and Zhai, 2005). Another policy that contributes both to the relative immobility of labour out of rural areas and to rural poverty is the unequal distribution of spending on education. In modelling the effects on China of full global trade liberalisation, Zhai and Hertel (2005) found that rising per capita spending on education in rural areas to the level in urban areas increased the overall benefits of liberalisation from 1.9% to 3.2% of GDP. This was shown to be particularly beneficial for rural areas, where the poverty rate falls by 1.9% from liberalisation alone, and by 5.2% when combined with the educational reform. These reforms take time to bring about their full benefits and, in the interim, it is important to take care of potential losers with better targeted interventions and safety nets.

> Box 1.2. **Trade reform and factor mobility in China: the long and the short of it**
> *(cont.)*
>
> In sum, there is a strong basis for concluding that while trade liberalisation both of China's own policies and of those of its trading partners is beneficial, the benefits will be much larger and more equitably distributed if combined with complementary reforms to enhance factor mobility. Notwithstanding Professor Keynes's bias, it is better not to get stuck in the short run.
>
> *Source:* Contributed by the World Bank.

1.3. Structural change in the agro-food sector

Agricultural production structures

China's agriculture is characterised by scarce land in relation to labour and small-scale production using little mechanisation. Currently, out of 248 million rural households, 200 million of these are farms having an average land allocation of just 0.65 ha.[5] Moreover, each farm typically consists of several non-contiguous tiny plots.[6] As in some other east Asian countries, agricultural output per unit of land is high by international standards but output per worker is low (Table 1.3). Limited arable land and a large rural labour force mean that, in general, China tends to have a comparative advantage in the production of labour intensive crops, such as fruits and vegetables, and a disadvantage in the production of land intensive crops such as grains and oilseeds. However, the situation is strongly diversified regionally and depends, among other things, on availability of land and water, climatic conditions, transportation costs and access to markets.

The dominance of small-scale farming is partly due to a high population density in provinces rich in arable land, but is also a direct result of the 1978 reforms of the commune

Table 1.3. **Arable land and yields in selected countries, 2000-2002 average**

	Arable land per capita (ha)	Arable land per employed in agriculture (ha)	Rice, paddy (tonne/ha)	Wheat (tonne/ha)
Russian Federation	0.86	15.5	3.6	1.9
United States of America	0.61	59.4	7.2	2.6
Brazil	0.34	4.5	3.2	1.7
France	0.31	21.5	5.6	7.1
Thailand	0.26	0.8	2.6	0.6
India	0.16	0.6	2.9	2.7
Italy	0.15	6.5	5.9	3.1
Germany	0.14	12.2	n.a.	7.4
China	**0.11**	**0.4**	**6.2**	**3.8**
United Kingdom	0.10	11.2	n.a.	7.7
Indonesia	0.10	0.4	4.4	n.a.
Vietnam	0.08	0.2	4.4	n.a.
Philippines	0.07	0.5	3.2	n.a.
Korea, Republic of	0.04	0.7	6.6	3.0
Japan	0.03	1.7	6.6	3.8

Note: For China, arable land per employed in agriculture has been calculated by dividing 130 million ha by 324 million agricultural workers.

Source: FAOSTAT, FAO (2004).

system of agriculture and the introduction of the Household Production Responsibility System (HPRS). The HPRS broke up commune production teams and made the farm household the basic unit of production in agriculture. To maintain egalitarian access to land, households were generally allocated use rights to agricultural land on a per capita basis (in most cases quality of land was also taken into consideration) (Lohmar *et al.*, 2002). In contrast to land reforms in transition economies of Central and Eastern Europe, there was no consideration in China for the amount of land owned before the collectivisation of the 1950s, and no restitution of land ownership rights nor privatisation of land. In fact, most of the ownership rights reside with village collectives and households lease land from them (see next section).

China's agricultural land and labour force are spread unevenly across the regions (Figure 1.4). Over 45% of agricultural labour is concentrated in just two regions of north and southwest China, while the largest share of cultivated area (based on revised Agricultural Census estimates) is in North, Northwest, and Southwest China.[7] The largest per farm availability of arable land is in Heilongjiang province (northeast) at 2.5 ha and the smallest in Zhejiang province (east) at 0.2 ha.[8]

China's crop and livestock production activities are scattered throughout the country, based to a large extent on the endowment in factors of production, climate and transportation costs. Nevertheless, certain regions in China account for a disproportionate share of the output of many key crop and livestock products, partly due to distorting policies applied in the past (Chapter 2). In general, labour intensive crops such as fruits and vegetables are produced mostly in densely populated coastal provinces and in areas close to cities. Land intensive crops such as wheat, corn, soybeans and cotton are produced mainly in the northern part of the country while most of the rice and sugar crops are produced in the southern half. Meat production is more evenly distributed and to a large extent carried out in small part-time "backyard" facilities.

Analysis of changes in farm structures is very difficult as there is no consistent time series data which would cover various categories of farms. The last comprehensive data on the distribution of farm size dates back to the first agricultural census in 1997 and clearly

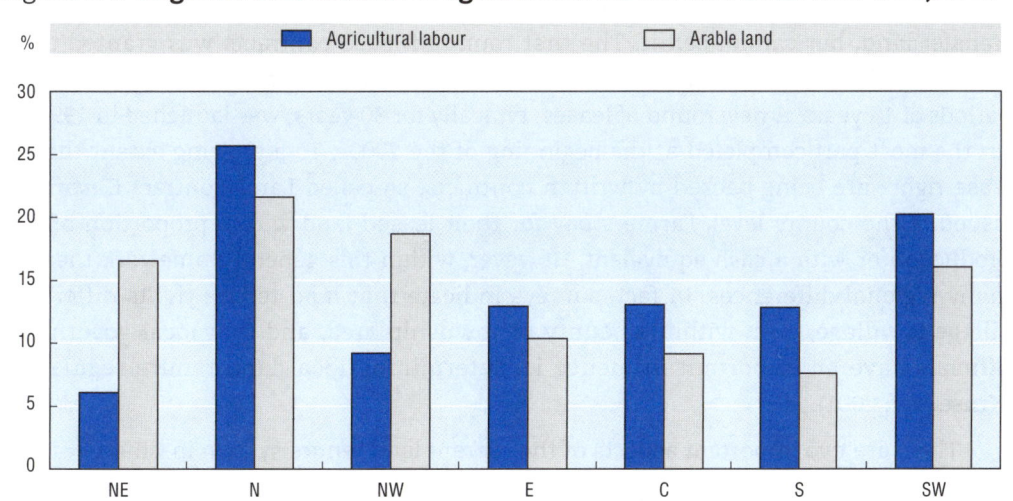

Figure 1.4. **Regional distribution of agricultural labour and cultivated area, 2003**

Source: China Statistical Yearbook, NBSC (2004).

confirms the dominance of small-scale farming. Farms of more than 2 ha accounted for just 1.9% of the total number and operated on 13.9% of cultivated land. More than 80% of farms operated on less than 0.6 ha (National Office for the Agricultural Census, 1998).

At the beginning of the 1990s, the Chinese government undertook several administrative and financial measures to support large-scale operations in agriculture, largely through collective forms of farming. These measures were imposed "from above" and have turned out to be very costly and unsustainable. According to the Ministry of Agriculture data, their share in total cultivated land increased to 6.5% in 1994, but then fell to below 3% in 1999 (Du et al., 2002). Another category includes around 2 000 state farms whose importance has also declined. In 2002, they averaged above 2 400 ha per farm and operated on 3.6% of China's total cultivated area. They produced 15 million tonnes of grains in 2002, which was 3.8% of China's total grain production (China Statistical Yearbook, 2003).

In recent years, there have been several attempts to stimulate household-based large scale farming (dahu). Quite often these farms are given access to so-called "waste land" which is contracted for a low fee for periods ranging from 50 to 100 years, and which can be adapted for agricultural production.[9] Another form of large-scale farming is the so-called fanzudaobao, developed mostly in the coastal area. Within this arrangement, the collective (for example village) rents land from farmers having long-term land lease contracts, sometimes makes infrastructural investments and then rents it out to external investors, mostly for vegetable and fruit production.

In 2003, changes to legislation (see next section) provided more secure rights to land plots and allowed the preservation of land use rights of families moving to towns. However, families moving to large cities still risk losing land use rights. It is too early to assess the impact of these changes on the functioning of the land market, but observations suggest that plots vacated are being merged into larger holdings, so that in many places, in particular in the East, subsistence farming is being replaced by more commercial types of farming (Browne, 2004).

Agricultural land tenure system and policies

The land tenure system in China is based on land lease contracts. Farmland is *de facto* owned by village collectives, which extend land lease contracts to individual farm households. Households have most of the property rights: they can use, sub-lease and transfer land, but cannot sell it. The first round of lease contracts was granted at the beginning of the 1980s within the framework of the HPRS and then extended in 1984 for periods of 15 years. A new round of leases, typically for 30 years, was launched in 1998 and for the most part completed at the beginning of the 2000s. To a growing extent the new lease rights are being backed by written contracts, so-called Land Contract Certificates issued at the county level. Farmers pay for their leased land with a proportion of their production or with a cash equivalent. However, within this general framework there are many regional differences. In fact, surveys indicate that land-tenure rights differ from village to village, even within a county or township area, and that local government officials have an important influence in determining local land tenure regulations (Krusekopf, 2001).

There are two important aspects of the current land tenure system in China:

- Land is considered a part of the social security system in rural areas. Therefore, even if large numbers of rural wage earners and migrants no longer rely on land, they keep their use rights to land as it is their only subsistence guarantee if they lose their jobs (Vermeer, 2001). Moreover, according to current regulations if agricultural households leave their village to take non-farm jobs in cities, they permanently forfeit their use rights.[10] Such problems are handled by leaving a family member in the village or by sub-leasing land to another farmer who takes on the obligation for payment of taxes and fees to local authorities.

- A large part of the rural population is still attached to the collective ownership of land, in particular in less developed areas of central and western China. There is an underlying principle that every member of the village should have access to a parcel of collectively owned land. Therefore, any change in the demographic situation in a village (birth, death, marriage, in- or out-migration) creates pressures for the administrative re-allocation of land. In fact, re-allocation of land (partial or involving all households) occurs on average every three to five years, which contrasts with the central government policy to extend the length of lease contracts between villages and households to 30 years.

Even if, officially, agricultural land is owned by village collectives, land ownership rights are subject to many levels of authority (Box 1.3). Moreover, within the current system there is scope for arbitrary decisions by local leaders, and for situations of conflict between village leaders and farmers, both in cases of administrative re-allocations of land and when local regulations limit land-transfer rights. There are also a large number of cases when local leaders at the township, village or *xiaozu* levels assuming the role of landowners decide to lease or sell land to external investors without the consensus of local farmers and without proper compensation for lost access to land.[11]

In 2002, the National People's Congress passed the "Rural Land Contract Law", effective from March 2003. The Law combines all existing regulations in one document and is intended to establish a comprehensive legal framework for land relations between farmers and collectives. The Law reinforces earlier regulations such as the regulation that contracted land cannot be taken back and reallocated by the collective (village or high level administration) during the period of the contract. It also specifies that migrants settling in towns (but not in cities) retain their land use rights in the village and that farmers should be sufficiently compensated for land transferred for other uses. Moreover, in the amendment to the Constitution passed in March 2004 the previous wording of Article 10 has been changed to indicate that compensation will be paid for land taken over by the State.[12]

While recent changes in legislation are intended to provide better protection of farmers' land use rights and some observers suggest that they have stimulated land renting transactions, ambiguous formulation of relevant articles may undermine effective protection in practice. The Law refers to farmers' rights to land contracts and operations (*tudi chengbao jingying quan*), but does not specify what those rights are precisely. Land ownership rights remain with "collective economic organisation", but it is not clear if it is village, *xiaozu* or township. It is also not clear who in fact represents the "collective": selected leaders, party representatives or all villagers. Migration to the cities is still discouraged as migrants settling in the cities are obliged to return contracted land to the village. On the other hand, it is very difficult for outsiders to enter the collective to obtain land contracts.

In the early stages of reform, land tenure policies based on the HPRS and the principle of egalitarian access to land brought a number of positive effects. Compared to the

> **Box 1.3. The distribution of land rights across levels of authority**
>
> **National government.** The central government provides national land laws and directives that give guidelines for sub-national policy makers.
>
> **Provinces.** The implementation of general laws differs across provinces. For example, while the central government's policy is to extend the length of lease contracts to 30 years, some provinces strongly support such policy and discourage reallocation activity, but others have been less supportive and allow for the frequent reallocation of plots of land among households. According to the current legislation, it is only at the province and the State Council level that land can be approved for conversion to non-agricultural uses, with the approval level required increasing progressively in accordance with the area of land being considered for conversion.
>
> **Counties.** County governments are responsible for the overall planning of land utilisation within their respective jurisdictions. Their duties also include issuing Land Contract Certificates to farmers.
>
> **Townships.** A township district contains approximately 10-20 villages. In some areas, townships may influence village land policies, including village-wide land re-allocations, and are perceived as the primary decision makers.
>
> **Villages.** Villages typically comprise between 300 and 500 households. Village leaders in general have ultimate authority on land allocation, but often delegate some or all of this authority to *xiaozu*. Village leaders sometimes impose compulsory planting requirements on some of the land allocated to farm households.
>
> **Xiaozu.** These are groups of 30 to 40 households (remnants of production teams of the collective era). *Xiaozu* are often the *de facto* collective owners of the land, but commonly work with village leaders on land allocation. Xiaozu leaders may periodically reallocate land among member households, usually to provide land for new households at marriage but occasionally based on births and deaths in households.
>
> **Households.** Households do not have ownership rights to land, but are allocated rights to use land, usually several non-contiguous small plots. Specific rights on each plot may vary, but are mainly the right to farm the land for a finite period and to keep or sell the produce. Farmers make most of the production decisions on their land, but the land must stay in agricultural production. Moreover, in the "Regulation on the Protection of Basic Farmland" issued by the State Council on 27 December 1998 (effective from 1st January 1999), Article 17 states that the change of "basic farmland" into orchards and fishing ponds is prohibited. As at that time China's grain markets were characterised by surpluses, the law was not strictly implemented and farmers could quite easily turn land to more profitable orchard and fisheries production. But the situation changed in the second half of 2003 when grain prices strongly increased. In mid-2004, a new decision was passed by the State Council (Suggestions on the further Deepening of the Reform of Grain Marketing System, *Guofa*, No. 17, May 2004), re-emphasising the prohibition and its implementation strengthened in the second half of 2004.
>
> Source: Lohmar et al. (2002); China's land use regulations.

communal system, the HPRS provided farmers with stronger rights to land and to production, which stimulated growth in agriculture and rural incomes. Through equal distribution of land use rights, China avoided having large numbers of rural landless workers vulnerable to famine or other economic shocks. It has also ensured that the vast majority of rural households are at a minimum, food self-sufficient. However, as China develops, such land tenure system shows increasingly its disadvantages. Land market transactions are limited. The survey by the Research Centre of Rural Economy (RCRE) of the Ministry of Agriculture suggests that land rental arrangements between farmers tend to be informal, short-term and most often between relatives (MOA, *China Agricultural Development Report*, 2003).

The government continues to make statements about the need to rationalise farms and achieve the (presumed) benefits of "scale farming", but further growth in land rental transactions may be constrained by the remaining ambiguity over land tenure rights, thus slowing the natural process of concentration of land resources in hands of most effective farmers. However, land concentration process is slow and does not necessarily lead to a significant increase in the average farm size. An acceleration of this process by arbitrary decisions could deprive many farmers of the access to land, which up to now, has helped contain absolute poverty in China. Therefore, "economies of scale" and "economies of specialisation" in farming cannot be imposed from above. The circumstances in which they exist have to be discovered and exploited by farmers themselves, but it is the government's role to allow land markets to operate (OECD, 2005a).

Structural change in the upstream and downstream sectors

Before the reform

Until the mid-1980s, both upstream and downstream agriculture-related enterprises were organised into state monopolies, structured vertically from Beijing to the local agencies, with each enterprise subordinated to a designated ministry or other central institution. The whole structure was strongly politicised with enterprise managers involved in the bargaining processes with bureaucrats over input supplies, production plans and state support. In practice almost all functions performed by enterprises in a market economy were handled by branch ministries.

Agricultural input supplies were monopolised by the Chinese Supply and Marketing Co-operative (SMC), a co-operative on paper but a government agency in reality. The SMC was also responsible for providing industrial goods to the rural population and for the procurement of the majority of agricultural products for the state. However, grain procurement was under the responsibility of Grain Bureaus, another government agency.

All foreign trade transactions were centrally controlled and monopolised by the State Trading Enterprises (STEs). In the agro-food sector the main STEs were: the CEREOILS (for cereals and oil products), China National Native Products & Animal By-product Import & Export Corporation, China National Textiles Import and Export Corporation (Chinatex, for cotton), China Machinery and Equipment Import & Export Corporation (including agricultural machinery), and SINOCHEM (for fertilisers and chemicals). They operated under the Ministry of Foreign Trade and Economic Co-operation (MOFTEC, currently Ministry of Commerce, MOFCOM) and subject to government licensing and import-export plans.

Food processing and retail trade enterprises were state- or collectively-owned. Until the end of the 1970s, private trading enterprises were not allowed. However, local farmers' markets were tolerated (except for a short period in the 1960s) and were the main outlets for private sales of agricultural products, mainly fresh fruits and vegetables.

Since the mid-1980s this system has been profoundly reformed. These reforms are briefly assessed in the following sections.

Reform in the input supply system

Despite some backward steps in the reform process, the input market in China had become highly competitive by the beginning of the 2000s. The state's share in input production also declined substantially. For example, in the first three quarters of 2003, private enterprises produced 61% of agricultural machinery, SOEs or state controlled enterprises 34%, and joint ventures 5% (*www.1agri.com/trade/news/detail/12-11/4818.html*).

While the SMCs have preserved an important role in the market, there are three other main channels through which farmers may purchase inputs: "three agricultural stations" (*nongye san zhan*),[13] direct sales by input producers and private traders. There are no nation-wide data to evaluate the relative importance of various channels, but observations suggest that prices have become competitive and farmers do not have difficulties in accessing a wide variety of inputs. China's WTO accession contributed to the lowering of tariffs on agricultural inputs, thus easing access to high quality imported inputs, and stronger competition led to a fall in input prices. For example, fertiliser and farm machinery prices fell in nominal terms by 25% and 17%, respectively, between 1996 and 2003 (*China Statistical Yearbook*, 2003).

However, the SMC and its nation-wide umbrella, the All-China Federation of Supply and Marketing Co-operative (ACFSMC), still need to undergo an important restructuring. The whole structure consists of 31 provincial, 337 prefecture, 2 365 county federations and 26 000 primary societies. It employs 4.3 million employees and has 0.5 million business outlets. The ACFSMC has 16 subordinate enterprises, 8 science and research institutes and 13 national-level specialised associations. For long periods, the SMCs were making losses and many of them are *de facto* bankrupt.

Chinese farmers continue to have difficulties in accessing credit. The two main formal channels of credit for farmers include the state-owned Agricultural Bank of China (ABC) and Rural Credit Co-operatives (RCCs).[14] In addition there is a wide range of informal rural finance institutions, part of them still considered illegal in China (OECD, 2005c). According to a recent national survey of rural families, only 16% of farmers have recourse to formal or informal credit. This is partly due to the little interest of farmers in borrowing money, but also to a lack of collateral (the land belongs to collectives) and to the high transaction costs involved in obtaining formal credit. Another factor is the closing of many local branches of financial institutions and the failure of new ones to emerge. As a result, more than 70% of loans are obtained through informal channels, while less than 30% are from formal financial institutions.[15] This partly results from the slow process of reform of rural finance institutions, but it can also be attributed to more general problems such as the continued channelling of financial resources to the state owned enterprises and growing financial fragility and management challenges facing Chinese banks (OECD, 2004d).

Marketing channels for outputs

With the exception of grains, for which the government agencies remain the main purchaser (see Chapter 2), Chinese farmers have an option of selling their products through diversified private channels. The most widespread are so called big marketing households (*yunxiao dahu*), which are family based traders specialising in the intermediation between farmers and wholesale markets. Quite often they have their own stands in agricultural wholesale markets in the big cities. Other main channels include direct sales by farmers to retail or wholesale purchasers, including supermarket chains, direct sales to local markets, direct contracting with food processors/exporters and sales through farmers' professional associations or co-operatives. When quality requirements are high (exports or high income domestic consumers), the intermediate traders, such as professional associations, provide guidance and support to farmers to improve quality and secure timely supplies.

As for the inputs market, there is no national level data on the diversification of marketing channels for agricultural commodities. Regional surveys, however, confirm that these channels are becoming increasingly diversified and competitive. For example, a survey conducted in six counties of Zhejiang province in 2003 indicates that while for paddy rice the state procurement system purchases more than two-thirds of total marketed production, for all other products there are private traders who dominate. In the case of vegetables, the most important channel is direct sale on wholesale and retail markets (43% of total marketed production), followed by sales to the processing enterprises (18%) and sales to the intermediate traders, including marketing households (16%). Fruits are mostly sold through intermediate traders and marketing households (90%). The same channel dominates for pigs (80%), eggs (60%) and poultry (80%) (Statistical Bureau of Zhejiang Province, 2004).

Studies on price transmission on China's agricultural markets suggest that these markets are "fairly well integrated into the rest of the economy" and confirm a competitive environment within which purchasing agents need to operate. This conclusion holds even for such products as rice, maize and soybeans purchased mainly by the state agents. As they need to compete with private marketing channels, markets for these products are "remarkably integrated", in particular for maize and soybeans (Huang and Rozelle, 2002a).

There is an important role for the government in helping farmers gain access to the emerging procurements systems. In addition to the more traditional role of providing appropriate infrastructure, the government could assist farmers to organise into marketing associations. This is confirmed by the experience of such countries as Japan and Korea. While Farmers' Professional Associations (FPAs) exist in China, they are still in a fairly early stage of emergence. Surveys indicate that only around 2-3% of China's rural households participate in any type of FPAs (Shen *et al.*, 2004). Similarly, China lacks an independent marketing co-operative system which, in the past, played an important role in integrating farmers with markets in such countries as France or the Netherlands. Although the initiative to meet and act as a group must be from the farmers themselves, the government can create a legal and regulatory framework to promote the creation of autonomous farmers' marketing organisations.

Food industry

The food industry is a fast growing sector with official statistics reporting average growth of 14% per year between 1999 and 2003.[16] The industry is still very fragmented with occasional reports indicating that the total number of food processing ventures is around 0.9 million in the mainland China (Taylor, 2005). This compares with about 20 thousand medium and large enterprises with annual sales in excess of CNY 5 million each (Table 1.4). Between 1999 and 2003, the sales value of these enterprises increased by 69%, the total number of enterprises fell by 4% due to consolidation, mergers, and business

Table 1.4. **Food industry in China between 1999 and 2003**

	1999	2003	Percentage change
Number of enterprises	20 125	19 277	-4.2
Number of employees (million)	3.2	3.9	23.6
Sales revenue (billion CNY)	733	1 235	68.6

Note: Data in this table relate to industrial enterprises "above designated size". These are all state-owned and non-state-owned enterprises whose value of annual sales exceeds CNY 5 million. Statistics for enterprises with sales below this mark are not available.

Source: *China Statistical Yearbook*, NBSC (2000) and (2004).

failures, and the total number of employees increased by 24%, indicating an almost 40% increase in labour productivity (Table 1.4).

The food industry is undergoing an ownership restructuring process with the shares of SOEs and collective enterprises falling and other enterprises' increasing. However, the SOEs still play an important role with the share in the total number of enterprises at 31% and in the gross value of production at 41% in 2002 (Table 1.5). In fact, the "privatisation" of food processing enterprises is largely due to the creation of new private enterprises as the restructuring of SOEs depends on national policy, which is still making slower progress than expected in this respect.

The food industry continues to attract foreign investors. Most often foreign companies set up joint ventures with local businesses. Many famous brand names, such as Cadbury, Campbell's Soup, Cargill, Coca Cola, Danone, Heinz, Hershey, Hormel, KFC, Kraft, M&M, Maxwell, McDonald's, Nabisco, Nestle, Ocean Spray, Pepsi Cola, Pizza Hut and Wrigley, have established enterprises in China. About 80% of products sold under these brand names are produced in local plants (Zhu, L.J., 2003).

It is estimated that only 30% of China's food consumption is processed as compared to 86% in more developed countries (Bean, 2003). This demonstrates that the industry has a large potential for future growth as China develops, consumers become wealthier and a larger share of the population lives in urban areas. With China's opening to international competition, the industry will continue to undergo merger and consolidation processes for greater profit and efficiency gains. Difficulties with the creation of a stable production base within small-scale farming systems may be a potential barrier to future development. As the vast majority of urban consumers and a large part of the modern food processing industry are located close to the eastern coast, growing imports may be an option to satisfy

Table 1.5. **The composition of food industry and tobacco enterprises in China by type of ownership, 2002**

	SOEs		Collective		Other	
	Number %	Production value %	Number %	Production value %	Number %	Production value %
Food sector, total	31	41	11	9	57	50
Food processing	30	23	12	14	58	63
Food manufacturing	28	23	11	9	61	67
Beverages	35	43	10	6	56	51
Tobacco	85	99	9	1	6	0

Source: *China Food Industry Yearbook* (2003).

expanding domestic demand if agricultural producers are not able to provide consistent supplies of high quality raw products.

Wholesale trade

As in other centrally planned economies, wholesale markets did not exist before the reforms started, having been replaced by the state distribution system. Under the reforms, wholesale markets expanded from 892 in 1986 to 4 387 in 2000 and the transaction value increased from CNY 2.8 billion to CNY 335 billion (Xu and Liu, 2003).

Until the mid-1990s, wholesale markets were usually owned by municipal governments or SOEs as a reflection of the government's perceived responsibility to provide urban citizens with stable, low-priced food. These entities also had the best access to finance, while private agents still faced constraints in legally establishing their businesses. The situation has changed and there has been a massive entry of private companies (Hu et al., 2004). Today, the government is focussing on the establishment of standardised management procedures and of market registration and monitoring systems.

The development of wholesale markets facilitated the overall market integration of the country. For example, the Shouguan Vegetables Wholesale Market in Shandong province (coast) supplies vegetables to more than 20 provinces. In turn, Jibu Vegetable Wholesale Market in Shenzhen (close to Hong Kong) acts as a link between producers and consumers in the southern provinces and even in some neighbouring countries. Since 2000, the number of wholesale markets has fallen, but the transaction value continues to increase as greater competition has led to greater concentration.

Retail trade

There are no reliable data on the overall retail food turnover, but it is estimated that the total amount of money spent on food by Chinese families was close to USD 200 billion in 2001, excluding expenditures on food in restaurants, hotels and canteens which could add more than USD 50 billion to this amount (Zhu, L.J., 2003). While SOEs involved in food retailing still exist, the sector is dominated currently by private enterprises.

Since the beginning of the 1990s growth in food retail sales has dominated by the development of supermarket chains. The first supermarket was established in 1990 and by 2002 the number had increased to more than 53 000. It is estimated that the share of such large stores (supermarkets, hypermarkets, and chain convenience stores) in total food retail sales in urban areas could be as high as 35% (Hu et al., 2004). The share in overall food sales in China is much lower, most likely below 20% (compared to 60-80% in the United States and the European Union), but the share will increase quickly as retail chains are spreading to all cities, including China's interior provinces.

The largest food retail chains are Chinese, but there is a growing number of foreign retailers such as Carrefour, Wal-Mart, Metro and Price Mart. Further inflow of foreign capital into the sector is expected in the near future with relaxation of the FDI rules. While current rules require foreign-investment businesses to be joint ventures, it will change in 2005 when full foreign ownership of retail businesses with no restrictions on the number of outlets will be permitted.

As China's retail market remains fragmented, mergers, consolidations and bankruptcies will and must be allowed to occur if the retail sector is to continue to increase its efficiency and responsiveness and if economies in the value chain are to be realised.

With the encouragement of China's government, the two largest retail food outlets, Linhua and Hualian, are now in the process of merging and have also expanded into the hypermarket format. This new chain is almost three times the size of its closest competitor and is the only chain to enjoy a nation-wide presence.[17]

Traditional food markets continue to be important outlets for farmers, in particular in rural areas and in small and medium-size towns. However, many of them, in particular in big cites, are being closed or consolidated. Local authorities in most cities consider them as unsanitary and they bring much less tax revenues than more regulated markets (Bean, 2003).

The newly emerging structure of food retailing induces changes in the whole food chain, including the food wholesaling system, which is farmers' closest contact with the market. Moreover, in an attempt to cut transaction costs, a growing number of supermarket chains create their own distribution centres replacing traditional wholesalers. These changes create new challenges in particular for small-scale and poorly educated farmers as they need to be able to understand and meet the specifications of product quality and safety standards. They also have to produce at low cost and on a regular basis a large quantity of consistent quality goods. On the other hand, farmers who manage to meet these requirements and get on the list of preferred suppliers may count on a stable customer, technical assistance and branding (Hu et al., 2004).

Foreign trade enterprises

The reform of state trading enterprises was relatively slow until the mid-1990s, but has since accelerated. The government changed the foreign trade regulatory framework, streamlined the existing STEs, allowed non-state entities to undertake foreign trade activities, and simplified registering procedures for entities wishing to become foreign trade operators.

In line with China's WTO accession process, its foreign trade regime was changed from the plan system to one based on tariffs, thus limiting the role of STEs. At the same time, China has substantially reduced import tariffs on agricultural products. The simple average tariff fell from 45% in 1992 to 20% in 2001 and then to 16% in 2004. After its accession to the WTO, imports of selected agricultural products such as maize, rice, wheat, cotton, edible oils are subject to the TRQ management. For in-quota imports, the quota is distributed between the STEs and private enterprises, with STEs continuing to control major shares for key products such as wheat, maize, rice, cotton, and sugar. The share for vegetable oils (palm, soybean and rapeseed) is relatively low.[18] Other products are subject to the regular customs duty (Chapter 2).

In the late 1990s, the government started to streamline the coverage of products under the unique responsibility of COFCO (China National Cereals, Oils and Foodstuffs Import & Export Corporation, earlier called CEREOILS), demonopolised trade in some other products and encouraged commercialisation of the COFCO network. For example, soybeans imports have been liberalised with a single tariff of 3% since 1999. For rice and maize, the Jilin Grain Group Import and Export Company (JGIEC), a provincial STE established in April 2001, took over responsibilities from COFCO for most maize and rice exports from north-eastern China. Moreover, some competition has also been introduced into the COFCO's provincial and municipal branches. Profit-based incentives have been given to managers and branch officials (Huang and Rozelle, 2004).

The reduction in the number of products controlled by single-desk state traders allowed other traders to emerge. However, any company willing to undertake foreign trade business had to apply to MOFCOM for examination and approval in terms of qualifications, available capital, business scope, etc. Nevertheless, non-state trading developed quickly as demonstrated by an increase in its share in the total agro-food exports from 42% in 2000 to 56% in 2003 (Cheng, 2004). The most important non-state enterprises are sino-foreign joint ventures (the share of 21% in 2003), fully foreign owned enterprises registered as joint ventures (15%) and private Chinese enterprises (12%).

The development of non-state trading will be enhanced by recent changes in the legislation. In the revised "Foreign Trade Law of the People's Republic of China", applied from 1 July 2004, the previous approval procedure has been replaced by a registration system. Under the new system, most limitations have been abolished and any company, enterprise or legal person can become a foreign trade operator registered by MOFCOM provided that they meet certain criteria (Chapter 2). As a result, in July 2004 in Nanjin Customs alone, the number of newly registered foreign-trade enterprises was 739. Among them 13 were state-owned, 14 collective, 250 foreign capital and 365 private Chinese enterprises.

Nevertheless, STEs remain important players in agro-food trade, especially for "strategic" agricultural commodities,[19] and the reforms undertaken so far are dominated by the devolution of COFCO's responsibilities to lower levels (i.e. provincial and municipal). The STEs share in agro-food exports was still high at 44% in 2003. COFCO is one of the largest STEs not only in Asia but also in the world. Over the past decade, COFCO imported as much as 16% of the world's traded wheat, and exported as much as 20% of the world's traded maize (Nyberg and Rozelle, 1999). In line with China's WTO accession protocol, COFCO continues to act as a key agent in the international grain trade for national and provincial grain trading companies and has preferential access to import quotas. In fact, in particular on the import side, quotas and allocations are still determined centrally with such institutions as MOFCOM (for sugar and vegetable oils), National Development and Reform Commission (NDRC; for grain and cotton) and COFCO playing a key role (Chapter 2).

1.4. The effects of economic reforms on China's agriculture

Evolution of market conditions: prices and costs

Both the economy-wide and sector-specific policies, discussed in detail in Chapter 2 of this report, contributed to a continued process of the agro-food sector's integration with the rest of the economy and with international markets. Markets have gradually emerged in rural China and have become the dominant mechanism for allocating agricultural products. According to Lardy (2001), the share of agricultural products sold at market prices was just 6% in 1978, but had risen to 40% by 1985 and then to 83% by 1999. Chinese official statistics indicate that in 2002, 94.5% of the total value of commercialised agricultural production was sold at market prices and only 5.5% at government-fixed administrative or guiding prices (*China Price Yearbook*, 2003). In fact, as grain markets and prices were liberalised in 2004, the government-fixed administrative prices have been discontinued (Chapter 2).

In most countries, growth in total productivity of all production factors (land, labour and capital) – total factor productivity (TFP) – and slow demand for agricultural products have led to declining agricultural terms of trade (ratio of prices received for outputs to prices paid for inputs) – despite various price support and subsidy programmes. In China, however, rising domestic demand for food, periodic strong rises in procurement prices fixed by the government, and price controls over input prices contributed to a substantial improvement in agricultural terms of trade between 1978 and 1995. The situation changed in the second half of the 1990s when China's commodity markets turned to surplus, input and output markets became more integrated, efficient, less exposed to government intervention and more open to both domestic and international competition. As a result, agricultural terms of trade started to decline (Figure 1.5).

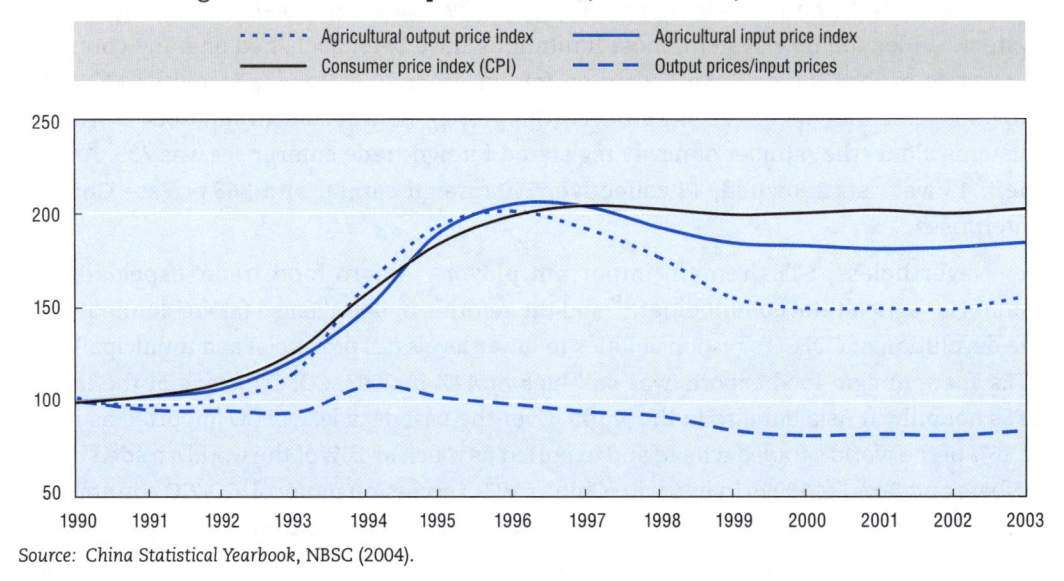

Figure 1.5. **Nominal price indices, 1990-2003, 1990 = 100**

Source: China Statistical Yearbook, NBSC (2004).

Changes in capital investment and input use

Growing demand for food and some outflow of labour from agriculture (see next section) stimulated investments and contributed to a rapid increase in purchases of agricultural inputs. The share of agriculture in total investment increased from 1.5% in 1990 to 5.6% in 2001. The use of various types of machinery such as ploughs, pumps, mini-tractors and harrows increased between two and four fold over the same period. However, the total number of even the most widely used machinery such as irrigation machinery (15.9 million) or small tractors (13.8 million) is much lower than the total number of rural households (248 million), meaning that in general Chinese agriculture remains labour intensive and the average level of technology is low (Table 1.6). This is rational within the dominant small-scale farming system and the existing endowment in factors of production, characterised by scarce capital and abundant labour resources.

Scarce land motivates farmers to maximise land productivity. Therefore, the use of variable inputs, in particular fertilisers, is very high and continues to increase. The average use of chemical fertilisers in active substance per hectare of sown area increased to 289 kg in 2003, 65% more than in 1990, including 141 kg of nitrogen (Figure 1.6). The increase in the use of potash has been particularly sharp (almost three-fold since 1990 from a low base relative to nitrogen), but still the use of this fertiliser is lower than recommended by the Ministry of Agriculture.[20]

1. THE POLICY CONTEXT

Table 1.6. **Agricultural machinery in China per 100 rural households, 1990-2003**

	1990	1995	2000	2003
Large and medium tractors	0.4	0.3	0.4	0.4
Small tractors	3.2	3.7	5.2	5.6
Irrigation machinery	3.8	4.4	6.1	6.4
Ploughs	0.1	0.2	0.4	0.4
Pumps	3.3	3.9	6.8	6.3
Combine harvests	0.0	0.0	0.1	0.1
Cereals drillings	n.a.	1.6	2.0	2.3
Harrows	n.a.	n.a.	1.5	2.0
Threshing machinery	2.2	2.6	3.6	3.6

Source: *China Agricultural Yearbook*, MOA, various editions.

Figure 1.6. **Chemical fertiliser use in China (active substance kg/ha of sown area), 1985-2003**

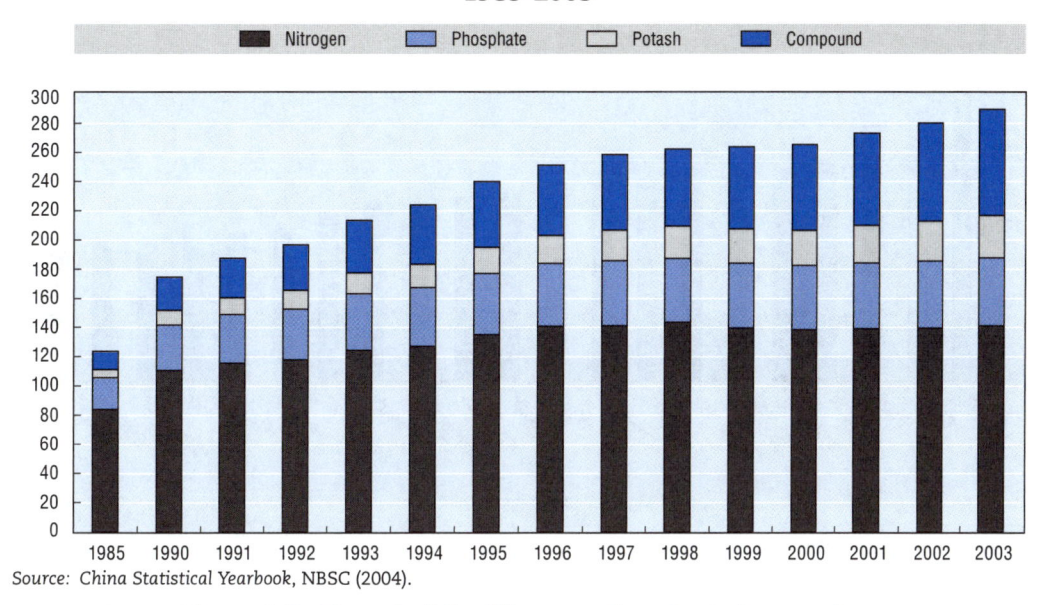

Source: *China Statistical Yearbook*, NBSC (2004).

Figure 1.7. **Chemical fertiliser use in selected countries (active substance kg/ha of sown area), 2002**

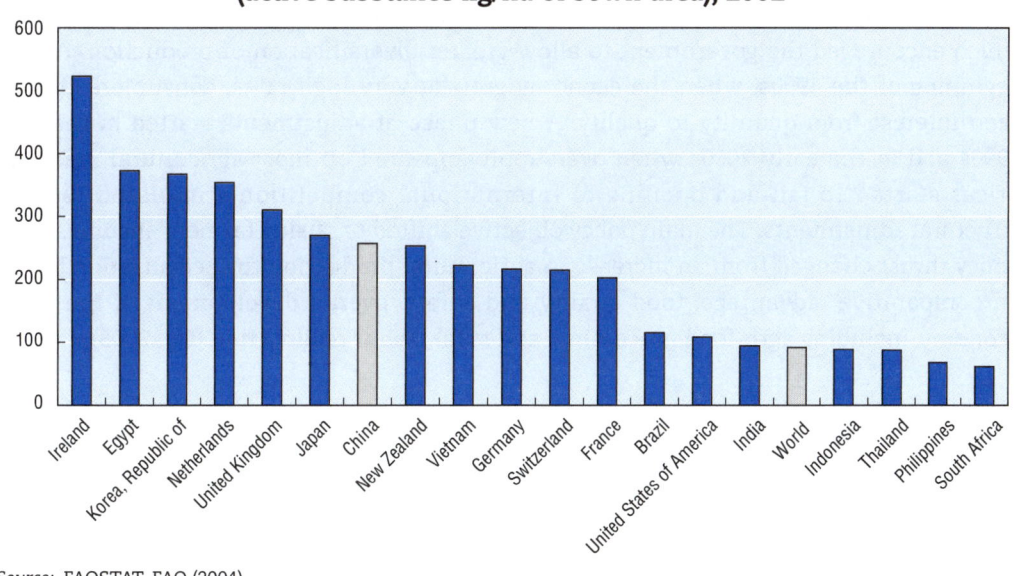

Source: FAOSTAT, FAO (2004).

China's use of fertilisers per hectare is almost three-fold higher than the global average and higher than in the majority of OECD countries (Figure 1.7). Such an intensive use of fertilisers creates a large number of environmental problems (see last section of this chapter).

Sectoral performance
Output

Volume and composition. While China's agricultural statistics still need to be improved substantially (Box 1.4), there is no doubt that during the reform period China's agriculture experienced phenomenal rates of growth. This is particularly true when compared with the performance of other countries (Figure 1.8). The HPRS boosted

Figure 1.8. **Growth in Gross Agricultural Output, 2003 (1989-1991 = 100)**

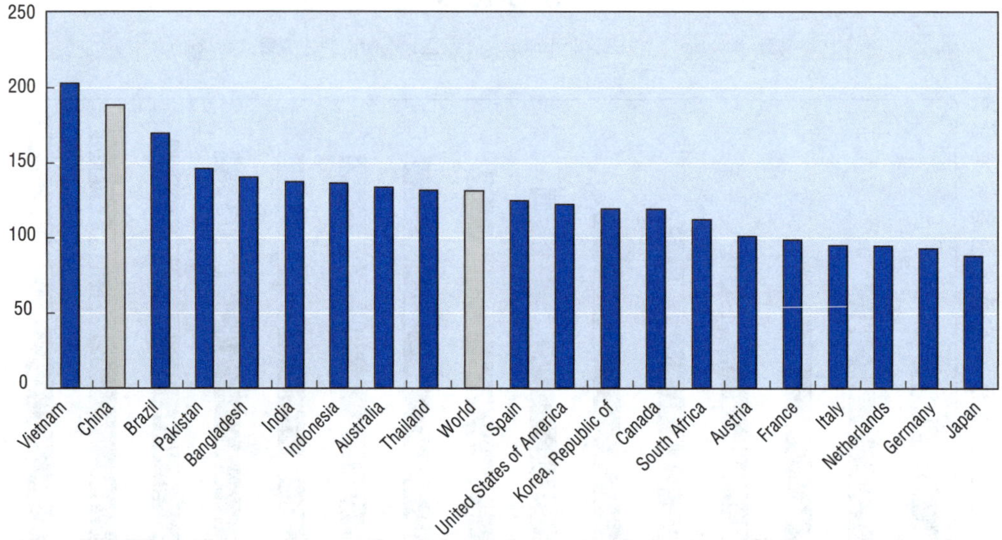

Source: FAOSTAT, FAO (2004).

production incentives, encouraged farmers to reduce costs, take risks, and enter new lines of production. By the mid-1980s, China's grain production had substantially improved which encouraged the government to allow greater diversification of production. At the beginning of the 1990s, when the economy grew at very high rates, consumers shifted their interest from quantity to quality. A new phase of adjustments started in the late 1990s and in the early 2000s when oversupply appeared on most agricultural markets, prices started to fall and opening to international competition stimulated further structural adjustments. The main policy objective shifted to raising farmers' incomes. The policy thrust changed from an increase in agricultural production to specialisation based on comparative advantage, food quality and safety, overall development of the rural economy, including agro-food processing, and rural labour reallocation (Du, 2003).

Box 1.4. **Problems with Chinese agricultural statistics**

Chinese statistics have come a long way from a pure reporting system in a centrally-planned economy to one that increasingly relies on surveys and modern statistical techniques. However, a large number of problems continue to make it difficult to assess economic processes in China with an adequate level of accuracy (Holz, 2005).

General institutional framework. Agricultural statistics suffer from the same weaknesses as China's statistical system as a whole. There is a legacy of reporting via ministries that limits the scope of the data collected, as well as the influence of the National Statistical Bureau of China (NBSC) on the conception and quality of data collection. There are also complications between the central and the sub-national level of the statistical system as sub-national statistical offices are closer to local governments than to the NBSC. Moreover, there is a rather sluggish move from enterprise reporting to survey techniques. User orientation of statistics is only beginning. Transparency of data collection methods is not sufficient and there are no clear rules concerning which data are available for free to the public and which data can be purchased from the NBSC (Holz, 2005).

Agricultural commodity data. Many agencies report their own agricultural supply, demand and price information. To some extent, this is the result of the fragmented structure of the government bureaucracy responsible for agriculture. Some ministries and agencies are concerned with marketing, others with production, and others with inputs. The sharing of statistics, information, and analysis among these agencies is relatively infrequent. NBSC and the Ministry of Agriculture have parallel reporting system and surveys.

Livestock production and consumption. Several studies have demonstrated that China's livestock supply and demand statistics are inconsistent. Supply does not match demand and the implied derived demand for feed does not harmonise with the actual supply of feed. While output data may be somewhat overestimated, consumption data are underestimated. By the end of the 1990s, China's livestock production series reached a level 2-3-fold higher than the consumption series. For example, while yearly per capita meat consumption in urban households was officially reported at 25.5 kg in 2000, various studies show that adjusted consumption could have been between 50 and 70 kg. The actual meat consumption in rural households could be also much higher than official data show. The analysis indicates that the single largest source of discrepancy is the adjustment to account for out-of-home consumption, which is not included in NBSC surveys (Tongeren and Huang, 2004; Aubert, 2004).

Grain stocks. Historically, Chinese authorities have made market information available only to privileged officials. For example, data on stocks of grain, oilseeds, cotton and other major crops are not publicly available and are considered to be a state secret (Gale, 2002a). In particular, the lack of data on grain stocks does not allow for any sound judgement on the short-term grain market prospects, let alone the medium term (see Section 3.3 in Chapter 3).

Farmland data. While the NBSC reports that China's arable land is 130 million ha, which is based on the first agricultural census conducted in 1997, the Ministry of Land Resources claims that by the end of 2003 the area had shrunk to 123.4 million ha due to the conversion of arable land to other uses. Moreover, the Ministry of Agriculture suggests that the area is 95 million ha which is based on the area covered by the second round of farmland leasing contracts and is the same as the NBSC reported until 1996. In turn, the State Administration of Taxation claims that the area is 84 million ha which corresponds to the area covered by the land tax. Mostly for tax reasons, many provinces did not

> **Box 1.4. Problems with Chinese agricultural statistics** *(cont.)*
>
> recognise census-based area expanded by 37% compared to previous estimates. Local authorities preferred to collect tax from the whole area and report tax revenues from a smaller area, thus allowing them to hide part of tax revenues. Now, the situation is reversing as direct payments for farmers (Chapter 2) are paid per unit of cultivated land and farmers claim payments for the total area they cultivate, but transfers to local governments to satisfy claims are based on the area covered by the land tax and part of farmers' claims risks to remain unsatisfied (Chang, H., 2004).
>
> *Farm structures.* While the 1997 agricultural census provided rich structural information on rural households, changes in land use patterns in recent years have not been sufficiently captured by the available data. In particular, there is evidence that with the development of supermarkets, there is a growing number of new external entrepreneurs leasing land from collectives or directly from farmers to create more viable, commercial-type of farming. However, this process is not yet reflected by the data.
>
> Until a more cohesive system of monitoring, analysing and reporting of China's agricultural sector emerges, it will continue to be difficult for analysts to obtain the data and information necessary to conduct rigorous research. In the absence of such information and analysis, it will also be difficult to assess policy programme and to make appropriate adjustments.

Between 1990 and 2003, the Gross Agricultural Output (GAO) increased by almost 90%, of which crop production 60% and livestock production 145%. However, in line with changes in demand, the rates of growth slowed from an impressive 6.2% per year in the first half of the 1990s to 3.6% per year between 2000 and 2003 (Figure 1.9).

Reflecting changes in consumers' demand, the composition of production continues to shift from crops to livestock and fish production. While crops accounted for 65% of the total value of primary production in 1990, the share fell to 50% in 2003. During the same period, the share of livestock production increased from 26% to 32%. The fishery sector's performance was even more spectacular with the share rising from 5% to 14%. The share of forestry remained stable at 4% (Table 1.7).

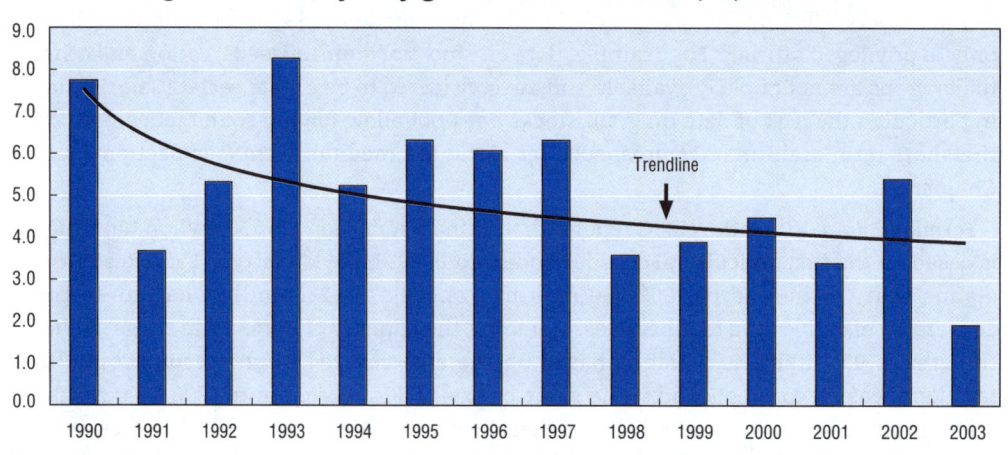

Figure 1.9. **GAO yearly growth rates in China, %, 1990-2003**

Source: FAOSTAT, FAO (2004).

Table 1.7. **Changes in the composition of the primary sector production, current prices, 1990-2003, %**

	1990	1995	2000	2003
Crops	64.7	58.4	55.7	50.1
Livestock	25.7	29.7	29.7	32.1
Forestry	4.3	3.5	3.8	4.2
Fishery	5.3	8.4	10.8	13.6
Total	100.0	100.0	100.0	100.0

Source: China Statistical Yearbook, NBSC (2004).

It is unlikely that China will sustain as high rates of agricultural production growth as over the last twenty five years of economic reforms. As consumers become richer, growth in demand for agricultural products slows down, even if accelerating urbanisation and growing demand for non-food uses and exports will provide new market opportunities. The amount of arable land will further shrink as there is a growing pressure to turn agricultural land back to forestry and to other non-agricultural uses. Moreover, China's land productivity is already very high and further intensification of land use through higher rates of chemical use risks an increase in environmental problems. Therefore, further growth in agricultural production is likely to be mostly through the wider application of new biotechnologies, the introduction of labour-augmenting technologies and innovation, and further restructuring to higher value, but less land intensive products such as fruits, vegetables and livestock.

Crops. While cereals remain the key crop in China, their share in total crop production and in area sown declined substantially between 1990 and 2003 as other crops had become more profitable and the government relaxed most of the policy measures which in the past forced farmers to produce grains. Impressive increases in vegetable and fruit production compared to performance of other crops (Figure 1.10) reflects adjustments in land use in response to changes in domestic demand but also to emerging export opportunities for selected products such as garlic, onions, apples, and pears (Chapter 2).

Figure 1.10. **Output indices for main crops, 1990-2003, 1990 = 100**

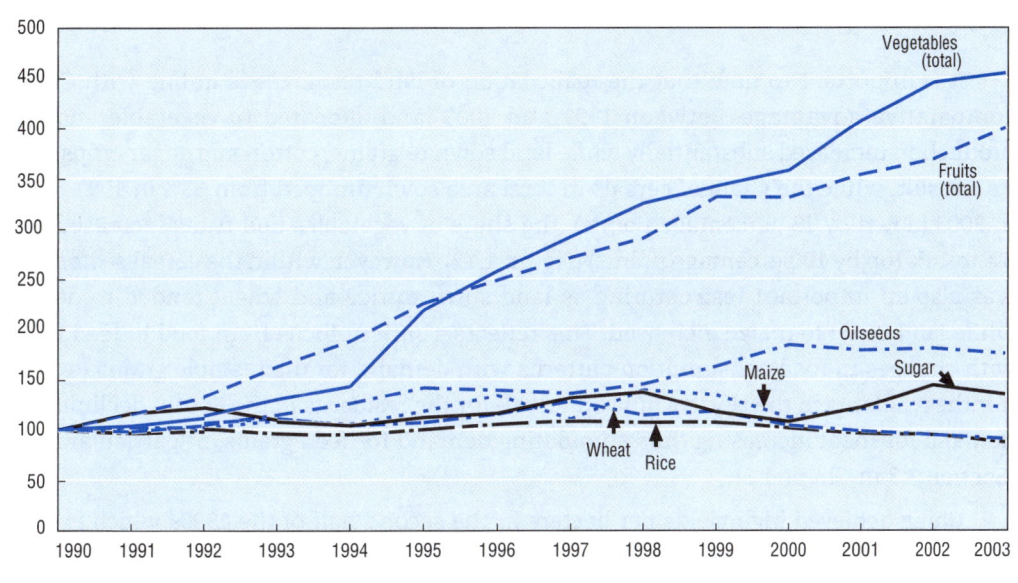

Note: Estimates for vegetables 1990-1993.
Sources: China Statistical Yearbook, NBSC (2004); China Rural Statistical Yearbook, NBSC, various editions.

1. THE POLICY CONTEXT

Cereal production (with rice in terms of paddy) at 374 million tonnes in 2003 was below the 1990 level, but such comparison hides quite important fluctuations in grain production in the 1990s. Due to government support policy, such as the "Governor's Grain-Bag Responsibility System" (GGBRS) implemented in 1995 and an increase in government procurement prices in 1994 and 1996, cereal output increased to 451 million tonnes in 1996 and remained at this level until 1999. This production substantially exceeded domestic consumption, and as domestic grain prices were in most cases higher than international prices and grain exports were still under strict government control, huge grain stocks were accumulated. Consequently, domestic grain prices fell (Figure 1.11). As a result, the government relaxed a number of policies forcing farmers to produce grains. For example the grain quota system was progressively discontinued between 1999 and 2002 (Chapter 2), thus allowing farmers to switch resources to other more profitable crops. Consequently, both area sown to cereals and the level of cereal production fell between 2000 and 2003.

Figure 1.11. **Total cereal production and nominal farm gate prices, 1993-2003**

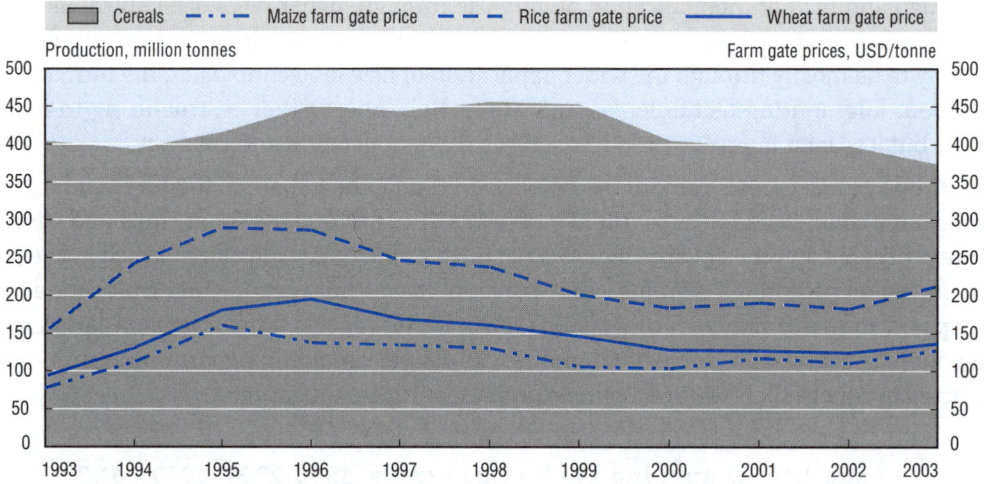

Sources: China Statistical Yearbook, NBSC (2004); OECD China PSE/CSE databases.

It is important to note that the reallocation of land resources is in line with China's comparative advantage. Between 1991 and 2003 land allocated to vegetable and fruit production increased substantially while land sown to grains, cotton and sugar crops fell.[21] As a result, while the share of cereals in total area sown dropped from 63% in 1991 to 50% in 2003 (down by 13 percentage points), the share of vegetables and fruits increased from 8% to 18% (up by 10 percentage points) (Figure 1.12). However, within the cereal sector there was also an important restructuring as land sown to rice and wheat tended to decline, while land sown to maize increased. This reflects a shift in focus from food to feed in line with changes in food consumption patterns with demand for main staple grains (wheat – for the residents in the North China and rice – for the residents in the South) declining and demand for meat increasing thus stimulating demand for feed grains, in particular maize (Section 3.3 in Chapter 3).

China achieved high yields per hectare in the second half of the 1990s which is partly linked with high rates of fertiliser use further stimulated by a strong increase in procurement prices for grains between 1994 and 1996 (Figure 1.11), wider application of scientific and technological progress and relatively favourable climatic conditions (Chen, 2002). For most

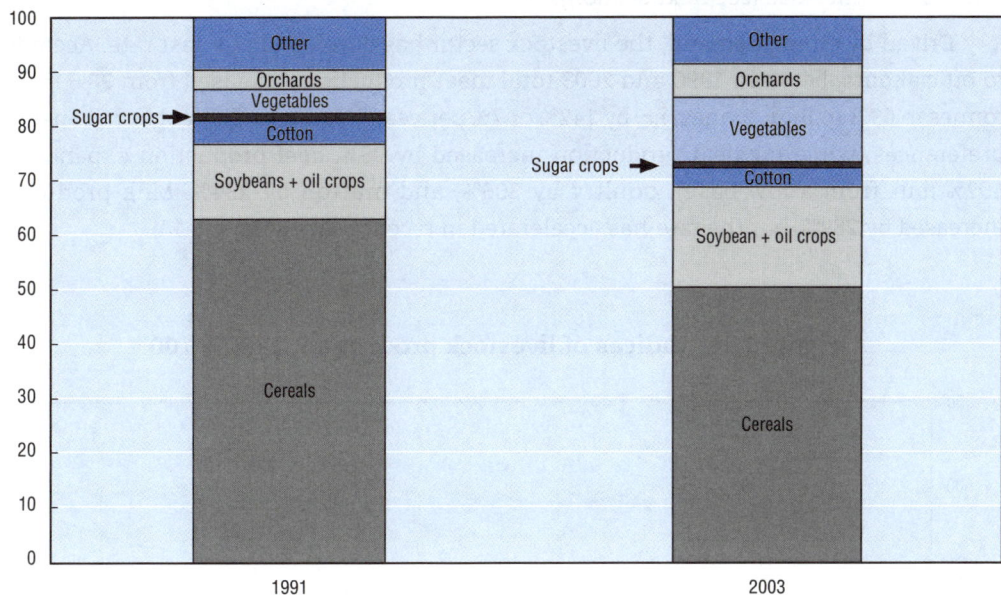

Figure 1.12. **Composition of the sown area, 1991 and 2003, %**

Source: China Statistical Yearbook, NBSC (2004).

crops, yields stabilised at the end of the 1990s or even fell, partly due to less favourable weather conditions (Figure 1.13).

Livestock. China is the number one livestock producer in the world. In 2002, its year-end stock of pigs accounted for 49% of the world's total, cattle 32%, sheep 14% and goats 23%. Total meat production accounted for 23% of the global output.

Pork is the largest component of China's diet of animal protein (50%), but poultrymeat, fish and eggs are growing rapidly in importance, accounting for approximately 12%, 16%, and 17%, respectively (Hsu, 1998). As consumer incomes rise, diets diversify. Accordingly,

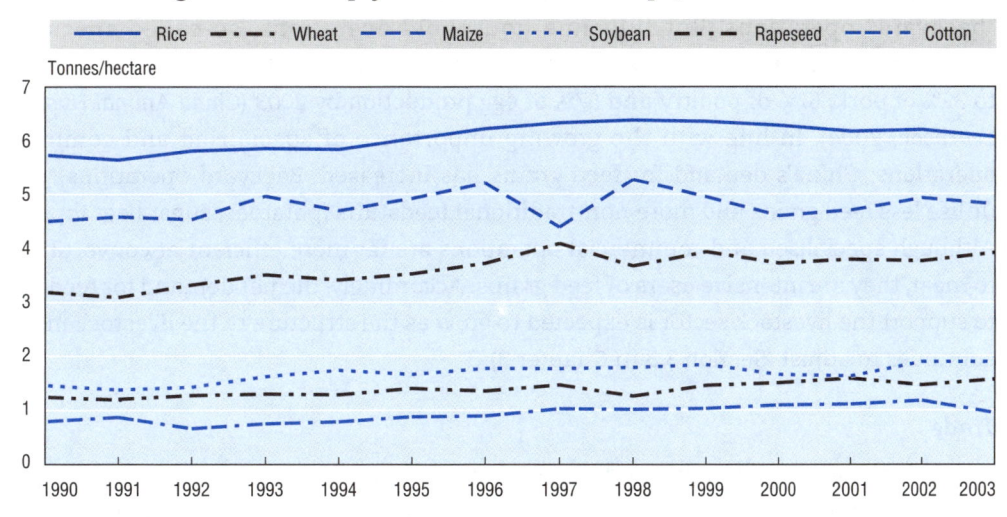

Figure 1.13. **Crop yields for selected crops, 1990-2003**

Source: China Statistical Yearbook, NBSC (2004).

the share of pork in the Chinese diet declines at the same time as total meat consumption continues to increase (see next section).

Driven by strong demand, the livestock sector has developed at a fast rate. According to official data, between 1990 and 2003 total meat production increased from 28.6 million tonnes to 69.3 million tonnes, *i.e.* by 142% or 7% per year.[22] Reflecting changes in consumer preferences, while pigmeat production increased by 98%, beef production expanded by 402% (but from a low base), poultry by 306%, and mutton by 234%. Milk production increased by 289%, but the rate has accelerated in recent years (Figure 1.14).

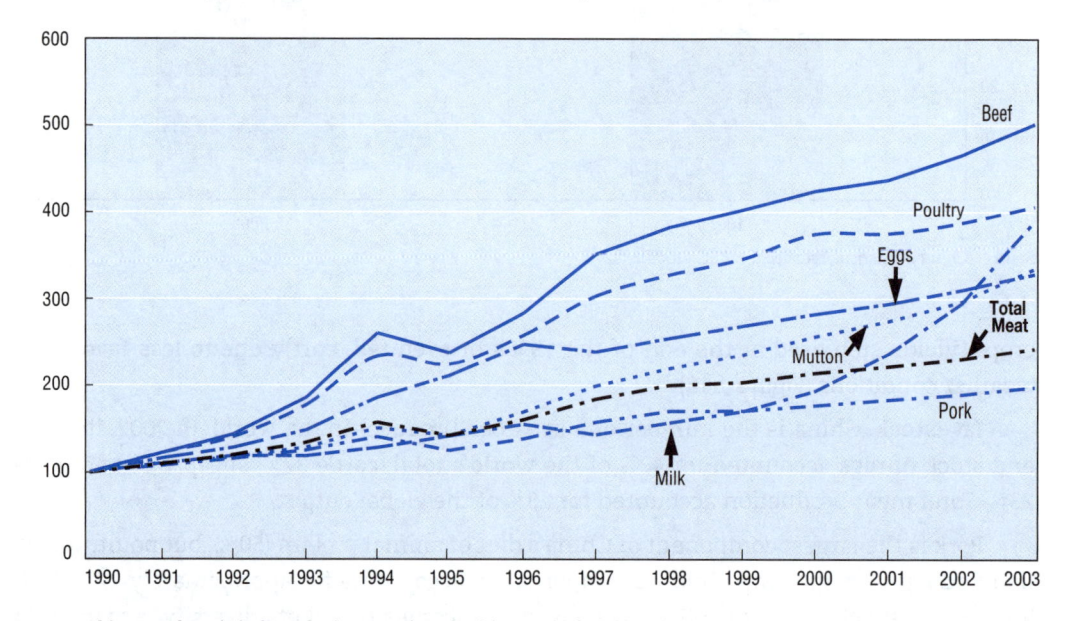

Figure 1.14. **Indices of livestock production, 1990 = 100**

Sources: China Rural Statistical Yearbook, NBSC (2004); *China Agricultural Yearbook*, MOA, various editions.

Traditionally, the vast majority of meat in China was produced by small, part-time "backyard" operations. But full-time household operations, so-called specialised households, and commercial operations have grown rapidly, and their share had increased to 32% of pork, 67% of poultry and 57% of egg production by 2003 (*China Animal Husbandry Yearbook*, 2004). In line with the growing importance of specialised and commercial operations, China's demand for feed grains has increased. Backyard operations tend to utilise less feed grains and more non-traditional feedstuffs (potatoes, sugar beet tops, etc.). Although specialised and commercial operations are far more efficient at converting feed to meat, they are intensive users of feed grains. Accordingly, the net demand for feed grains to support the livestock sector is expected to grow as the structure of the livestock industry continues to adjust (Section 3.3 in Chapter 3).

Trade

China's agro-food imports and exports have developed in line with the overall value of agricultural production but slower than the whole economy. Therefore, while the ratio of China's agro-food trade to the Agricultural Gross Value Added (AGVA) remained relatively stable, the share of agro-food products in the total value of exports and imports more than halved between 1992 and 2003. Despite growing exposure to international competition and a remarkable economic growth stimulating domestic demand for agro-food products, China's agro-food sector (including fishery) remained a net export earner up to 2003 (Table 1.8; see also Section 2.3 in Chapter 2).

Despite trade liberalisation, China's agro-food sector is still less linked with international markets than other sectors of China's economy and less than in many other countries. For example, while the sum of agro-food exports and imports to the AGVA is about 19%, the ratio of total exports and imports to China's GDP is much higher at 55%. Moreover, the share of China's agro-food exports to the AGVA is about 10% compared to 34% for the EU-15 (intra EU trade excluded) and 77% for the USA (WB/OECD, 2004). A comprehensive analysis of China's trade performance is provided in Section 2.3 in Chapter 2.

Table 1.8. **China's agricultural trade, 1992-2003**

		1992	1993	1994	1995	1996	1997	1998	1999	2000	2001	2002	2003
Agricultural Gross Value Added (AGVA; including forestry and fishery), current prices	USD billion	105.3	119.5	109.7	143.6	166.6	171.4	175.7	174.8	176.7	186.1	194.6	211.5
Agro-food exports (including fishery)	USD billion	11.3	11.4	14.3	14.4	14.3	15.0	13.8	13.5	15.6	16.1	18.1	21.4
Agro-food imports (including fishery)	USD billion	5.3	4.0	7.1	12.2	10.8	10.0	8.3	8.2	11.2	11.8	12.4	18.9
Balance	USD billion	6.0	7.5	7.2	2.2	3.4	5.0	5.5	5.3	4.4	4.3	5.7	2.5
The coverage degree of imports by exports	%	214.0	289.6	200.3	118.2	131.7	150.3	166.1	164.9	139.3	136.3	145.8	113.2
Share of agro-food trade in total trade:													
Exports	%	13.3	12.5	11.8	9.7	9.4	8.2	7.5	6.9	6.3	6.1	5.6	4.9
Imports	%	6.6	3.8	6.2	9.2	7.8	7.0	5.9	5.0	5.0	4.9	4.2	4.6
Ratio of agro-food exports to AGVA	%	10.8	9.6	13.0	10.0	8.6	8.7	7.9	7.7	8.8	8.7	9.3	10.1
Ratio of agro-food imports to AGVA	%	5.0	3.3	6.5	8.5	6.5	5.8	4.7	4.7	6.3	6.4	6.4	8.9
Ratio of agro-food exports and imports to AGVA	%	15.8	12.9	19.5	18.5	15.1	14.5	12.6	12.4	15.2	15.0	15.7	19.0
Ratio of net agro-food imports to AGVA	%	5.7	6.3	6.5	1.5	2.1	2.9	3.1	3.0	2.5	2.3	2.9	1.2

Sources: OECD Secretariat on the basis of *China Customs Administration Statistics* (2004); *China Agricultural Development Report*, MOA (2004); *China Statistical Yearbook*, NSBC (2004).

Employment

Total employment in agriculture increased until the beginning of the 1990s when it reached almost 350 million before dropping to below 330 million in the mid-1990s.[23] Then it increased again to 335 million in 1999, mostly due to the overall cooling of the Chinese economy and to a fall in employment in township and village enterprises (TVEs), until the mid-1990s the major employer of redundant farm labour (see Annex A). Stronger overall economic growth and somewhat better performance of TVEs since 2000 allowed a fall in agricultural employment to 313 million in 2003.[24] As employment in other sectors of the Chinese economy increased at high rates, the share of agricultural labour in the total declined from 71% at the end of the 1970s to 42% in 2003. Agriculture's share in rural employment was driven by the performance of TVEs and fell from above 90% at the end of the 1970s to around 65% in the mid-1990s and since then has remained almost unchanged (Figure 1.15).

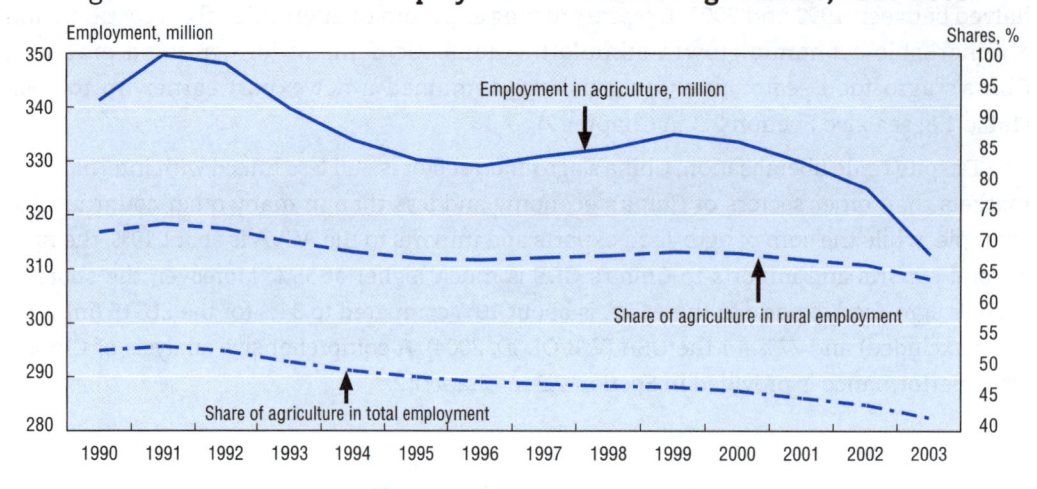

Figure 1.15. **Evolution of employment in Chinese agriculture, 1990-2003**

Sources: *China Statistical Yearbook*, NBSC, various editions; *China Labour Statistical Yearbook*, NBSC, various editions.

As discussed above, a large part of the current agricultural workforce is not productively employed. The high level of excess labour resulted in a high ratio of labour per unit of land, low labour productivity in agriculture, low agricultural incomes, and hidden unemployment in rural areas and has become a driving force behind the ongoing rural workforce migration to urban areas. The income gap between rural and urban households is wide and growing (see below). Therefore, the greatest challenge for the Chinese authorities will be providing alternative employment opportunities for redundant farm labour (Annex A).

Productivity

During the reform period, agriculture enjoyed tremendous productivity gains. It is estimated that the average rate of total factor productivity (TFP) growth was 6.6% between 1978 and 1996 (Carter *et al.*, 2003). While the rate was 8.4% between 1978-1987 it slowed to 4.9% between 1987 and 1996, mostly due to the slower, but still impressive, growth in the gross value of agricultural production despite a stronger growth in the use of factors of production, in particular current inputs.[25]

As employment in agriculture was higher in 2003 than in 1978, gains in agricultural labour productivity were below the rate in agricultural production growth. Therefore, while productivity growth allowed for a substantial increase in farmers' real incomes it was not sufficient to close the gap with other sectors of the economy enjoying similar or even higher progress in labour productivity. Persistent differences in labour productivity are, in turn, the main reasons for a large income gap between agriculture and the rest of the economy (see next section).

Table 1.9 provides a comparison of labour productivity in the primary sector (mostly agriculture) with that in the secondary and tertiary sectors. It shows that by the mid-1990s the gap contracted to around a four-fold difference, but it expanded again in the second half of the 1990s, and in 2001 was close to or even larger (above seven-fold *vis-à-vis* secondary sector) than at the beginning of the reform process.[26]

The above analysis confirms that to close the labour productivity gap between agriculture and other sectors of the economy, labour productivity in agriculture must

Table 1.9. **Relative labour productivity by sector, 1978-2001**

	Relative labour productivity (%)			Primary sector = 1		
	Primary	Secondary	Tertiary	Primary	Secondary	Tertiary
1978	39.9	278.6	194.3	1.0	7.0	4.9
1980	43.8	266.5	163.4	1.0	6.1	3.7
1985	45.5	207.2	169.6	1.0	4.6	3.7
1990	45.1	194.4	169.2	1.0	4.3	3.8
1995	39.3	212.2	123.8	1.0	5.4	3.2
2000	32.8	223.1	121.5	1.0	6.8	3.7
2001	30.4	229.1	121.3	1.0	7.5	4.0

Note: The relative labour productivity is calculated as the ratio of the share of GDP of the sector/share of employment of that sector multiplied by 100.
Source: CSLS (2003).

increase at much higher rates than the volume of agricultural production. Moreover, it is unlikely that China will sustain as high rates of agricultural production growth as were obtained over the first two and a half decades of the reform period. Therefore, the prospects for substantial gains in agricultural labour productivity reside in the outflow of labour from agriculture to other sectors of the economy. This will be the only way to secure higher incomes for those depending on agriculture.

Rural incomes in China

The evolution of rural incomes. One of the remarkable achievements of the reform period has been the strong growth of real per capita rural incomes, which rose more than three-fold between 1980 and 2000, representing an annual rate of about 6%. The rate of increase, however, has not been the same across time or among all regions in China. In the first half of the 1980s the annual rate was 14%, but in the late 1980s it slowed to below 3% on average. Subsequently, income growth rates accelerated to above 10% in the mid-1990s, but have since slowed again to 2-3% at the end of the 1990s (Figure 1.16).

Figure 1.16. **Rural household income per person, 1981-2004**

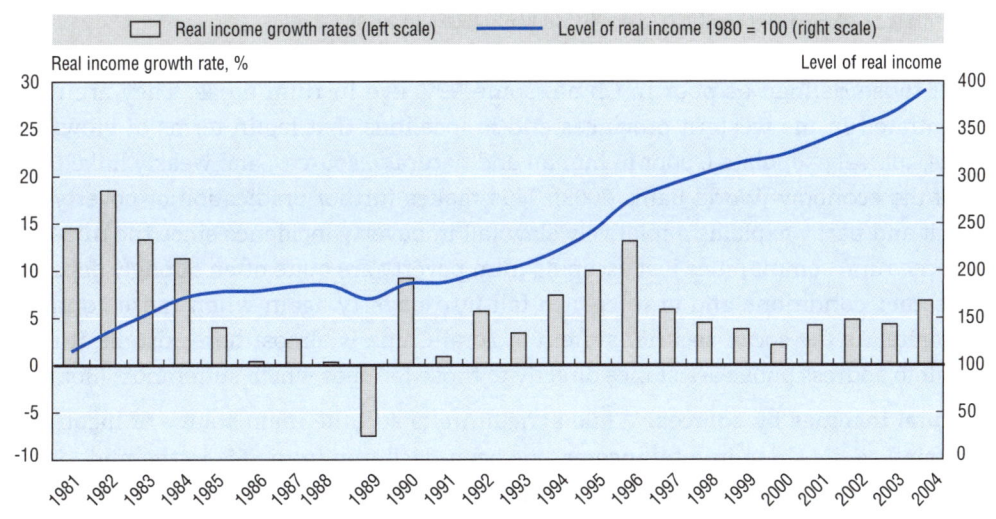

Note: Rural CPI used as deflator.
Sources: Lu and Wang (2001); *China Statistical Yearbook*, NBSC, various editions.

The income distribution across rural areas in different parts of China depends increasingly on the availability of non-agricultural job opportunities in rural areas. As the availability of jobs is highly uneven, the variation in per capita income by province is much greater for rural than for urban areas (OECD, 2002b). This is reflected in the Gini coefficient for rural incomes, which rose from 0.24 in the early 1980s to 0.37 in 2003 (Rural Survey Organisation of NBSC, 2004).

No matter how uneven the distribution of rural income may be, the overall rise of incomes in rural areas, including those in poor areas, has led to a dramatic fall in the incidence of poverty in China since the beginning of reforms in 1978. Using China's standard of defining poverty (income below USD 0.2 per person per day at current exchange rate; below USD 0.6-0.7 at PPP), the number of people in poverty fell from 250 million in 1978 to 29 million in 2003 and the proportion of the rural population affected by poverty fell from 31% to 3% over the same period. However, China's poverty line is lower than the line applied by the World Bank to measure poverty in other countries (income USD 1 per day at PPP). According to World Bank estimates, the number of people living in poverty in rural China was still high at 88 million in 2002, but the progress in eliminating poverty is still remarkable as about 400 million people rose above the poverty line between 1979 and 2002 (Table 1.10).

Table 1.10. **Rural poverty in China, 1978-2003**

	Chinese definition of absolute poverty line			World Bank definition of absolute poverty line		
	Poverty line	Incidence	Number affected	USD/day income	Incidence	Number affected
	(CNY per year, current prices)	(%)	(million)		(%)	(million)
1978	100.0	30.7	250.0	1.0	62.0e (1980)	490.0 (1979)
1985	206.0	14.8	125.0	1.0	n.a.	n.a.
1990	300.0	9.4	85.0	1.0	31.3	280.0
1995	530.0	7.1	65.0	1.0	21.8	200.0
2000	625.0	3.4	32.0	1.0	13.7	111.0
2002	627.0	3.0	28.0	1.0	11.3e	88
2003	637.0	3.1	29.0	1.0	n.a.	n.a.

e: estimated by the OECD Secretariat on the basis of World Bank data.
Sources: *China Development Report*, NBSC (2004); World Bank (2000 and 2003).

Of those defined as poor in China, some 99% live in rural areas. They are mostly concentrated in the western provinces and in localities that lag in terms of growth, are remote, sparsely populated, poor in human and natural resources, and weakly linked to the rest of the economy (World Bank, 2003). This makes further eradication of poverty more difficult and partly explains a relatively slow fall in poverty incidence since the mid-1990s. Moreover, rural communities just escaping from poverty are quite often strongly dependent on weather conditions and are likely to fall into poverty again when these conditions deteriorate. As the social security system in rural China is almost non-existent, it is very difficult to address public assistance directly to those families which suffer most (Box 1.5).

Rural incomes by sources. While agriculture is still the main source of incomes for rural families, its share in total income has been declining from 66% in the mid-1980s to 46% in 2003. In contrast, the share of wages, mostly from rural industry jobs, and of incomes from non-agricultural household businesses, predominantly in the service sector, have been increasing during the reform period (Table 1.11).

> **Box 1.5. Social security in rural China**
>
> In rural China, land and family are the traditional pillars of social security. For example, according to a household survey, 92% of the rural elderly population is supported by families and only 8% by rural pension schemes (Cai and Zhang, 2000). In urban areas the state operated social insurance system is much more active, covering some 40% of the labour force, including civil servants (OECD, 2002b).
>
> The rural co-operative medical system has largely been dismantled and replaced by a paid system. As a result, about 10% of the rural population has access to subsidised medical care system compared to 90% in 1978.
>
> A programme targeting poor farmers and providing compensation for natural disasters, *i.e.* crop failures, is the most important rural social programme in terms of beneficiary numbers. It covered 16.6 million persons in 1999 and 16.8 million in 2002 (*China Financial Yearbook*, 2003). Another programme called "five guarantees" (food, clothing, housing, funeral and education) covered 3 million persons in 1999, but only 1.6 million in 2002 and is essentially financed locally.
>
> There are several arguments to suggest that the family-based social security system in rural China will not be sustainable in the long term. First, as rural households become more and more dependent on off-farm incomes, access to land plots is becoming less satisfactory as a substitute for social insurance. Second, declining birth rates diminish the likelihood that children will be able to take care of three or four generations, including elderly family members. Third, the spread of urban value systems can have an impact on traditional family responsibilities. Since social policy reforms in rural areas could potentially be very costly and are rather unlikely to be on the government's immediate policy agenda, there is a need to consider possible ways for more equitable distribution of access to various social policy schemes for the rural population, in particular to ensure acceptable old-age incomes for the rural population.
>
> *Source:* OECD (2002b).

Table 1.11. **Rural household incomes by source, 1985-2003**

	Net income[1] CNY/person	Per cent distribution				
		Agriculture[2]	Household non-agricultural business	Wages	Transfers and property income	Total
1985	398	66.4	8.1	18.1	7.4	100.0
1990	686	66.4	9.1	20.2	4.2	100.0
1995	1 578	60.6	10.7	22.4	6.2	100.0
1998	2 162	52.6	15.3	26.5	5.7	100.0
1999	2 210	51.5	14.0	28.5	6.0	100.0
2000	2 253	48.4	14.9	31.2	5.5	100.0
2001	2 366	47.6	14.1	32.6	5.7	100.0
2002	2 476	45.8	14.2	33.9	6.0	100.0
2003	2 622	45.6	13.2	35.0	6.2	100.0

1. All incomes, monetary and in-kind; current prices.
2. Net income from household based farming and related activities, including crop production, animal husbandry, forestry and fishing.

Source: China Statistical Yearbook, NBSC (2000 and 2004).

The level and the share of non-agricultural incomes in total net incomes in peasant families is strongly diversified across provinces, in particular between those located in western and eastern China. In Guizhou, the poorest province, wages (including the remittances of the migrants) amounted only to CNY 460 out of the total net yearly income of CNY 1 565 per person in 2003. In Zhejiang, the richest province, wages amounted to CNY 2 575 out of the total of CNY 5 390 (Figure 1.17).

Figure 1.17. **Wages and net incomes per person in peasant families across provinces, CNY, 2003**

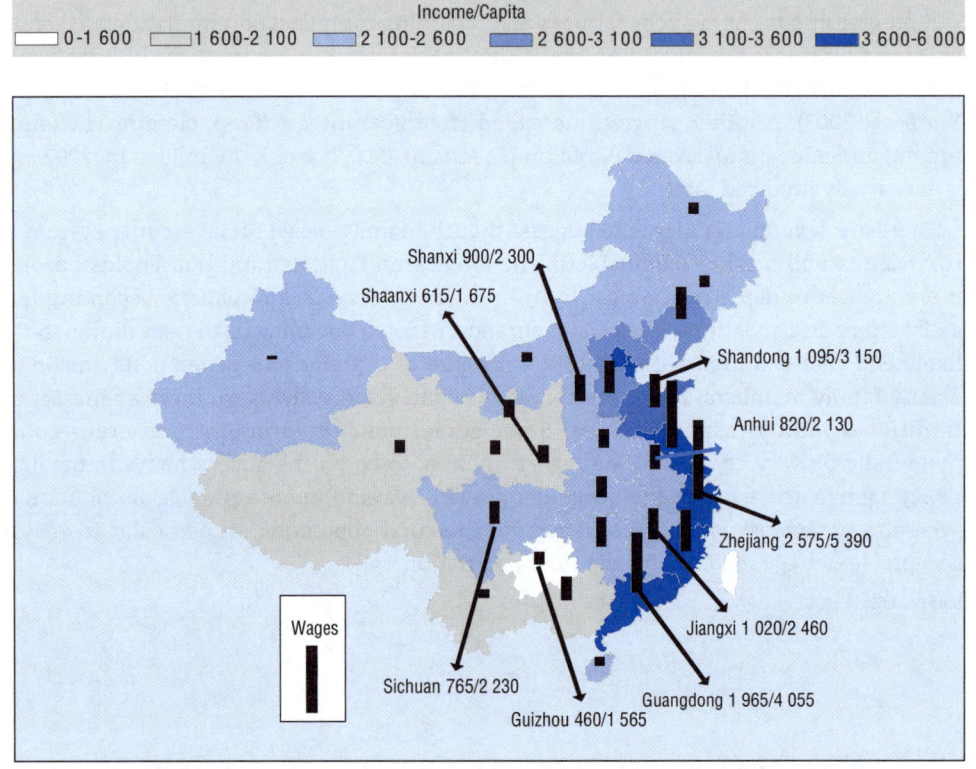

Income/Capita
0-1 600 | 1 600-2 100 | 2 100-2 600 | 2 600-3 100 | 3 100-3 600 | 3 600-6 000

Shanxi 900/2 300
Shaanxi 615/1 675
Shandong 1 095/3 150
Anhui 820/2 130
Zhejiang 2 575/5 390
Jiangxi 1 020/2 460
Sichuan 765/2 230
Guizhou 460/1 565
Guangdong 1 965/4 055

Wages

Notes: Wages include remittances from migrants. Very high wages in the municipalities of Beijing, Tianjin, and Shanghai are not displayed. Chongqing is included in Sichuan.
Sources: China Yearbook of Rural Household Survey, NBSC (2004); Aubert (2005).

Rural-urban income gap. One of the most striking features of China's development in the reform period has been a large and, in recent years, growing income disparity between the rural and urban populations, nationally, regionally and locally. Notwithstanding some methodological difficulties in measuring this disparity,[27] all available indicators confirm an income gap between rural and urban areas, which is of growing concern for the government. Some of these indicators are provided in Figure 1.18.

While the overall level of inequality is not particularly high (Section 1.1 and OECD, 2004b), China's rural-urban income gap is one of the largest in the world. Yang and Zhou (1996) demonstrated, on the basis of urban-rural income ratios for 36 countries over the 1985-1995 decade, that urban incomes are rarely more than twice rural incomes. In China, the urban-rural per capita income ratio fell to 1.85 in the mid-1980s, but since then has almost constantly been growing with the exception of the mid-1990s, when an increase in grain prices and an expansion of TVEs accelerated growth in rural incomes.

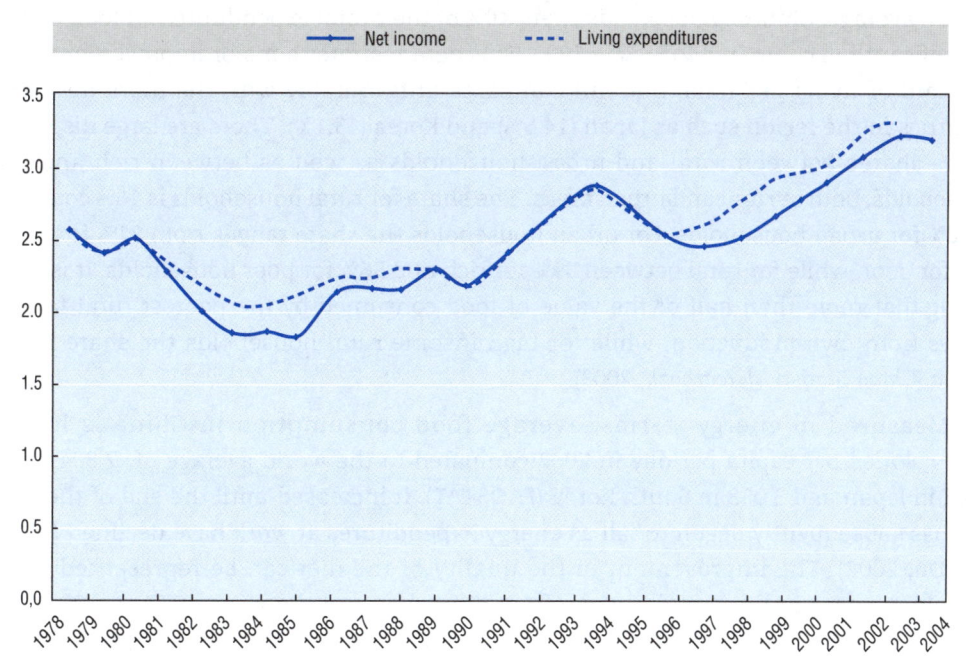

Figure 1.18. **Urban to rural per capita income and living expenditures ratios, 1978-2004**

Source: China Statistical Yearbook, NBSC, various editions.

Since the mid-1990s, growing competitive pressures and financial difficulties of TVEs as well as falling purchasing power of agricultural products, in particular grains, contributed to a further widening of the income gap. The ratio was 3.2 in 2003 and 2004, the highest over the whole reform period (Figure 1.18).

The large rural-urban income disparity stems from a considerable gap in labour productivity between agriculture and other sectors of the economy and, more generally, between rural and urban areas. This is largely due to constrained factor mobility, especially of labour and capital, but also to differences in access to education, health care and other social services. Although housing and enterprise reforms are mitigating urban population's advantages, as urban workers now pay higher rents and contribute more to their pension and medical benefits, the gap in access to social services remains large (Box 1.5).

As mentioned above, rural-urban income disparities have tended to worsen during the reform period despite rapid growth in rural industrial employment, output and wages, and despite the substantial direct and indirect contributions that rural industries have made to the rural economy. Unquestionably, rapid rural industry expansion until the mid-1990s (Annex A) prevented the income ratio from being even more unfavourable, but it was unable to reverse the worsening trend (Rozelle, 2001; Kwiecinski and Li, 2002).

Food consumption[28]

Fuelled by growing real incomes, urbanisation, changes in lifestyles, development of new marketing channels, and expansion of mass media and communication, food consumption patterns in China during the last two decades have gone through significant changes both in terms of quantity and structure (Table 1.12).

On average, Chinese households spent 40% of their total expenditures on food in 2002, a significant fall compared to 55% in 1990. This demonstrates a major improvement in the standard of living in China, but still compares unfavourably with the more developed countries in the region such as Japan (14.5%) and Korea (15.1%). There are large disparities in the shares between rural and urban households as well as between rich and poor households, both in rural and urban areas. The share for rural households is 46% compared to 38% for urban households. For urban households the share ranges from 31% for rich to 47% for poor, while for rural between 39% for rich and 56% for poor households. It is worth noting that more than half of the value of food consumed by the poorest rural families comes from own production, while for high income rural households the share is one-fourth (*China Statistical Yearbook*, 2003).

Measured in energy terms, average food consumption in China is high at 2 951 calories per capita per day in 2002 compared to the world average of 2 804 and to 2 761 in Japan and 3 058 in South Korea (FAOSTAT). It increased until the end of the 1980s and has subsequently begun to fall as energy expenditures at work have declined (Popkin and Du, 2002). The improvement in the quality of the diet can be represented by the 4.4-fold increase in the protein intake from animal sources between 1977 and 2001. The level of protein consumption per capita is higher than the world average and just below the levels in Japan and South Korea (FAOSTAT).

Even if on average food consumption in China is high, the problem of undernutrition still exists. But, according to the FAO estimates, the number of undernourished people fell from 193 million at the beginning of the 1990s to 135 million in 1999-2001 and the incidence from almost 17% to below 11% over the same period. The still existing undernutrition results from the remaining niches of poverty rather than from a physical lack of food.

Dietary changes affected both urban and rural households even if they still demonstrate very different food consumption patterns. Urban households consume on average almost 50% less grains (in comparable terms), but much more meat and milk products as well as fruits. Fresh vegetable consumption is roughly at the same level. However, China's official statistics do not accurately reflect changes in food consumption patterns. For example, meat consumption in urban households is reported to remain almost unchanged between 1990

Table 1.12. **Food consumption in China, 1990-2002 (kg/person/year)**[1]

	1990	1991	1992	1993	1994	1995	1996	1997	1998	1999	2000	2001	2002
Cereals – excluding beer	210.0	206.2	206.9	204.9	198.7	196.4	199.9	197.6	194.0	190.4	184.2	177.3	168.2
Vegetables	99.0	99.6	110.3	128.4	139.9	148.9	165.1	175.2	179.9	198.4	226.9	242.2	257.1
Fruits – excluding wine	14.6	16.6	18.1	22.2	25.4	30.1	33.3	36.0	37.0	42.2	41.5	44.4	45.7
Oil crops	5.2	4.6	4.9	6.5	7.4	6.8	6.7	7.7	7.5	6.6	7.5	7.3	6.5
Sugar, raw equivalent	7.0	7.4	6.6	4.8	5.4	6.3	7.2	8.1	7.9	6.0	5.5	5.8	6.1
Meat	24.8	26.8	29.4	32.4	35.5	38.1	37.5	42.7	46.1	47.0	49.1	50.1	51.7
Milk – excluding butter	5.2	5.5	5.7	5.6	6.2	6.6	7.1	6.9	7.2	7.8	8.7	10.2	12.5
Eggs	6.3	7.3	8.0	9.1	11.3	12.7	14.8	14.1	14.9	15.6	16.2	16.8	17.5

1. The figures in the table are derived from the FAO commodity balances and reflect the gross availability of food products per capita and do not necessarily indicate the actual amounts of food consumed by individuals.

Source: FAOSTAT, FAO (2004).

and 2001 and only in the last two years data show an increase by about one-third (*China Statistical Yearbook*, 2003). The main reason for the underestimation is the out-of-home food consumption which is not captured by statistics (Box 1.4).

China Health and Nutrition Survey provides more detailed information on dietary changes and confirms a shift away from carbohydrates to fat in the diet in all groups of the population and in all income groups (Table 1.13). In fact, China has moved to the stage where the prevalence of obesity is greater than that of undernutrition, and the diet based on fat creates more health problems than does undernutrition and infectious diseases (Popkin and Du, 2002).

Table 1.13. **Shifts in energy sources in the Chinese diet, ages 20-45**

	Energy from fat (%)			Energy from carbohydrates (%)		
	1989	1993	1997	1989	1993	1997
Urban	21.4	32.0	32.8	65.8	55.0	53.3
Rural	18.2	22.7	25.4	70.0	65.2	62.1
Low income	16.0	19.7	23.0	72.9	68.6	64.5
Mid-income	20.3	25.5	27.1	67.5	62.2	60.3
High income	21.5	31.5	31.6	65.4	55.4	54.8
Total	19.3	25.5	27.3	68.7	62.1	59.8

Source: Popkin and Du (2002).

The consequences of these long-term shifts in diet and energy requirements is giving rise to a double burden whereby obesity and undernutrition increasingly coexist, even within the same households. According to recent research, 8% of households have one member who is underweight and one member who is overweight. Among households that include at least one underweight member, 23% have at least one member who is overweight.[29] Typically, an underweight child lives with an overweight non-elderly adult member. While intra-household distribution of calories may contribute to such pattern, the higher propensity of stunted children of becoming overweight later in their development has also been documented. Obesity in China is still comparatively low, with a prevalence of 2.5% and 2.4% among males and females respectively. Similarly, just 12% and 13.8% of males and females respectively are overweight. On the other hand, obesity and overweight trends are quite high by international standards: prevalence in obesity and overweight among adult males is increasing respectively at 0.3 and 0.8 per cent annually, while among adult females the figures are 0.2 and 0.4 respectively (Popkin, 2004).

The implications of the nutritional transition will become increasingly relevant in order to understand the future of China's agricultural sector. First, changes in dietary patterns will have an impact on the composition of the domestic demand for food. As almost 90% of China's production is sold domestically, long term shifts in food consumption will continue to influence the production structure of agriculture. Second, as in other countries that have undergone the nutrition transition (*e.g.* Brazil), China's epidemiological profile will also be affected by the nutrition transition as a diet richer in saturated fats and sugars will lead to a higher incidence of dietary-related non communicable diseases (DR-NCD) at the expense of communicable diseases. As a result, it is estimated that the economic costs of undernutrition will be overcome within the next two decades by the economic costs related to obesity (Popkin, 2004). In particular, public health costs will increase substantially as a result of the growth in DR-NCD. Labour productivity will also be affected as obesity and overweight become more prevalent. Consequently, food and nutrition policies need to be reconsidered in view of the challenge posed by the westernisation of the Chinese diet. In particular, nutrition education and public policies aimed at influencing life styles will become increasingly important. Overall, a key challenge from a policy point of view is posed by the coexistence of overweight and underweight family members, which might limit the effectiveness of policies that target households rather than individuals.

Agri-environmental situation

General

Historically, increased food production was achieved by expanding agricultural land, but in recent years cultivated land has been decreasing and increasing output has been due to higher productivity as a result of greater applications of fertiliser, pesticide and mechanical inputs. This has been associated with a move away from traditional farm practices towards more intensive chemical-using farming systems, as well as some shift from arable crop to livestock and horticultural production. Pressure on utilisable agricultural land and water resources has also arisen from urban and industrial growth, while agriculture is itself affected by pollution from other human activities, especially industrial production. As a result, these developments have led to severe problems of land degradation, desertification, soil erosion, water pollution and loss of biodiversity.

About 55% of China's land area is classified as agricultural but about three-quarters is grassland and only one quarter is cultivated annually. China's cultivated land area in per capita terms on average is only 0.11 ha, or one-third of the world average. It is estimated that cultivated land disappeared at an average of 0.5 million ha annually in the decade since 1986, due to industrial and urban development and road building in rural areas. Much of this land contains highly fertile soils (Ye et al., 2002).

In many respects, the pressures on the agri-environment in China follow the same trends in the structural adjustment of agriculture observed in other industrialising economies, in particular where policies of increasing food self-sufficiency have been pursued. However, given the rapid rate of past and projected economic growth, the sheer size of the country, and the vulnerability of much of the land and water resources, including from climate variability, the pressures on the environment pose a potential constraint on future agricultural growth, with international trade implications.

Soil and land

Almost all cultivated land is farmed very intensively within a small-scale farming system (with high fertiliser and pesticide use and multiple cropping). Soil organic matter is a good indicator of soil fertility, but the traditional system of ploughing in straw is being abandoned, along with the use of animal manure as a major source of soil nutrients. The use of chemical fertilisers has become dominant during the last two decades, with potential long-term adverse effects on soil organic matter, as well as on water pollution. According to the national census of soil in over 1 400 counties, 53% of farmland is phosphorous deficient, 23% is potassium deficient and 14% is deficient in both.

Land salinisation actually affects 82 million ha, accounting for 8.5% of the total land area, with an additional risk of land salinisation on 17.3 million ha. Pesticide pollution of soils affects 16 million ha. Including pollution by industry, the total polluted agricultural area is almost 22 million ha. Soils are also subject to the effects of acid rain, which currently affects 25% of Chinese territory. In 1993, the acidified land area covered 2.8 million ha, mainly in south China (Guangdong, Guangxi), southwest (Sichuan basin and part of Guizhou), and currently affecting Central China (Changsa in Hunan province, Nanchang in Jiangxi province), East China (Xiamen in Fujian province, Shanghai) and north China.

Soil erosion is widespread, affecting 356 million ha (agricultural and non-agricultural) in 2003 and is particularly severe in mountainous and hilly areas, to a large extent resulting from over-pasturing and deforestation. Of the total amount of soil erosion, water erosion accounts for 165 million ha and wind erosion for 191 million ha, and the area of erosion has more than doubled since the 1950s. Erosion occurs on 38% of the total land area or almost 2/3 of agricultural land. It is estimated that erosion currently results in an accumulated loss of 2.6 million ha of arable land (Jing and Jing, 2004). Soil erosion from deforestation and poor land reclamation practices has led to land degradation, a reduction in river discharge capacity, which has resulted in increased flooding. It has been estimated that the annual loss caused by floods from all sources averaged more than CNY 80 billion (about USD 9.5 billion) in the 1990s (Liu and Zhang, 2004). Classified by the rate of soil erosion, about one-half is "low", one-quarter is "moderate" and one-quarter is "high or severe" (*China Environment Communiqué, 2003*).

Desertification is another problem, which is mainly located in the northwest and north China. The total land area already desertified covers 837 000 km^2, or nearly 9% of the total land area. If the land at risk of desertification is included, the total area would be as high as 2.25 million km^2, or 23% of the total land area. Over time, desertification has increased. From the 1950s to the 1970s, the annual rate of desertification was 1 560 km^2, increasing to 2 100 km^2 in the 1980s and to 2 460 km^2 in recent years.

Water

Water is very scarce in China. The annual shortage of water is estimated at 36 billion m^3, of which agriculture accounts for 30 billion m^3, or about 8% of China's total agricultural use of 400 billion m^3. Per capita water resources in the country are only between one-third and one-quarter of the world average. Agriculture is the main user of water accounting for over 70% of total consumption in the country. Water resources are very unevenly distributed across regions. It is estimated that underground water resources are only 290 billion m^3 of which 90 billion m^3 is already exploited for human consumption and irrigation. While 65% of cultivated land is to the north of the Yangtze River, this part of China holds just 20% of total water resources.

About 27 million ha of agricultural land are affected by drought, mainly in the northwest and north China (Wang and Li, 2004a). Currently, as a result around 3.1 million km^2 of land (mainly grassland) suffer from desertification, accounting for about one-third of the national territory in which one-third of the population lives. Over exploitation of water resources by farmers has led to a decrease in cultivable land. The protection offered by low-growing shrubs is also diminishing, leading to an encroachment of the desert.

The serious shortage of water is exacerbated by the poor efficiency of water use in agriculture, partly due to the under pricing of water (Chapter 2). Currently, only about 40-60% of water flowing through irrigation systems is effectively used. In some places in the north, the water flow per unit of land is much higher (by between 50 and 150%) than optimally needed for the crop planted, resulting in the wasteful use of water and increasing stress on ecosystems.

The serious problem of scarce water resources is compounded by widespread water pollution and eutrophication. Waste from industrial and agricultural production – especially livestock waste – increased four-fold in the three decades since 1970. Many lakes and water courses contain an excess of nutrients and need treatment before they are suitable as freshwater sources. According to the data from 752 monitoring stations located in the seven big river drainage basins, 44% of rivers are polluted to a varying degree (*China Environment Communiqué, 2000*), including from agricultural nutrients.

The intensity and overuse of chemical fertilisers and pesticides and the growing intensity of livestock production resulting in manure runoffs are the main causes of water pollution and loss of biodiversity. As such, agriculture is the main source of nitrogen pollution. It is estimated that between 1985 and 2000 about 141 million tonnes (or about 9 million tonnes per year) of nitrogen fertiliser flowed into the water system, resulting in the pollution of both surface and ground water (Liu and Zhang, 2004). While in developed countries 50-60% of the pesticides applied are effectively absorbed by crops, in China this ratio is estimated at only 30-40% (Sun *et al.*, 2004).

In 2000, a survey conducted by the State Environmental Protection Administration (SEPA) highlighted three serious environmental problems emanating from the livestock sector. First, the amount of animal excrement was 1.9 billion tonnes in 1999, 2.4 times higher than industrial solid wastes. Second, livestock pollutants in animal excrement reached 71 million tonnes, much higher than the emission from industry. Third, livestock pollutants place heavy pressure on the ecosystem. Livestock pollutants are major factors in the non-point source pollution in the countryside (Zhu, 2004).

The majority of the livestock excrement flows into the rivers and lakes without treatment or with very simple physical treatments. The livestock excrement contains high levels of nitrogen and phosphorous and these pollute the water either through direct emission, or by rain flow. Nitrates in excrement penetrate into the soil, which heavily pollutes the underground water (Zhu, 2000). The lack of wastewater piping systems in the interface between rural and urban areas causes downstream pollution in cities.

Since the early 1980s, the area of vegetables, fruits and flowers has been increased more than four-fold. Due to their high profitability, it is common for farmers to use very high levels of N and P fertilisers on these crops. The average fertiliser application rate is 569-2 000 kg of active nutrients per hectare, about 10 times higher than for grains (Zhang *et al.*, 2004). Increasing vegetable production with high fertiliser input is one of the highest risks for eutrophication of water bodies in China.

Air

While agriculture is not a major contributor to overall air pollution in China, livestock and rice farming is the major source of emissions of methane, nitrogen oxide and ammonium nitrate. Livestock farming as well as land application and storage of manure are responsible for emissions of methane and ammonia. While methane is a potent greenhouse gas, ammonia causes acidification of soils and water where it deposits and is harmful to forests.

Biodiversity

China is ranked 8th in the world (1st in the northern hemisphere) in terms of biodiversity abundance. Nearly all of China's unique and globally significant biodiversity resources are under stress. Many species are seriously threatened, with up to one fifth endangered. Nearly one fourth of all species listed in the *Convention on International Trade in Endangered Species* (CITES) are found in China (World Bank, 2001). In the past 50 years, about 200 plants have become extinct and 4 600 plants and 433 animals are endangered (Liu and Zhang, 2004). These problems are mainly caused by environmental destruction, pollution from the heavy application of chemicals, and inappropriate exploitation and utilisation of agricultural resources, including land (*e.g.* loss of wetlands to agricultural use) and water (*e.g.* reduction of water flows that damage aquatic ecosystems). Another serious problem in agriculture is the invasion of harmful alien species which threaten biodiversity in rural areas (Wang and Li, 2004).

Notes

1. This section draws on *China in the World Economy: The Domestic Policy Challenges*, OECD, 2002.
2. In this report, China's currency is called Yuan which is equal to Yuan renminbi. The currency code applied is CNY which is in line with ISO classification.
3. Agriculture is defined here as the primary sector, which in China's statistics includes not only crop and livestock production but also fisheries and forestry production.
4. In China, grains are defined as including cereals, oilseeds, beans (including soybeans), and tubers. In this report, grains refer to cereal grains and the terms are used interchangeably. Discussions with Chinese officials reveal that China is considering altering the official definition of grains to include only cereal grains.
5. Various sources provide in general a smaller average of around 0.5 ha which is a result of dividing the total arable land by the total number of rural households. However, as not all rural households are engaged in agricultural production, the average is probably slightly higher at 0.65 ha.
6. The average number of plots per household fell from 8.4 in 1986 to 6.1 in 1999, and the average size of plot increased marginally, but in 1999 it was still very small at just around 0.1 ha (Tan *et al.*, 2005).
7. In this report, China's regions are classified as follows: Northeast: Heilongjiang, Liaoning, Jilin; North: Shandong, Hebei, Beijing, Tianjin, Henan, Shanxi; Northwest: Shaanxi, Gansu, Inner Mongolia, Ningxia, Xinjiang, Qinghai; East: Zhejiang, Jiangsu, Shanghai, Anhui; Centre: Hubei, Hunan, Jiangxi; South: Guangdong, Guangxi, Fujian, Hainan; Southwest: Chongqing, Sichuan, Guizhou, Yunnan, and Tibet.
8. Expressed in terms of sown area, per farm availability of land would be even more homogeneous throughout China. Excluding the vast areas of Inner Mongolia, Heilongjiang and Jilin, where there is only one crop a year, the farm sizes, taking into account the multi-cropping ratios, varied from 0.5 ha of sown area in Zhejiang province to 0.8 ha in Henan province in 2003.
9. This is so called "four wastes" land: mountain, slope, marsh and gully and ravine land. It can also include other than arable land.
10. Before legislation changes introduced in March 2003, agricultural households risked losing their land use rights if they moved to urban areas, not only to cities.
11. The most striking examples of confiscation arise from the transfer of land from agricultural production to other uses. Between 1997 and 2003, China's farmland was reduced by 6.7 million ha. More than half of this land was transferred for the creation of "development zones" used mostly for industrial investments. According to China's Ministry of Land Resources (MLR), only 30% of these zones were approved by the central and provincial governments and all other zones were illegal. Local governments and other intermediate institutions appropriated land from farmers with no or very low compensation at about 20-30% of actual market price and then sold the land to external investors with high profits. The MLR estimated that farmers were owed at least CNY 9.9 billion. The same problem arises when farmers are resettled due to large infrastructural

constructions. For example, over the last five decades 16 million farmers were resettled due to construction of dams and other hydropower projects. It is reported that 10 million of them live in poverty as the compensation is usually far from enough to make up for their income losses in the resettlement process (*China Daily*, 30 July 2004).

12. In October 2004, the State Council issued a new directive on land management (*Guofa*, No. 28: Decision on Deepening Reform and Strengthening the Land Management) which again stressed that farmland should be strictly protected; the conversion of agricultural land to other uses should be approved by the State Council and/or the provincial government; and farmers should be fully compensated for the lost land; the living standard of farmers losing land can not be lowered.

13. These are: soil and fertiliser station, seed station and plant protection station. They are independent agencies (all under the direction of the local Agricultural Bureau) providing agricultural extension services to farmers.

14. For a long period of time, the unclear ownership structure and complex management system of RCCs resulted in their very poor performance. Their non-performing loan ratio is much higher than for the state-owned commercial banks. In 2004, the government launched a pilot reform of RCCs in eight provinces. In selected counties within these provinces, part of RCCs have been transformed into Rural Commercial Banks (RCBs); the regulation of the interest rate has been loosened, allowing the rate to float within a range of 100%; the supervision over RCCs has been handed over to the provincial government and the central financial authorities are to provide a regulatory framework. This reform will be extended to 29 provinces in 2005.

15. Usually, informal channels of credit are based on personal networks cultivated by farmers among relatives and friends. While informal channels is often the only option available for farmers (Hu, 2004), they may involve substantial transaction costs. The maintenance of these networks, for example through mutual gifts, can cost rural families up to 25% of their monetary living expenses (Li, 2005).

16. According to the Chinese classification, the food industry consists of the food processing sector, the food manufacturing sector, the tobacco and the beverage sectors. The food processing sector comprises rice milling, flour milling, oil refining, sugar refining, slaughtering, salt processing, feed processing and aquatic products processing. The food manufacturing sector mainly covers the making of pastries and confections, dairy products, canned foods, fermented products and condiments. The beverage sector includes the making of alcoholic beverages, soft drinks and tea.

17. Industry analysts suggest, however, that the new chain must overhaul its management structure as redundancies may undermine its competitiveness. Although the two chains were merged, it was a bureaucratic and not a commercial decision and both appear to be operating independently and continue to report revenues as separate entities.

18. For grain imports within TRQ, the sole STE is COFCO; for the vegetable oil it is COFCO and five other STEs (China National Native Products and Animal By-products Import & Export Co., China Resources Co., China Nam Kwong National Import & Export Co., China Liangfeng Cereals Import & Export Co., China Cereals, Oil & Foodstuff Co.), see MOFTEC, 2002.

19. In China, "strategic" agricultural commodities are: cereals, oilseeds and vegetable oils, sugar, cotton, tobacco, tea and silk.

20. While the Ministry recommends that the proportions between the three fertilisers: nitrogen/phosphate/potash should be 1/0.37/0.25, the actual proportions are 1:0.33:0.20 (*China Chemical Industry Newsletter*, 21 April 2004). More than 90% of potash used in China comes from imports.

21. Due to multiple cropping, in particular in the southern China, area sown in China is larger than arable area. For example in 2003, sown area was 152.4 million ha. If compared to 130.0 million ha of arable area (1997 Agricultural Census figure, quoted in *China Statistical Yearbook*, 2004), the multi-cropping ratio would be about 1.2. However, if compared to 95 million ha of cultivated area, as suggested by the Ministry of Agriculture (see Box 1.4), the ratio would be 1.6, which seems more compatible with the cropping systems in China (Aubert, 2005).

22. It is very likely that official data on livestock production in China are overestimated (see Box 1.4). Since the downward correction of the total meat production figures, following the Agricultural Census of 1997, the official rates of growth of meat production, strongly inflated in the first half of the 1990s, have been considerably lowered in more recent years (Aubert, 2004).

23. This is an official estimate covering the whole primary sector, thus including agriculture, forestry and fisheries.

24. Agricultural labour figures may be overestimated as they most likely include 100 million "floating" migrants employed in urban areas for at least part of the year. It may be roughly estimated that probably not more that half of them can be counted as "agricultural workers". With this adjustment, the actual number of employed in agriculture would be somewhat around 265 million, or about 35% of the total active population (Aubert, 2005).

25. In the analysis, the agricultural inputs were classified into 4 categories: labour, land (sown area, adjusted for both multi-cropping and irrigation), capital (sum of the value of draft animals, non draft animals, poultry and farm machinery), and current inputs (seed and feeds, organic and chemical fertiliser, electricity and insecticides).

26. Other studies confirm this conclusion. Measured by the value added per worker in PPP (constant 1995 USD), in 2000 the labour productivity in industry was 7.2 times higher and in services 3.4 times higher than in agriculture (Fogel, 2004).

27. These difficulties include: substantial additions to urban residents' incomes in terms of direct provisions of goods and services from state employers and, until the recent past, price subsidies for many food products; the large share of food consumed in rural areas from a family's own production; different quality of products and services consumed in rural and urban areas, etc. For a discussion on this issue, see *e.g.* Johnson, 2000.

28. This section has been prepared in co-operation with the FAO.

29. In the case of Brazil, these figures are 11% and 45%, respectively.

ISBN 92-64-01260-5
OECD Review of Agricultural Policies
China
© OECD 2005

Chapter 2

Policy Trends

China has implemented substantial economic policy reforms since 1978. A fundamental element has been reform of the raft of agricultural and agriculture-related policies contained in China's governance framework. While the general direction of reforms has been consistent, there have also been numerous small policy shifts and changes in the detail of policies. In this chapter, trends in agricultural and related policies during the period 1990-2005 are highlighted, followed by an evaluation of the support provided to producers.

In Section 2.1, the framework of agricultural policy will be provided. This framework is examined with regard to key policy objectives, the national and sub-national institutional arrangements for administering agricultural policy, and the major instruments employed to implement policy.

Section 2.2 contains an overview of domestic agriculture-related policies. This section is arranged in seven sub-sections devoted to: price and income support measures; reduction of input costs; agricultural taxation; rural public services infrastructure; consumer measures; environmental measures; and overall budgetary outlays on agro-food policies.

Trade policies related to the agro-food sector are examined in Section 2.3. This section contains six sub-sections: overall reforms of the trade system; the objectives of Chinese trade policy in the agro-food sector; agro-food import and export policy measures applied by China; trade relations; trade policy measures applied by partners; and agro-food trade flows.

Finally, Section 2.4 quantifies the extent of support provided to agriculture and the burden that this imposes on Chinese consumers and taxpayers.

2.1. Agricultural policy framework

China has carried out fundamental reforms of its economic policies since 1978, resulting in a gradual transition from a centrally planned economy towards a socialist market economy. In line with this transition, the agricultural policy framework has also been evolving. While the principal economic objective of creating a market-based economy appears to have consistently under-pinned agricultural policy trends, it is clear that the reform process has not been completely smooth – occasionally, reforms appear to have been set aside in order to allow measures dealing with unforeseen events to be implemented. Like all national governments, the Chinese leadership has a multi-faceted set of objectives and is faced with the task of developing policies to be applied in an uncertain future. Policies sometimes have unintended consequences as a result of changing circumstances, leading policy makers to adjust policies to better accommodate the environment in which they are applied.

China's policy-implementation process has been relatively flexible, in the sense that broadly-defined central government policies have been implemented in a variety of ways, according to the capacity and needs of the sub-national government bodies responsible for policy implementation. Given this reality, it is difficult to describe trends in agricultural policies with complete accuracy. This chapter focuses on major developments in agricultural policies during the period 1990-2005. Policies from earlier periods may be described when it is necessary in order to properly present the evolutionary process of policies.

Agricultural policy objectives

The basic set of central government policy objectives has been stable for the period 1990-2005, although the government has varied its priorities according to changing socioeconomic conditions. A wide range of government reports, statements and planning documents make it clear that the agricultural sector is viewed as very important in terms of the Chinese economy in general, and has a high profile in policy making. In broad terms, agricultural policy has addressed the provision of adequate supplies of food at stable prices, as well as maintenance and improvement of rural incomes (Tuan and Ke, 1999). In 2004 the central government confirmed that food security remained at the forefront of policy considerations, along with the objective of increasing incomes in rural areas (NDRC, 2004). The main objectives of agricultural policy have been:

- Food security.
- Farm household income.
- Food safety.
- Environmental protection.
- Agricultural competitiveness.

Food security

Ensuring an adequate supply of affordable food is a top priority of China's policy makers. China is the most populous country in the world but its land and water resources are relatively limited. Traditionally, grain self-sufficiency has been regarded as the key to achieving food security. Although food self-sufficiency may not have been clearly defined in government policy statements, it is widely interpreted to mean that China should produce 95% of its own grain requirements (Fang, Tuan and Zhong, 2002; Felloni et al., 2003). A Chinese government White Paper issued in 1996 also proposes that China achieve 95% self-sufficiency in grains (State Council, 1996).[1] This self-sufficiency objective is a key factor in understanding how agricultural policies have evolved.

There are two different elements to food self-sufficiency in China. One element is the market supply of food (grains especially), and the other is the non-market supply – farm produce which is consumed directly by the farm household. Given the fact that most farmers derive their grain supply from their own production, food security for the agricultural population has generally been considered in terms of mitigating the grain shortage effects of natural disasters such as floods and droughts. The perceived balance in the non-agricultural market (essentially the urban market) has had a primary influence on agricultural policies. The availability of grains for supplying the commercial market, including grain reserves controlled by the State Grain Authority often triggers changes to policies related to grain production, procurement and trade.

In order to prevent widespread hunger, it is not sufficient to ensure that enough food is available in a country or region, the food must also be accessible to the population (Sen, 1981). A major element of entitlement to food is a consumer's capacity to afford the food at the prices it is being sold. Thus, not only did policy makers seek to ensure that sufficient food was supplied, but also sought to ensure that marketed supplies were available at affordable prices. This policy focus was particularly evident in the earlier part of the 1990s.

Farm income

During the latter part of the 1990s and early in the 2000s, the growing income gap between urban and rural populations, and between developed and underdeveloped rural areas, became important policy issues. Policy makers began to address the integration of urban and rural development and devised regional development programmes to accelerate economic growth in less developed regions. Policies aimed at raising agricultural incomes nationwide were also adopted, with agriculture-related taxation policy reform embodying a fundamental shift from taxing agriculture to supporting agriculture. These agricultural-income supporting policies were strengthened in 2004 through the adoption of "The Suggestions of the Central Committee of the Communist Party of China and the State Council on Policies for Boosting Growth in Farmers' Income", the "No. 1 Document" of the Chinese central authorities for 2004 (State Council, 2004a).[2]

Food safety

In an environment in which urban food shortages are not of immediate concern to consumers, and coincide with their increasing affluence, urban consumers have become increasingly concerned with food safety issues. Especially since 1997, governments at various levels in China have implemented a range of measures to improve food safety.

Environmental protection

Government policy responses to the impact of human activity on the environment have been perceived as increasingly important. Agriculture has the potential to impact negatively on the environment in a number of ways: through land clearing, land use techniques, agro-chemical applications and run-off, and water usage. Sustainable use of land and water resources has been given a high profile, especially since the late 1990s, with significant levels of funding being made available to projects such as the so called "grain for green" project.

Agricultural competitiveness

Starting from the early 1990s, the process of transition towards a market economy was accelerated. The negotiations for WTO accession introduced external pressures to carry out policy and institutional reforms. In order to be well prepared for the increased competition after the opening of its domestic markets, the government placed a strong emphasis on improving the efficiency of agricultural production. Along with education programmes targeted at improving farmers' take-up of technology, measures to help raise the competitiveness of Chinese products in both the domestic and international markets were put into practice. These measures were implemented in response to the impact of cheaper imports on domestic prices.

Evolution of policy objectives

In general, the reforms of agricultural policies and institutions were directed towards increasing the role of markets. However, changes in domestic circumstances and in world market conditions led to reprioritising of measures to achieve the broad reform objectives. Moreover, the interventions that still exist and weak market institutions, such as enforceable contracts, transparent information and open bargaining among several buyers and sellers, indicate that China has a long way to go. Generally the 1990-2005 period can be divided into two sub-stages in terms of the major policy initiatives and measures implemented.

1990 to 1997. In this period, the principal agricultural policy objective was to increase agricultural production, especially that of food grains. In line with the general economic policy initiative towards a market-oriented economy, this period was also characterised by substantial deregulation of agricultural marketing and a significant lessening of controls on the prices of agricultural products and on marketing channels.

Food security was a leading government priority throughout this period. Initially, the central government was responsible for food security, but some responsibility was devolved to provincial governments. The Governor's Grain-Bag Responsibility System (GGBRS) was introduced in 1995. Under the GGBRS, provincial governments were to ensure the availability of adequate supplies of food grains within provincial boundaries.

Macroeconomic influences, as outlined in Chapter 1, resulted in fluctuations in the level of regulation of agricultural product prices. Following the liberalisation of price controls in the early 1990s, inflation and rising food prices in 1994 and 1995 resulted in a strengthening of government controls on prices and marketing channels, followed by a more gradual easing of regulation as prices stabilised. While affordable food and stable prices were a policy objective throughout this period, the objective became a higher priority in times of rising prices.

China was engaged in WTO accession negotiations prior to 1990. Until accession in 2001, China undertook incremental adjustments to agriculture and agricultural trade policies with a view to becoming more competitive in international markets.

1998 to 2005. This period was characterised by the adoption of policies supporting rural income, representing a fundamental shift in the government's agricultural policy agenda. The new policy direction was clearly spelt out in the document issued in 1998 by the Central Committee of the Communist Party of China (CCCPC) "The decision of the CPC Central Committee on several major issues in agriculture and rural work". The decision firmly made the reduction of taxation of farmers and the improvement of their incomes as the guiding principles of government policy until 2010.

Agriculture began to be supported with the aim of maintaining and improving the incomes of those dependent on agriculture. Food security remained an important policy objective, while policies addressing food safety achieved a higher profile in this period. As a result of growing urban affluence and a lack of strong growth in food prices, food marketing and price controls became less important.

The competitiveness of China's agricultural production also became a higher priority than previously as accession became imminent. Since WTO accession in December 2001, the competitiveness of China's agricultural sector has continued to be an important policy objective.

Major floods in the southern parts of China in 1998 gave renewed impetus to agro-environmental policies. The floods highlighted land clearing and land usage practices which contributed to the severity of the floods, prompting the "grain for green" policy response. At the same time, continuing pressure on the main northern river systems was increasing concerns over falling water tables and increasing desertification. These concerns underpin the continuation of the Comprehensive Agricultural Development Programme which provides funding for soil and water conservancy projects throughout China.

Currently, improving farmers' incomes and narrowing the urban–rural income gap are top priorities for the Chinese government, while food security, or food-grain self-sufficiency, remains a principal policy focus. High level policy documents, including the "No. 1 Documents" from 2004 and 2005, clearly outline these policy objectives, while proposing policy measures that include reduced taxation of farmers, direct subsidies to grain farmers, measures to maintain farm land in agricultural production, and measures to improve the up-take of technology in the agricultural sector.

Institutional arrangements for administering agricultural policy[3]

General framework

China has a multi-layered policy development and implementation environment. Many government bodies and statutory institutions at both national and local levels (province, prefecture, county, and township/village levels) are involved in various ways in the policy process. Normally, sub-national institutions have their own array of objectives, some of which may deviate from national government policy priorities. Based on observed trends over the past twenty years, the institutional framework for administering agricultural policy can be characterised in the following way:

- Broad policy direction is provided by the CCCPC (Chapter 1).

- Nominally, the policy framework is formulated by the National People's Congress (NPC) and the Standing Committee of the NPC. The NPC has the power to create laws.
- The state administration is governed by the State Council. Policy is operationalised by the ministries and other elements of the national bureaucracy or state administration.
- National policy operationalisation often involves inter-departmental liaison and negotiation through the "leading groups" mechanism. Leading groups oversee policy activation which involves the area of responsibility of more than one ministry or commission. The leading groups typically deal with a single broad area of policy, and draw their membership from the senior levels of relevant ministries and commissions and of the CPC.
- Responsibility for implementing regulations in the agricultural sector is generally devolved to sub-national bureaucracies (at the province, prefecture, and county levels), and to the administrators of townships and villages.
- Policy is often implemented in a variety of ways across the country, and is not always implemented effectively. This is partly due to the responsibility to implement policy being devolved to sub-national levels of government, partly to inadequate monitoring of policy implementation, and partly due to policy initiatives not being fully funded.

The CPC occupies the dominant position in Chinese political life. In providing core leadership for China, the CCCPC sets the broad policy directions for the country. Then, the CCCPC and administrative elements of the national government work jointly to develop the policy framework and specific instruments within that framework. Sub-national levels of government have some influence in policy development, but are mainly involved in implementing national policy and have no specific rights to develop their own policies.

Typically, the CCCPC and the State Council develop a number of policy suggestions for the NPC, such as those in the "No. 1 Document". The senior leadership is advised in their policy development and decision-making by CPC "Leading Groups". These groups are normally headed by the premier, a vice premier or state councillor, with members consisting of ministers or their deputies from relevant ministries, plus other relevant personnel. These groups draw input from a range of specialist sources, including government think-tanks, academia, and specialists within the public service. Upon acceptance of the suggestions, the NPC requests the State Council to determine how to implement the policy suggestions and to develop a budget for the proposals. When the budget is approved, the relevant Ministries put appropriate regulations in place and responsibility for implementing the regulations is passed on to the appropriate elements of sub-national governments.

Nominally, the NPC is the peak legislative body. The CCCPC exercises political power and the NPC wields legislative power on behalf of the Chinese people. The Congress plays a key role in setting up the framework for governance, enacting legislation, electing top government officials, and examining and approving fiscal budgets. In practice, as the full NPC is in session for a limited time each year, the Standing Committee of the NPC exercises power on behalf of the Congress. Within the NPC there are nine special committees which take major responsibility for certain important sectors of the work of the NPC. The Agricultural Sub-Committee is one of the nine special committees, taking responsibility for agricultural and rural affairs. The Financial and Economic Sub-Committee and Environment and Resources Protection Sub-Committee also have an interest in agricultural and rural issues as they have an impact on their specific activities.

National government

The State Council is the top administrative body, exercising overall management and administration of government business on behalf of the NPC in accordance with the Constitution and existing laws. Important policy and administrative decisions, which have a general impact on the agricultural sector, can be made by the State Council. In the policy making process, each ministry or national bureau can make policy suggestions to the State Council on issues for which it has responsibility and can make comments on proposals of other government bodies from its own perspective. Inter-ministerial discussions are held whenever necessary to discuss issues for which there are shared responsibilities. The so-called "leading groups" play an important role under such circumstances. The decisions made by the State Council are then implemented by relevant ministries and bureaus. Clearly, processes are in place to enable extensive consultation, discussion and negotiation on key policy decisions.

Under the State Council, the Ministry of Agriculture (MOA) takes primary responsibility for issues directly involving agriculture. The MOA is authorised to manage mainly agricultural production, while pre- and post-production activities and a wide range of supportive services are under the authority of other ministries. For instance, the purchase and marketing of major cereals is the duty of the State Grain Administration.[4] Similarly, supply of manufactured farm inputs (such as chemical fertilisers and pesticides) is also largely beyond the direct authority of the MOA. International agricultural trade was once monopolised by state trading enterprises[5] (STEs) under what is now the Ministry of Commerce (MOFCOM);[6] most state owned enterprises, including those engaged in agricultural activities, are now under the supervision of the State-owned Assets Supervision Administration Commission (SASAC). Given that different ministries often have divergent priorities and interests, such an administrative separation of authority has increased the time required to develop policy. However, this framework ensures the agricultural policy is not developed in isolation, but reflects the policy framework of the whole of the Chinese government. An outline of the major central institutions involved in agriculture policy making and implementation, is provided in Figure 2.1.

Sub-national government

During the past twenty years, government power has been gradually decentralised. During this period, sub-national governments have become increasingly more influential in the policy making process, especially with respect to those policies which have major impacts on regional and local economies or on local government budgets. Sub-national governments have also often had the freedom to decide how to implement national government policies, resulting in some variation in the way national policies have been implemented.

The reforms of the fiscal system in 1994 installed a revenue sharing arrangement between central and sub-national governments. Sub-national governments have since been required to fund certain elements of policy-related costs from their own budgets. The economically prosperous provinces have reportedly been able to increase expenditures on government investments that either contribute to the growth of local economies or improve the image of the local region – this includes supporting the agricultural sector and maintaining stable prices for staple foods in urban markets. The poorer provinces often lack the fiscal resources to support agriculture to the same extent as the more affluent provinces. As a result of the vast diversity of conditions in the agricultural sector, and the differences in

2. POLICY TRENDS

Figure 2.1. **Central institutions with oversight over China's agro-food sector**

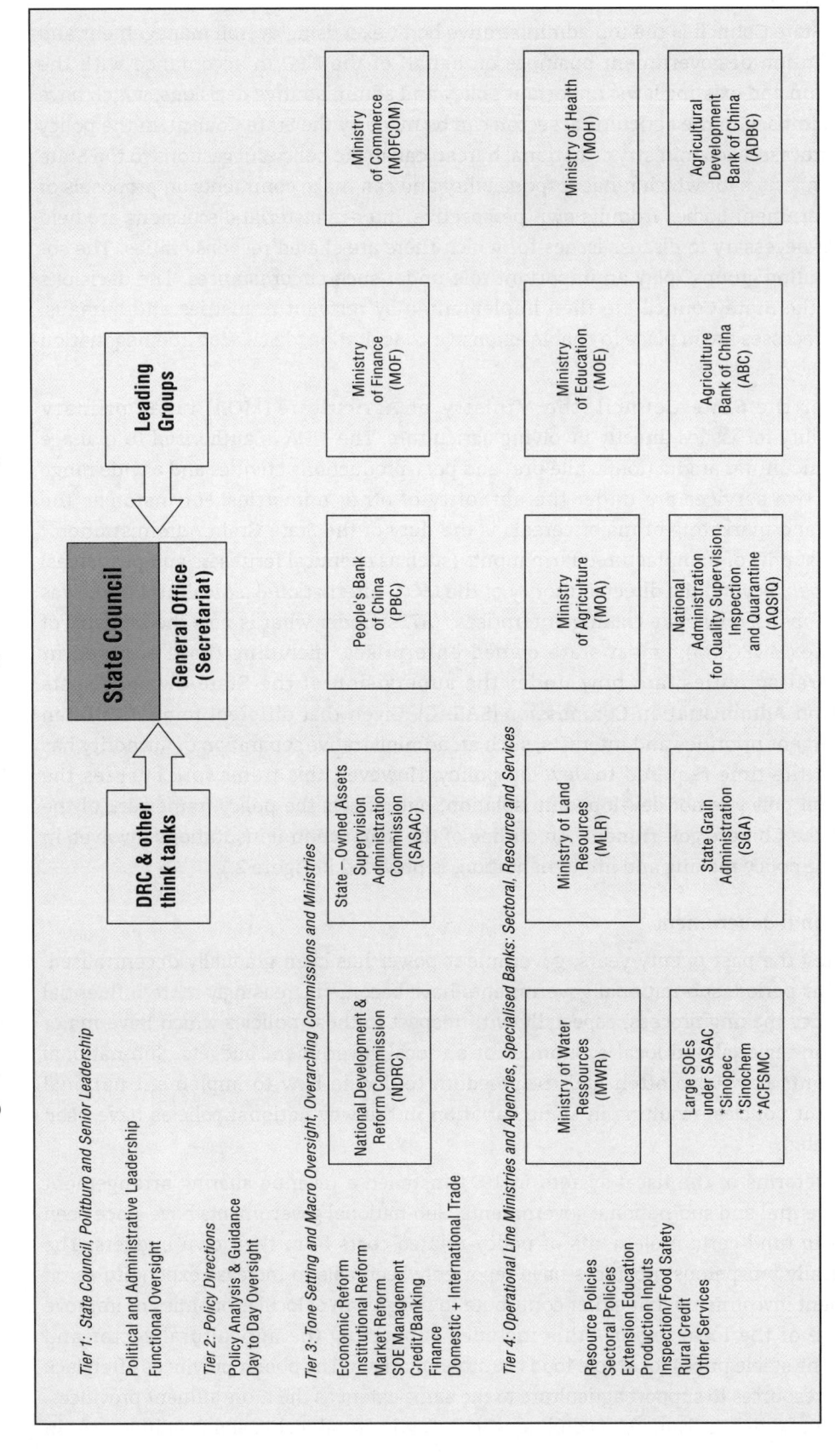

Tier 1: State Council, Politburo and Senior Leadership

Political and Administrative Leadership
Functional Oversight

Tier 2: Policy Directions

Policy Analysis & Guidance
Day to Day Oversight

Tier 3: Tone Setting and Macro Oversight, Overarching Commissions and Ministries

Economic Reform
Institutional Reform
Market Reform
SOE Management
Credit/Banking
Finance
Domestic + International Trade

Tier 4: Operational Line Ministries and Agencies, Specialised Banks: Sectoral, Resource and Services

Resource Policies
Sectoral Policies
Extension Education
Production Inputs
Inspection/Food Safety
Rural Credit
Other Services

State Council — General Office (Secretariat)

Leading Groups

DRC & other think tanks

- National Development & Reform Commission (NDRC)
- State – Owned Assets Supervision Administration Commission (SASAC)
- People's Bank of China (PBC)
- Ministry of Finance (MOF)
- Ministry of Commerce (MOFCOM)

- Ministry of Water Ressources (MWR)
- Ministry of Land Resources (MLR)
- Ministry of Agriculture (MOA)
- Ministry of Education (MOE)
- Ministry of Health (MOH)

- Large SOEs under SASAC
  * Sinopec
  * Sinochem
  * ACFSMC
- State Grain Administration (SGA)
- National Administration for Quality Supervision, Inspection and Quarantine (AQSIQ)
- Agriculture Bank of China (ABC)
- Agricultural Development Bank of China (ADBC)

Source: OECD (2005a).

financial capacity of sub-national governments across China, the implementation of some national policy programmes is tailored by local governments to suit local conditions. Although they have no specific policy creation role, sub-national governments have significant control over how policy is implemented within their jurisdiction.

Sub-national governments also have an important role in providing feedback to higher levels of government. This feedback contributes to the information set on which policy and government budgets are based. The feedback received has been misleading in some instances. It is reported that officials of some sub-national governments have purposefully created a perception of their locality as being less prosperous than in reality in order to obtain additional funding from higher levels of government.[7]

Basic policy instruments

In line with the multi-layered policy development and implementation environment, policy and policy instruments also consist of multiple layers, often differentiated in terms of timeframes. As in most countries, in China, laws embody policy directions which generally have no perceived end-point. Within the legal framework, medium-term policy direction is provided in plans which are in place for a set number of years – such as the plans developed under the five year plan system. Annual policy direction, consistent with the broad goals of the five year plans, is provided by the annual plans and relevant regulations or decisions issued by the CCCPC and the State Council, such as the "No. 1 Document". In 2004 and 2005, these documents focused on agricultural and rural issues, in particular on increasing farmers' incomes.

The notable changes in central government policy and apparent shifts in policy priorities led to frequent adjustments of policy instruments and measures during 1990-2005. Both indirect interventions and direct administrative controls over certain economic activities were used in a variety of combinations, depending on the issue being addressed and the environment in which the measures were to be implemented.

Under the traditional central planning system, the government managed the economy primarily through quantitative control over all economic activities, including production, domestic marketing and pricing, processing, foreign trade, supply of inputs and allocation of primary resources. This situation has been gradually changing since the economic policy reforms of 1978. The introduction of market mechanisms has required that government move towards a greater reliance on indirect measures to achieve policy objectives. There has also been an increased imperative to exercise effective governance through more transparent legislation and regulations.

China has a number of laws and regulations which have a significant impact on the agricultural sector. These include general laws and regulations having an economic impact, such as taxation rules; and laws and regulations which are specifically intended to govern economic activities in the agricultural sector. The latter group of laws and regulations are summarised in Table 2.1. Most of these regulatory instruments have been enacted or amended since 1990, establishing a legal framework for the governance of agriculture-related activities. The principal piece of legislation in this framework is the Agricultural Law. The Agricultural Law establishes the broad direction of policy for agriculture as an economic sector (Box 2.1); other laws, such as the Grassland Law, deal with specific aspects of the agricultural sector. In general, these laws outline government statements of intent and provide guiding principles, rather than committing the government to specific actions and binding obligations.

> **Box 2.1. The Agricultural Law of China**
>
> First introduced in 1993, and then amended in 2003, the Agricultural Law highlights the major role played by agriculture in the Chinese economy and contains the primary principles for development of agriculture and the rural economy.
>
> The Agricultural Law emphasises that the government places agricultural development at the top of the national economic development agenda. It also declares that the government will adhere, in the long term, to the existing arrangement of the HPRS (Chapter 1) and collective management of communal resources. It commits the government to provide necessary assistance and protection to agricultural production and producers' incomes, and also commits the government to take appropriate measures to ensure food security.
>
> The Agricultural Law also outlines government commitments to structural adjustment of the rural economy, industrialisation of farm businesses, food safety, improving farm input supply, development of agricultural sciences and technologies, and improvement of agricultural sector training and education. It also reflects the necessity for institutional reforms, in order to comply with WTO membership requirements. Trade of agricultural products, rural labour issues, farmer's rights, and agricultural taxation are also addressed in this law.
>
> In terms of policy, the Agricultural Law contains both objectives and policy measures. The issue of protecting agricultural production and producer's incomes, for example, is addressed both in terms of the objective of producing sufficient food and broadly defined measures, including maintaining farmland in agricultural production, establishing grain risk funds (budgetary allocations used to ensure grain market stability), and establishing national and local grain reserves or buffer stocks.

More specific guidance and instruction for policy implementation is generally contained in regulations issued by the State Council and various administrative elements of the central government. The regulations operationalise the higher level policy statements, generally detailing explicit measures and commitments to be implemented by the operational elements of government.

In the past, China's system of governance contained significant scope for sub-national elements of government to implement government policy according to local conditions. The policy framework was developed and promulgated to provide broad direction within which policy measures could be adjusted by provincial authorities and township and village administrators. Additionally, there are many constraints to the enforcement of regulations. These include lack of technical, financial and administrative capacity in local government institutions, goal conflict in the implementing institution(s), under-development of the legal system, and lack of knowledge of the legal system by both farmers and administrators.

The main domestic policy measures employed by the national government, discussed in greater detail in the following sections, cover producer support measures, general services and consumer support measures. In turn, producer support measures cover both domestic and trade policy measures.

Domestic policy measures include:

- *State pricing*: in place for major agricultural commodities for much of the period 1990-2004. From 2004, centrally set state pricing only applies to tobacco (under a state

Table 2.1. **Major laws and regulations in the agro-food sector**

Title	Passed/amended	Implemented
Laws		
Grassland law	June 1985/Dec. 2002	Oct. 1985/March 2003
Law of land management	June 1986/Aug. 1998	Jan. 1987/Jan. 1999
Fishery law	Jan. 1986/Oct. 2000	July 1986/Dec. 2000
Forestry law	Sept. 1984/April 1998	Jan. 1985/April 1998
Water law	Jan. 1988/Aug. 2002	July 1988/Oct. 2002
Water and soil conservation law	June 1991	June 1991
Quarantine law	Oct. 1991	April 1992
Law of monopolised sale of tobacco and products	June 1991	Jan. 1992
Agricultural law	June 1991/Dec. 2002	July 1993/March 2003
Agricultural technical extension law	July 1993	July 1993
Food hygiene law	Oct. 1995	Oct. 1995
Law of township and village enterprises	Oct. 1996	Jan. 1997
Animal diseases prevention law	July 1997	Jan. 1998
Flood prevention law	Aug. 1997	Jan. 1998
Law for organisation of village committee	Nov. 1998	Nov. 1998
Seed law	July 2000	Dec. 2000
Rural land contract law	Aug. 2002	March 2003
Law for promoting agricultural mechanisation	June 2004	Nov. 2004
Regulations		
Regulation on plant quarantine	Jan. 1983/May 1992	Jan. 1983/May 1992
Regulation on preventing animal diseases	Feb. 1985	July 1985
Regulation on animal medicines	May 1987	Jan. 1988
Regulation on management of farmers' burden and labour service	Dec. 1991	Dec. 1991
Regulation on breeding animals	April 1994	July 1994
Regulation on management of pesticides	May 1997	May 1997
Regulation on protecting new plant varieties	March 1997	Oct. 1997
Regulation on pig slaughtering	Dec. 1997	Jan. 1998
Regulation on grain procurement(abolished in June 2004)	June 1998	June 1998
Regulation on protecting basic farmland	Dec. 1998	Jan. 1999
Agricultural GMO safety regulation	May 2001	May 2001
Regulation on retirement of land for forest	Dec. 2002	Jan. 2003
Regulation on management of national grain reserves	Aug. 2003	Aug. 2003
Regulation on grain marketing	May 2004	June 2004

Source: Various Chinese government sources.

monopoly). For most of the period 1990–2004, state pricing was accompanied by state procurement.

- *Input subsidies*: charges for water, electricity and transport tend to be lower for farmers, but the level of subsidy is difficult to assess as the cost of provision is different across various users. To lower prices of fertilisers, fertiliser producers have been given access to lower priced inputs, such as electricity. Since 2002, farmers have been subsidised for the cost of purchasing improved quality soy seed. In 2004, this scheme was extended to include subsidies for purchasing improved seed for production of wheat, corn and rice, as well as soybeans.

- *Credit subsidies*: until the end of the 1990s, preferential loans were provided mostly to state marketing organisations to fund purchase and storage of key agricultural products. In the 2000s, most of these programmes were discontinued, but are still applied for grains.

- *Direct payments*: initiated as a trial in 2002 and implemented nationally in 2004. Farmers engaged in growing grains have received a direct subsidy based on the area of land they sow to rice, wheat or corn.

- *Payments for returning farmland to forests*: also known as the "grain for green" programme, commenced in 1999. Farmers cultivating ecologically vulnerable land received a cash subsidy and a grain allocation for each *mu* (1/15 hectare) they retired from agricultural production. Subsidised seedlings were also available for afforestation. In 2004, the grain allocation was converted to a cash equivalent.

- *Agricultural taxes*: between 1990-2004, farmers were required to pay agricultural taxes either in cash or in kind. In addition, they also paid various fees to local governments and collectives and provided "labour accumulation" for the construction of communal facilities. Agricultural tax reform was initiated as a trial in 2000 and is being phased in across rural China from 2004.

Trade policy measures include:

- *Tariffs*: the simple average tariff for agricultural products fell from 45.4% in 1992 to 15.3% in 2005, remaining at that level under the agreed terms of China's accession to the WTO.

- *Tariff rate quotas*: under the terms of its WTO accession, China can apply TRQs to wheat, rice, corn, sugar, cotton, wool and some vegetable oils. China's TRQ system includes criteria for allocating the import quotas to state trading enterprises (STEs) and non-STEs.

- *State trading*: dominating until the mid-1990s. Its role has been diminishing since then, but still important for key commodities.

- *Export subsidies*: prior to joining the WTO, China provided export subsidies for corn and rice. In line with its WTO accession commitments, China is not allowed to apply export subsidies.

General services provided to the agricultural sector as a whole include:

- *Research and development*: government funding for this element of agricultural support is relatively small and tended to decrease.

- *Agricultural schools*: government funding for agricultural schools is also a small expenditure item, but unlike research funding, agricultural school funding has been increasing.

- *Inspection services*: while China has funded food inspection services through-out the period 1990-2005, in the latter part of that period, food safety has become a higher priority concern of policy-makers. Not only has expenditure on inspection services increased, China has also undertaken significant work to upgrade food safety standards.

- *Public stockholding*: China mainly engages in public stockholding of food grains. In line with China's food security policies, the government at national and sub-national levels is active in maintaining buffer stocks of food grains.

- *Agricultural infrastructure*: investment in agriculture-related projects is a major tool for the government to achieve development targets and is by far the largest component in government's budgetary support for agriculture. Government has continued to accept primary responsibility for such projects as: pollution control, land rehabilitation, transport and irrigation infrastructure maintenance and development. Large-scale state-owned farms have been one avenue of state investment. However, state farms are only a small part of the Chinese agricultural sector.

Consumer support measures include:

- *Food price subsidies*: since 1992, China has paid subsidies to urban consumers to offset the price increases of staple food products. Although some of the subsidies are still paid, there has been a significant decline in the level of budgetary expenditure on them.

2.2. Domestic policies[8]

China's domestic agricultural policies have undergone significant changes in the period 1990-2005, as previously mentioned. In this section, the trends in those policies are discussed in more detail.

Price and income support measures

Price and income support measures can be discussed in terms of several periods, according to major changes in the measures applied.

1990-1996

In the early 1990s, rice, wheat, corn, soybeans and cotton were subject to state pricing and state procurement measures. The state purchased a set quantity at a state set price under a contract procurement system, farmers could sell any extra production (above their contracted quota amount) at negotiated prices[9] to the state or, with the exception of cotton, at free market prices to other buyers (Figure 2.2); supply marketing co-operatives (SMCs) retained a monopoly on purchasing cotton from producers throughout this period.

This state procurement arrangement was not initially intended to support rural incomes, but to secure food supply to urban residents at reasonable prices and to ensure an adequate supply of raw cotton to the state textile industry. For most of the period preceding 1997, the state prices of cereals and soybeans were significantly lower than Chinese free market prices (Figure 2.2). The largest gaps were observed in 1995 when the rate of inflation increased. During this period, the price gaps between producing and consuming regions were also relatively large due to inter-provincial controls on grain movements. Many grain-producing regions instituted controls on grain movements under the GGBRS in an attempt to ensure adequate supplies within provincial borders. Negotiated prices were in all cases much higher than the state-set prices, but lower than the free market prices.

Domestic market prices often deviated from border prices during the period 1990-1996, partly due to centrally planned import and export arrangements reducing domestic responsiveness to world prices. However, domestic price developments were surprisingly much in parallel (though not identical) with the development of the respective border prices (Figure 2.2; see also section 2.4 and Annex B).

Funding for state procurement was provided through China's central bank. The People's Bank of China (PBC) was the main source of funds to the Agricultural Development Bank of China, which in turn provided "soft" loans[10] to the State Grain Enterprises (SGEs) charged with making procurement purchases. This system was still in place in 2005.

Tobacco and sugar are important cash crops, subject to government control throughout the 1990-2003 period. Tobacco was, and remains, subject to a state monopoly under which the state purchases all raw tobacco grown in China, at prices determined by the National Tobacco Bureau. Similarly, sugar cane and sugar beets were subject to state pricing, with these crops being purchased by a state owned sugar processing sector.

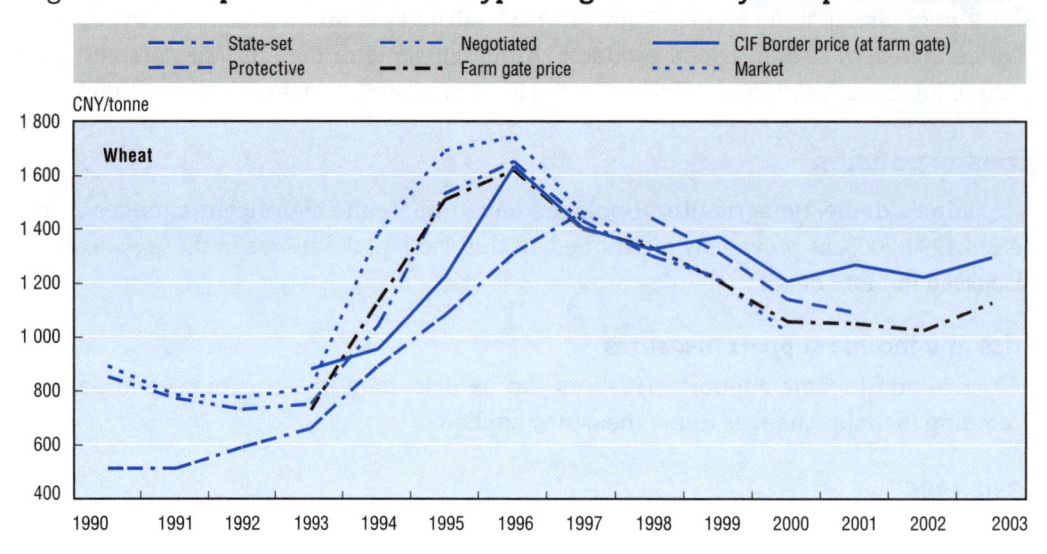

Figure 2.2. **Comparison of different types of grain and soybean prices in China**

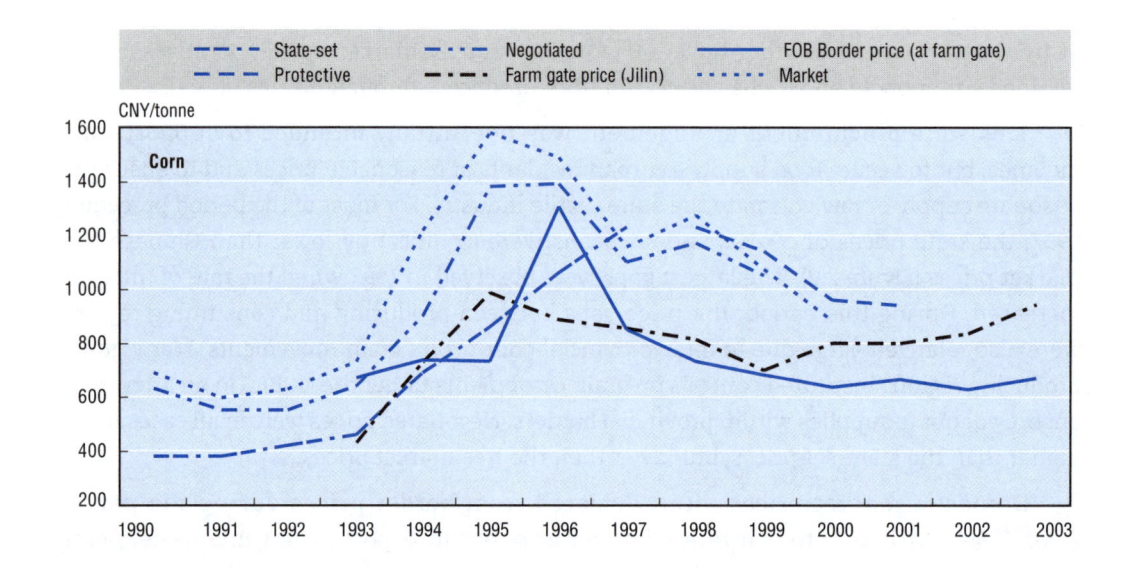

1997-2003

In late 1997, the national government reinstalled a scheme of so-called protective prices for grains in order to protect grain producers. Protective prices were prices set by the government, at which the state guaranteed to purchase all output of specific commodities offered for sale to the state. The protective prices were generally above the market price (except for soybean; Figure 2.2). There were two major problems with the implementation of this measure. First, the SGEs were expected to undertake state procurement at protective prices and to operate as a commercial enterprise marketing grains at a profit. It is reported that many SGEs sought to depress prices paid to producers or even rejected purchases of grains they believed could not be easily on-sold – profit-seeking behaviour not in accordance with the income supporting aims of state procurement at guaranteed prices. Second, budgetary constraints for many sub-national

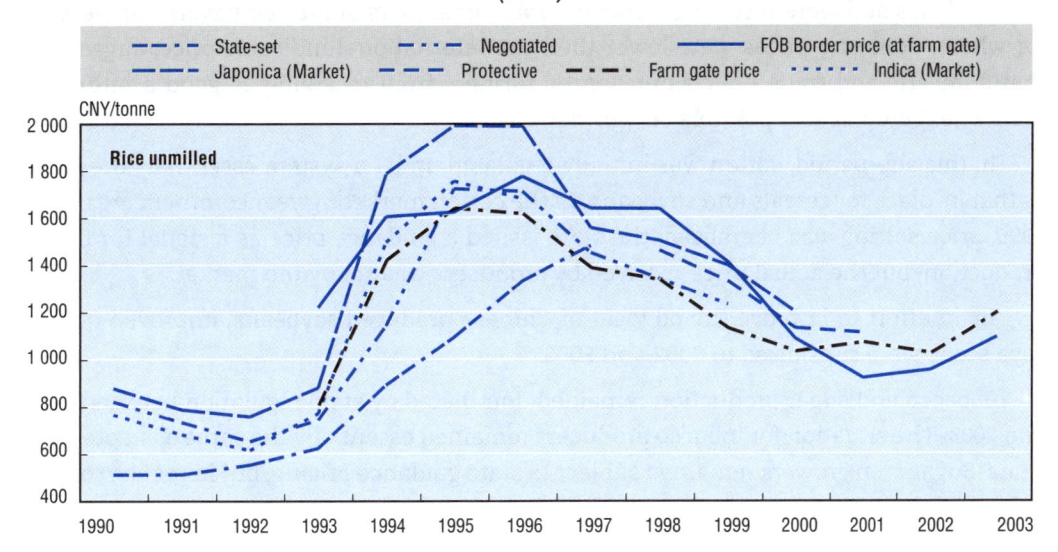

Figure 2.2. **Comparison of different types of grain and soybean prices in China**
(cont.)

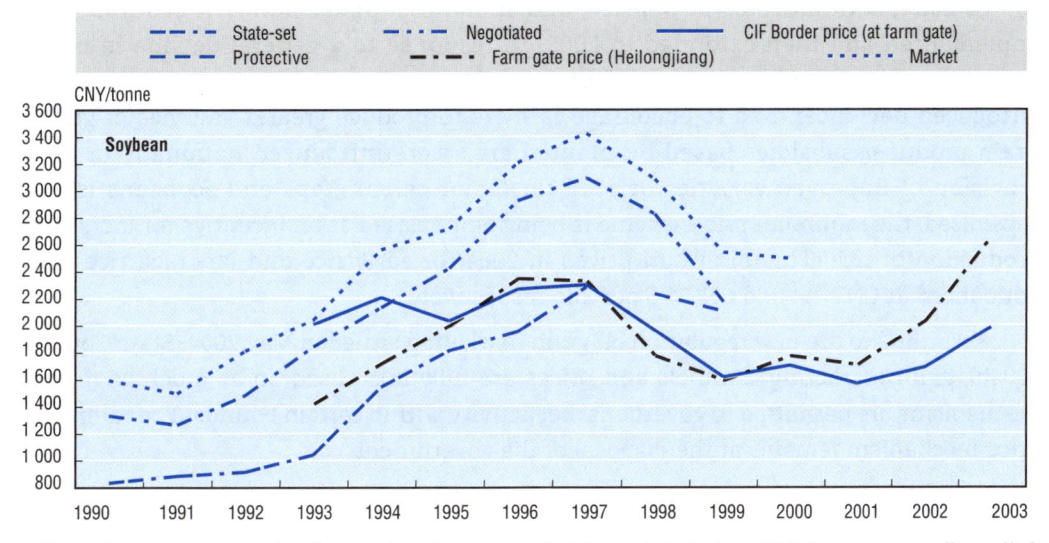

Notes: State-set, negotiated and protective prices are marked for periods during which they were actually applied and are quoted after *China Agricultural Development Report*, various editions; market prices are from the Ministry of Agriculture; border prices adjusted to the farm gate level and farm gate prices are from the OECD PSE/CSE databases 2005.

governments in major grain producing regions resulted in the policy being underfunded; payment for state purchased grains could not always be made in cash or in a timely manner.

Following the initial implementation problems with the protective price system, the national government decided to reduce the protective prices as well as coverage of this scheme. In 2000, indica rice in the whole of China, spring wheat in northern China, wheat in southern China, and corn south from the Yangtze River were removed from the scheme and protective prices for cereals that were still covered were reduced.

Between 1997 and 2003, corn, soybean and, in the 2000s, rice prices were generally above the relevant border prices. This suggests that Chinese farmers were protected by trade policies and were receiving transfers from consumers and/or taxpayers. Conversely, for wheat, domestic prices were lower than the international indicator price, suggesting that domestic and trade policies prevented farmers from receiving as good a return on their wheat crop as was possible on world markets.

In this sub-period, cotton was initially regulated under a system essentially the same as that in-place for cereals and soybeans. In the cotton marketing year commencing in late 1999, price setting was liberalised; the state issued a guidance price as a signal for future production, but the actual price received by producers was set by the market.

In an effort to improve the oil yield of Chinese produced soybeans, improved quality soya seeds were subsidised in 2002 and 2003.

Tobacco and sugar production remained dominated by state regulation between 1997 and 2003. The situation for tobacco producers remained essentially the same as in previous years. Sugar crop growers remained subject to state guidance pricing, but in reality, this set a floor price which was generally significantly less than the actual price growers received from sugar mills.

2004-2005

In 2004, a number of new price and income support policy measures were implemented and then extended to 2005. In response to a general decline in cereal production since 1998 and sharp rises in grain price in late 2003, the government introduced new incentives to encourage farmers to produce greater volumes of grains. Grain producer subsidies based on planted area were introduced nationally in 2004. Provision of improved varieties of seed for major cereal crops and soybeans is also subsidised, the minimum price scheme is being emphasised as an incentive for increasing production of rice. The minimum prices in 2004 for early rice and japonica rice were announced at CNY 1.4 and CNY 1.5 per kilo respectively.

According to the new regulation on grain marketing issued in May 2004 (State Council, 2004b), grain marketing channels and prices are fully liberalised, in as much as market mechanisms are permitted to govern market activity within certain bounds. The minimum price mechanism remains at the disposal of the government.

While this reform ends the monopolistic position of SGEs in purchasing grains, the new regulations do set some conditions on enterprises wishing to operate on grain markets. Only companies which meet certain criteria are permitted to engage in trading grains. The specific criteria are set by provincial governments. In Anhui, for example, corporations must have registered capital of at least CNY 500 000, user rights for a storage facility with a minimum capacity of 1 000 tonnes, employees qualified to test and manage the stored grain, and adequate equipment for testing and weighing grain. Household businesses must have registered capital of at least CNY 30 000 and user rights to a storage facility with a minimum capacity of at least 50 tonnes.

The government planned in 2004 to appropriate CNY 10 billion from its grain risk fund to directly subsidise producers. By mid-2004, actual payments had reached CNY 11.6 billion and had contributed to the livelihoods of about 600 million rural people (Jing, 2004). Thirteen provinces were provided with special subsidies for purchasing improved seeds of rice, wheat, corn and soybean with a total of CNY 1.2 billion expended.

Meanwhile, the government also provided subsidies for the purchase of farm machinery. This subsidy will mainly assist a small number of large farms which are of sufficient scale to warrant mechanisation. Benefits from this subsidy may accrue indirectly to the majority of farmers through reduced costs faced by farm service providers, such as harvesting companies.

The new grain subsidies are not expected to have a major financial impact on farmers. While the direct subsidies are an important symbolic change from the traditional taxing of agriculture, initial analysis indicates that the new subsidies have contributed only modestly (in the order of 5%) to increases in grain farmer's incomes in 2003-04 (Gale *et al.* 2005). A more detailed presentation of the most recent changes in grain policies is provided in Annex B.

For tobacco, sugar and cotton producers, the price and income support measures since 2004 are unchanged from the immediately preceding period.

Reduction of input costs

Throughout the period 1990-2005, China has sought to reduce the input costs faced by agricultural producers. Measures have included tax relief for input manufactures and service providers, inputs wholesale price controls, and price regulation of essential services.

Manufactured inputs supply

At the beginning of the 1980s, manufactured farm inputs were channelled predominantly through the Supply and Marketing Co-operatives (SMCs) which were quasi-statutory bodies at that time. During the 1990s, the government gradually relaxed restrictions on marketing channels, maintaining a reduced range of interventions (Chapter 1).

Until the late 1990s, supply of manufactured agricultural inputs was generally insufficient to meet the level of demand at the state-set prices. The government used controlled supply of high quality farm inputs at state prices as an instrument to induce farmers to deliver more grains, oilseeds and cotton to the State.

In 1987, the "three links scheme" (*san gua gou*), a scheme of linking farmers' deliveries of cereals, soybean and cotton with advanced payment and supplying chemical fertilisers and diesel oil at state-set prices, was introduced. The advanced payment was made before planting at a rate of 20% of the state procurement purchase from the expected crop. Supply of fertilisers and diesel oil varied depending on the crop, region and availability of inputs in that year. However, given the reality that individual households delivered only small amounts of products, the application of the scheme was administratively burdensome and largely ineffective.

In February 1993, the State Council declared a series of measures to reform the grain marketing system, one of which was replacement of the in-kind supply of fertilisers and diesel oil by input price subsidies in cash. The government set reference standards for subsidies paid by both national and provincial governments from their budgets respectively. Provincial governments were allowed to supplement the rate of subsidies. This arrangement was intended to induce farmers to sell more grains to the State. In 1994, the central government implemented increased procurement prices for grains, and ceased paying cash input subsidies – however, provincial governments retained the option of paying subsidies for manufactured agricultural inputs. Between 1994 and 1998-99, this scheme was phased out, being wound-up in different provinces at different times.

Fertilisers

The Chinese government devoted substantial resources to the establishment of modern farm input industries and distribution system during the 1970s and 1980s. As a result, supply and application of chemical fertilisers increased steadily and rapidly from the mid-1970s. Under the central planning regime, production, sale and distribution of chemical fertilisers were subject to strict state planning and pricing.

Beginning in 1985, large state-owned chemical fertiliser factories were permitted to sell outputs in excess of government planned requirements at market prices as an incentive measure. Meanwhile, the government continued to impose a ceiling on factory sales prices in order to control the price level. Development of small fertiliser factories using local resources was strongly encouraged during the mid-1980s. Fertiliser factories were also exempted from paying value-added tax (VAT).

In 1993, the regime for importing farm inputs changed from a strict STE monopoly to the agent system, in which the STEs could act as import/export agents for private individuals or companies. Along with this measure, the delivery prices of imported products under state plans were charged as CIF prices plus handling costs.[11] The agent fee for importing was set by the national government, initially as a proportion of the CIF price, being replaced by a set absolute fee in late 1994. These pricing changes ensured that Chinese agriculture was exposed to international input prices. A governor responsibility system was also installed for chemical fertilisers in 1995. A system of state reserves of chemical fertilisers and farm chemicals at national and provincial levels was subsequently established.

The policies towards small factories changed in the late 1990s. Recognising the inefficient resource utilisation and high pollution levels of many of the small fertiliser factories, government regulations were adopted which forced many of them to close.

In 1998, the State Council decided to broaden the reforms of the fertiliser supply system. Commencing in January 1999, the factory prices of fertilisers produced by large factories (with an annual turnover larger than 300 000 tonnes) changed from being a state-set price, to a state-guidance price. This reform permitted large factories to adjust fertiliser prices within a range of 10% either side of the state reference price, allowing some adjustment for fluctuations in production costs and market demand. The reference prices and price ranges of fertilisers produced by medium and small factories are determined by sub-national price bureaux, with reference to the guidance prices applied to large factories' output. State production and procurement plans were abolished, and wholesale and retail firms were permitted to buy, sell and set retail prices for domestically produced fertilisers. However, provincial price bureaux retained the right to set price ceilings for certain types of fertilisers. The intent is that state intervention would protect both producers and consumers of fertilisers from excessive price fluctuations.

Under the 1998 reforms, the State retained control of chemical fertiliser imports, with prices still determined on a full cost recovery basis – being the CIF price of goods plus VAT of 13% plus the operational costs of the importer plus the importer's commission fee. The rates for operational costs and commission fees for planned imports by the designated STEs remained subject to state regulations.

In 2004, the government continued to use administrative measures to retard the increase of farm input prices. For instance, at the time of the fertiliser price increases in early 2004, the NDRC despatched supervision teams to several provinces to check whether fertiliser manufacturing businesses were adhering to the permitted wholesale pricing regime. The government also adopted a provisional one-year measure in 2004 to offer subsidies to diammonium orthophosphate producers and importers at a rate of CNY 100 per tonne (*People's Daily*, 6 April 2004). In addition, fertiliser producers pay lower rates for electricity supplies (Table 2.2) and are exempt from contributing to the so-called agricultural power grid loan-payback fund, a fund to pay for rural electricity transmission grid improvements.

Pesticides and herbicides

During the past two decades, the government has not applied stringent planning controls to production and marketing of pesticides and herbicides. In this environment, the farm chemical industry has developed rapidly. High returns induced large capital inflows to the industry. Foreign firms have a significant involvement in this industry; developing joint ventures, transferring knowledge, and selling their branded products in the Chinese market.

Electricity

Throughout the period 1990–2004, the wholesale prices of electricity within provinces were determined by provincial governments. In general, preferential treatment is given to agricultural production and farm input industries (Table 2.2). However, the transmission systems in rural areas tends to be of a poor standard, leading to significant "leakage" of electricity – as a result, the cost per unit of electricity actually consumed by rural users has frequently been higher than for urban users.[12]

The State Council made the decision to unify domestic rural and urban electricity prices from 1998, and the government invested CNY 288.5 billion for renovation of its rural grids. Special attention was given to western China. By April 2004, all provinces had implemented the measure. It is estimated by NDRC that unifying the rural and urban electricity prices may help rural people save CNY 42 billion every year (Xinhuanet, 22 April 2004). However, it has been reported that this programme is troubled by revenue failing to cover the operational costs of supplying electricity to rural areas.

While the State Council decided to implement unified electricity prices for urban and rural domestic consumption in 1998, this measure was already incorporated in the Electric Power Law of the People's Republic of China (promulgated in 1995, effective from April 1996). Article 50 provides the legal basis for unification of urban and rural supply charges, and also stipulates that fees for electricity for use in agriculture should be set to reflect the cost of electricity supply, plus a reasonable marginal profit.

This suggests that, while rural domestic electricity consumption may have transmission costs subsidised by urban users, use of electricity in agricultural production receives no preferential treatment in national law. This does, however, leave the possibility that provincial governments may subsidise electricity use in rural areas for specific purposes. According to the data in Table 2.2 below, this appears to be the case – irrigation and drainage activities in poor counties appear to receive subsidised electricity.

Chemical fertiliser production also receives subsidised electricity, in an attempt to suppress agricultural input prices. The subsidy is received by the fertiliser producer, who

Table 2.2. **Prices of electricity by different users in selected provinces in 2002**

Unit: Yuan/kilowatt hour

Type of user	Hubei	Ningxia	Sichuan	Fujian
Urban residence	0.498	0.447	0.453	0.417
Commercial businesses	0.912	0.736	0.742	0.773
Ordinary industrial and business firm	0.564	0.494	0.549	0.592
Chemical fertiliser factory	0.393	0.420	0.402	..
Large industrial firm	0.399	0.319	0.442	0.484
Chemical fertiliser factory	0.283	0.287	0.312	0.230
Agricultural production	0.445	0.362	0.472	0.195
Irrigation and drainage in poor counties	0.215	0.215	0.157	..

Source: NDRC Web site.

then delivers product to the marketing and distribution network. As mentioned before, the fertiliser retail market has had no explicit price controls since 1998, with increasing levels of commercial competition in supply to end-users, and prices determined by market forces. Consequently, it is difficult to determine the benefit flowing to farmers from the electricity subsidy.

Transport

During the 1990s, the government regulated the prices of rail and water transportation. The charges for shipment were determined according to the type of goods being transported, and preferential rates were given to state planned shipments of agricultural products and inputs. Since May 2001, water transport rates have been deregulated, but rail freight is still subject to state-set rates.

In July 2000, agricultural products were reclassified to different categories for rail transport, generally resulting in rail freight charges for agricultural products increasing, with the exception of the rate for fresh fruit, which declined marginally. Freight rates applied since July 2000 are listed in Table 2.3. In reality, the state-set rates are not always binding given the excess demand for rail transport services and the monopolistic position of the railway authority. The scarcity of freight capacity creates a situation in which market forces are likely to bid up freight rates, especially in the absence of a viable competitor to the state railway. The government has also provided preferential treatment to shipment of certain agricultural products and inputs with respect to the railway construction fee. In April 2002, the government exempted grains and cotton from this fee, in order to assist shipment of these products. Early in 2002, the fee was CNY 0.033 per tonne-kilometre. The exemption ceased in May 2004.

Table 2.3. **Rates of railway shipment for selected goods (since July 2000)**

Commodity	Freight charges	
	Fixed (CNY/tonne)	Variable (CNY/tonne/km)
Grains, cotton, live animals, chemical fertilisers, pesticides.	7.0	0.0319
Fresh vegetables, fresh fruits.	7.9	0.0360
Frozen meats and fishery products; tobacco, vegetable oils.	8.5	0.0390

Source: Ministry of Railway.

Water

While agriculture has a special position in the Chinese economy, and water is a key input in agriculture, the water market is not well developed. Under the central planning regime, the national government managed water utilisation by issuing water allocation plans to major river-basin authorities and organising large water conservation and management projects. Local governments managed water utilisation in their localities in a similar manner. Water has been used largely as a free resource, with water prices mainly reflecting the financial costs of distributing the resource. The productive value of water is not considered in formulating water prices.

Nominally, water for agriculture has been available at a low price in order to assist agricultural production (Table 2.4). However, in practice, the water fee has often been collected on the basis of land area farmed, rather than on actual water usage, and some local leaders roll other charges into the water fee. These problems not only result in inefficient use of water in agricultural production, but also increase production costs in some areas. The central government has begun to consider reforms to water supply management and pricing, but there is currently no firm indication regarding what those reforms will comprise.

While farmers pay much lower prices for water than other users, there is no data which would provide a basis for assessing the degree of support for farmers. The price differentials evident in Table 2.4 may reflect a quality differential as well as differences in delivery costs. Moreover, many northern farmers access ground-water at their own expense and supplement it with surface water provided by irrigation authorities. In addition, different costs are likely to be encountered in producing and supplying treated urban drinking water, water for industrial use, and agricultural irrigation water.

Table 2.4. **Comparison of water prices among different usages**
Unit: CNY/m^3

Year	Agriculture	Industry	Living
1994	0.0163	0.0826	0.0868
1995	0.0221	0.0847	0.1007
1996	0.0273	0.0964	0.1211
1997	0.0295	0.1476	0.1373
1998	0.0323	0.1519	0.1509
1999	0.0332	0.1676	0.1935
2000	0.0349	0.1917	0.2037
2001	0.0361	0.2284	0.2395

Source: Survey in 2002 by the National Development and Reform Commission.

Rural credit

China established the Rural Credit Co-operatives (RCCs) during the socialist transformation of agriculture in the mid-1950s. However, during the 1960s and 1970s, the RCCs were largely converted to branches of the Agricultural Bank of China (ABC), with all major activities of the RCCs being controlled by the ABC. The programme of reform of rural financing in the early 1980s focused on turning the RCCs back into genuine co-operative financial institutions, under the guidance and supervision of the ABC.

The reforms of the early 1990s transformed the ABC into a state-owned commercial bank and its policy operations were taken over by Agricultural Development Bank of

China (ADBC). The policy-based operations of the ABC, and then the ADBC, have mainly been to provide loans to state marketing organisations to fund the purchase and storage of agricultural and side-line products. These loans may also be provided to fund forestry construction projects and water conservancy developments. As a result of the ADBC policy-based loans being made to marketing and distribution firms, rather than to farm production enterprises, it is difficult to determine to what extent the loans provided a benefit to agricultural producers.

In 1996, the responsibility for supervising RCCs was transferred to the People's Bank of China (PBC), China's central bank. Thus, the ABC ceased to be the monopoly provider of rural banking services. With these reforms, a new rural financial system consisting of the central bank, commercial banks and co-operative financial institutions has emerged.

Given the predominance of smallholdings in China's agricultural sector, banking institutions in rural areas have a large number of business clients who are limited in their production scales and are asset-poor. Farmers are often behind schedule in making mortgage payments and their incomes are uncertain as they are reliant on local weather conditions. These characteristics result in relatively higher costs and greater risks in providing credit to farmers than many other business sectors. Consequently, banks with a commercial outlook are generally not enthusiastic about providing rural credit services.

In a bid to increase their competitiveness in the face of fierce foreign competition, the state-owned commercial banks have generally withdrawn credit services from most counties and rural areas to focus on more profitable opportunities in the larger cities. The main task of their rural branches has been to attract rural deposits. A large quantity of deposits drawn from rural areas provides credit to profitable urban businesses. As a result, the burden of financing agriculture is left mainly to RCCs. As indicated previously, the RCCs operate in a risky sector for credit provision. At the end of 2002, the non-performing loans of RCCs stood at CNY 515 billion, accounting for 37% of their total outstanding loans.

The ADBC provided a wide range of preferential loans in line with government policy programmes. Loan quotas and loan eligibility criteria vary between programmes, as does the level of interest rate subsidy (paid by central government budget allocation). This included loans to major grain and cotton producing counties and to high-yield, high-quality and high-profitability agriculture demonstration counties. Between 1994 and 1998, CNY 36.57 billion in loans was used to assist the government policy programmes (*Farmers' Daily*, 16 April 1999).

Preferential loan rates were below the prevailing commercial rates (Table 2.5). The interest rates of loans to poor regions and for poverty alleviation are the lowest. However, the availability of preferential loans was limited and allocation of loans, not being part of an entitlement programme, was frequently made according to a process of negotiation between the lender and supporters of the loan application.

Significant problems existed in the targeting of these preferential, poverty alleviation loans. As the loans were initially targeted at selected counties, loan funds did not necessarily end up in the hands of the poor, and the uses to which the loans were put did not necessarily benefit the poor. There is some evidence that loan funds were diverted to supplement sub-provincial government budgets and that loans were often made to industrial enterprises rather than to support agricultural production development (Wu, 1997; Rozelle, Zhang & Huang, 2000; Park, Wang & Wu, 2002).

Table 2.5. **Annual interest rates of selected types of loans (%)**

For loan approved at ...	1 Jan. 1990	21 April 1991	15 May 1993	23 Aug. 1996	1 July 1998	7 Dec. 1998	10 June 1999	June 2002
Commercial loans								
Loan period								
Less than 6 months	11.34	8.10	8.82	9.18	6.57	6.12	5.58	5.04
7-12 months	11.34	8.64	9.36	10.08	6.93	6.39	5.85	5.31
1-3 years	12.78	9.00	10.8	10.98	7.11	6.66	5.94	5.49
Preferential loans								
Various loan periods								
Purchase of grains and cotton	10.08	7.74	8.46	9.18	n.a.	n.a.	n.a.	n.a.
Rural region development	7.02	5.76	6.48	7.20	4.05	3.60	2.97	
Loan for poverty alleviation	2.88	2.88	2.88	2.88	2.88	2.88	2.88	

n.a.: not applicable.
Note: The loan interest rates were adjusted frequently; this table covers only those for selected periods.
Source: People's Bank of China.

According to the People's Bank of China, short-term agricultural loans issued by the banking sector in 2003 totalled CNY 841.2 billion, accounting for 10% of banking sector short-term loan activity. In the same year, the banking sector made loans worth CNY 766.2 billion to village and township enterprises. As the primary industry sector (farming, forestry, animal husbandry, and fisheries) contributed around 15% of China's GDP in 2003, a loan activity rate of around 10% seems to indicate that the agricultural sector is under-represented in the bank loans market. The gap has partly been covered by the rapid development of informal financial institutions (Chapter 1).

Agricultural taxation

China has a dual tax system, with the agricultural sector being taxed according to a different system to that applied to the industrial and tertiary sectors. Sub-provincial level fees and charges levied on individuals in rural areas are also different to those applied to urban residents.

In concert with China's economic liberalisation, the taxation system underwent major changes. Within the period 1990-2004, the most wide-ranging taxation reforms occurred in 1994. The main reforms relating to agriculture were the implementation of a value added tax (VAT) and the devolution of responsibility for collecting the Animal Slaughtering Tax to provincial governments. The tax revenue sharing arrangement introduced at that time underlined the emerging economic responsibilities at the provincial government level and paved the way for devolution of other agricultural based taxes.

The main taxes levied on agriculture are the Agricultural Tax, the Animal Slaughtering Tax (prior to 2004), the Special Agricultural Products Tax, and the VAT.

Agricultural Tax

Under the central planning regime, agricultural tax was used as an instrument for the State to acquire grains by collecting tax in the form of grains. The tax base used was a so-called "constant yield" base, which resulted in variability in the effective rate of taxation between regions and between years. The tax calculation was set in 1958, calculated as a certain percentage of the normal yield of the taxable cultivated land. The tax rate varied according to local conditions, but was to be no more than 25%. The national average tax rate for this tax was 15.5%.

In 1983, the government adopted a policy to fix the amounts of the Agricultural Tax as well as implementing state procurement of grains. With this reform, the Agricultural Tax became a lump-sum tax in terms of volume of grain turned over to the State. Over time, the effective rate of tax tended to differ between localities due to different rates of production growth.

Reform of agriculture-related taxes was proposed in 2000, leading to the trial of a system in which the Agricultural Tax was set at 7% of the annual grain-equivalent value of agricultural output for the years 1993-1997, with an additional 1.4% replacing a range of local administrative fees and charges, the so-called peasant burden (Box 2.2). This new system was due to be implemented nationally in 2001, but budgetary constraints prevented this occurring, although the experimental implementation of these reforms from 2002 took place in counties which constitute about three quarters of the rural population. These reforms were implemented nationally in rural China in 2003. In 2004, the national government announced the progressive phasing out (over a five-year period) of the Agricultural Tax, commencing in 2004. A number of provinces voluntarily reduced the levels of Agricultural Tax levied, prior to this national policy being announced (see sub-section below on current tax reform).

Animal Slaughtering Tax

The Animal Slaughtering Tax was collected by sub-national governments. In principle, this was a lump-sum tax imposed on the number of animals bought and sold. The tax was a set fee determined by governments below the provincial level. The applied tax rate varied for different species and in different regions. It is reported that some local governments imposed this tax on all animals raised or even allocated planned tax revenues to all farm households based on their land areas. During the 1990s, the national government issued instructions several times to stop such practices.

Under the general reforms to agriculture-related taxation developed and implemented in the early 2000s, this tax was to be gradually abolished as the reforms were implemented across China. Liu and Liu (2004) note that by 2004, this tax was no longer collected in any province.

Special Agricultural Products Tax

Special Agricultural Products Tax was formally introduced in 1983, shortly after the household responsibility system was adopted nationally. This tax was calculated on the output value of certain taxable agricultural products (*e.g.* tobacco, horticultural products, wool and animal skins, timber and rubber, fishery products). The tax was applied to high value products in an attempt to dissuade farmers from abandoning grain production. This tax was retained in the 1994 taxation system reforms, with only minor revisions. During the late 1990s, the tax rates were frequently revised (Table 2.6). This tax was criticised by Chinese economists and other scholars as hindering the adjustment of Chinese agricultural production to align with China's comparative advantages. In the reforms of the rural taxation system since 2002, the central government allowed provincial governments to decide how to reform this tax for application at the provincial level. The only stipulation was that the provinces bear any impacts on their tax revenue. In 2004, the central government announced that, henceforth, the Special Agricultural Tax would apply only to tobacco (Table 2.6).

Table 2.6. **The rates of tax on special agricultural products – % of value**
Selected products

Taxable commodity	1990	1994	1997	1999	2004
Citrus	15	12	12	12	0
Apple	15	12	12	12	0
Melons	10	8	8	8	0
Mushrooms	..	8	8	8	0
Tea	..	16	12	12	0
Tobacco	..	31	31	20	20
Aquaculture	10	8	8	8	0

Source: State Administration of Taxation.

Value Added Tax

Value Added Tax was introduced in 1994. Prior to 1994, marketing and processing of agricultural products was subject to the business (product) tax. This tax was replaced by the Value Added Tax under the 1994 taxation reforms. The VAT rate for agricultural products is 13%, 4 percentage points lower than the VAT rate generally applied to other products.

VAT is not collected from the primary producers of agricultural products, but is collected from the primary purchasers of agricultural products, when they on-sell. Primary handlers can claim 10% of the original purchase price as a deduction from their VAT liability. VAT liability is calculated as "output VAT" less "input VAT" – for agricultural products "input VAT" is taken to be 10% of the input purchase price. This is intended to provide an incentive for purchasers of farm products to buy more, and to prevent downward pressure on prices received by farmers.

In addition to VAT not being applied to the primary producers of agricultural products, VAT exemptions or reduced rates have also been applied to some important agricultural inputs, such as fertilisers. The benefit of this to farmers is difficult to determine as these exemptions are generally applied in the production of the inputs rather than in the sale to farmers. The intention was that farmers would pay lower prices for inputs than would otherwise be the case.

Business Tax

Business Tax has existed in China since prior to the tax reforms of 1994. Like VAT, it is not collected from farmers. Moreover, agricultural services, such as agricultural machinery services, irrigation and drainage services, pest and disease control, and animal breeding services, are generally exempt from paying the Business Tax. This is intended to reduce these input costs for farmers.

The agricultural taxation burden

Agriculture-related taxation represented 1.7% of agricultural GDP in 1990, growing to account for 4.2% of agricultural GDP in 2002. The growth in agriculture-related taxation as a proportion of agricultural GDP is partly associated with the structural adjustment of agriculture. The rapid growth of horticulture, and livestock and fishery production, resulted in revenue from the Animal Slaughtering Tax and the Special Agricultural Products Tax rising significantly during the period 1990-2003. Moreover, the significant rise in the combined Agricultural Tax and Animal Husbandry Tax in 2002 was largely due to the

Table 2.7. **Agriculture-related taxes**

Unit: CNY million – current prices

Year	Total	Agricultural tax and animal husbandry tax[1]	Contract tax[2]	Tax on special agricultural products	Tax on use of cultivated land[2]
1990	8 786	5 962	118	1 249	1 457
1991	9 065	5 665	189	1 425	1 786
1992	11 917	7 010	361	1 624	2 922
1993	12 574	7 265	621	1 753	2 935
1994	23 149	11 951	1 182	6 369	3 647
1995	27 809	12 812	1 826	9 717	3 454
1996	36 946	18 206	2 520	13 100	3 120
1997	39 748	18 238	3 234	15 027	3 249
1998	39 880	17 867	5 899	12 779	3 335
1999	42 350	16 308	9 596	13 143	3 303
2000	46 531	16 817	13 108	13 074	3 532
2001	48 170	16 432	15 708	12 197	3 833
2002	71 785	32 149	23 907	9 995	5 734
2003	87 177	33 422	35 805	8 960	8 990

1. Animal slaughtering tax is not classified as an agriculture-related tax under the current Chinese taxation system, but it is likely to have an impact on meat production.
2. Contract tax and tax on use of cultivated land are both related to conversion of cultivated land to non-agricultural usages, contract tax has minimal impact on the agricultural sector.

Source: China Financial Yearbook 2004, Ministry of Finance.

impact of the trial or partial implementation of the Agricultural Tax reforms. Notably, township and village fees and charges not previously categorised as "agricultural taxes" became part of the Agricultural Tax, boosting the tax revenue collected under that category (Table 2.7).

Due to the diversity and complexity of the dual tax system operating in China, it is difficult to determine whether farmers benefit from or are disadvantaged by the system, compared to urban residents and non-agricultural businesses.

In the period 1990-2003, each year the Chinese agricultural sector contributed on average 4% of China's official taxation revenue and generated 19% of China's GDP. This seems to indicate that the agricultural sector receives preferential taxation treatment. Moreover, farmers engaged only in agricultural production pay the Agricultural Tax, and possibly the Animal Slaughtering Tax (prior to 2004) and Special Agricultural Products Taxes, but they do not pay business taxes, personal incomes taxes, or business income taxes unless they are engaged in non-agricultural economic pursuits from which they derive some income. Urban residents are also subject to a range of taxes such as the House Property tax and City Maintenance and Construction Tax, which rural residents do not pay.

However, such comparison does not take account of the generally lower incomes in the rural sector and farmer's relatively lower ability to pay taxes, the non-tax government revenue collected from the rural sector, and the implicit taxation involved in the state procurement system. Arguably, until the late 1990s, farmers made a contribution to the national treasury through implicit transfers associated with state pricing and procurement. Mandatory state purchases of grains at prices lower than those prevailing in the grains market meant that the state denied farmers a proportion of the return they could have expected if their grains had been sold at market prices.

Apart from formal taxation discussed above, the Chinese farmers had to pay various fees and charges imposed by township governments and village leaders, which supposedly were used for communal social services (Box 2.2). These township and village fees and charges were paid by rural-classified residents whether or not they were engaged in agriculture-related activities (the relatively small number of urban-classified people dwelling in rural areas do not pay these fees). Aubert and Li (2002) estimate that the value of payments made by farmers (including agricultural and related taxes, fees for townships and the villages, legally and illegally collected funds, apportionments and fines) could be between CNY 180 billion and 220 billion in 2000. This is 4-5 fold more than officially collected agriculture-related taxes, but also between 40% and 50% more than the officially estimated "peasant's burden", which in addition to agriculture-related taxes includes official township and village levies (five *tongchu* and three *tiliu*; Box 2.2).

> **Box 2.2. A brief history of the "peasant burden"**
>
> Until 2000, farmers were subject to four major types of taxes:
>
> – Government taxes – such as Agricultural Tax and Animal Slaughtering Tax.
> – Township levies – the five *tongchou*, for education, public transport, military expenses, family planning, and social expenses.
> – Village levies – the three *tiliu*, for the public accumulation fund, public welfare fund, and other administrative expenses.
> – Miscellaneous fees, levies and fines – paid to various government institutions at different hierarchical levels.
>
> In 1991, the national government set a limit on the fees and charges levied by township and village administrations (the five *Tongchou* and three *tiliu*). Under the Regulation on Fees and Labouring, the burden imposed on peasants at the township and village levels was capped at a maximum of 5% of the previous year's net income. It is reported that many fees and charges were still often determined arbitrarily by local township and village authorities, and classified as miscellaneous fees, levies and fines, in an attempt to cover expenses at the township and village level. This resulted in officially reported township and village fees and charges being within the regulated limit, but with payments of surcharges and penalties, and payments to other institutions being largely unregulated additional payments at the township and village level.
>
> During the 1990s, the cost of "official" township and village fees and charges almost doubled the cost of "normal" taxes. In addition, farmers were also obliged to provide 10–20 unpaid workdays for local development projects, such as constructing roads and repairing ditches. This practice should come to an end in line with recent reforms of the agricultural taxation system.
>
> Source: Aubert and Li (2002).

Current tax reform

As described above, the rural taxation reforms, tried in 2000, and progressively implemented in rural China from 2003, attempt to address the issue of fees and charges for farmers, incorporating most agricultural taxes, fees and charges in one tax, and capping the tax at a maximum rate (8.4%) relative to the annual grain-equivalent value of agricultural output for the previous years. Reforms include the removal of the Animal Slaughter Tax and of the Special Agricultural Tax on all products except tobacco.

In addition, the Chinese central government announced in 2004 that the Agricultural Tax would be phased out over five years, beginning in 2004. The initial 1 percentage point reduction in Agricultural tax is expected to cost the Chinese bureaucracy around CNY 11-12 billion in 2004. Trial implementation of the Agricultural Tax abolition commenced in Jilin and Heilongjiang, two important grain producing provinces, in 2002. In 2002 and 2003, the central government transferred around CNY 24 billion and CNY 31 billion respectively, to those two provinces to fund this reform. The appropriation rose further, to CNY 39.6 billion in 2004.

While the national government set the maximum level at which agriculture-related taxes may be applied, it also permits provincial governments to collect agriculture-related taxes at a lower tax rate than the maximum rate. Shanghai decided not to collect the Agricultural Tax in 2003 and continued this measure in 2004. Beijing, Tianjin and Xiamen City took this step in 2004.[13] The government of Zhejiang announced that in 2004 it would not collect the Agricultural Tax from farmers producing grain and oil crops. Hebei, Inner Mongolia, Liaoning, Shandong, Jiangsu, Jiangxi, Anhui, Henan, Hubei, Sichuan and Guangdong reduced the Agricultural Tax rate by 3 percentage points. Sub-provincial measures have been applied in some areas, for example Guangdong province will collect no agricultural taxes in 2004 in Pearl River delta (Jing, 2004). At the beginning of March 2005, the government announced that agricultural tax reform should be further accelerated with the aim of phasing out all national farm taxes in 2006. An additional amount of CNY 14 billion will be transferred to provinces undertaking rural tax reform (MOF, 2005).

While rural tax reform is designed to provide more transparency, diminish abuses, eliminate illegal fees and charges, and to unify the rural and urban tax systems, there are two main threats to its sustainability. First, the success of reform depends on continued and growing (during the implementation period) tax revenue transfers from central government to provinces and counties as compensation for lower sub-national tax revenues. This will necessitate a more general tax reform in China to increase central government tax revenues and to institutionalise the system of distributing tax revenues across various levels of administration. Second, as discussed above, the official value of the "peasant burden" seems to be underestimated. Therefore, even if some taxes are successfully phased out and some illegal practices discontinued, the reform is unlikely to result in the non-collection of miscellaneous fees, levies and fines (paid to various government institutions at different hierarchical levels), as they seem not to be sufficiently accounted for in the costs of the reform.

Rural public services infrastructure

Rural public services infrastructure policies include policies on agricultural research and development, agricultural education services, food safety inspection services, physical infrastructure, and public stockholding.

Agricultural research and development

Research and development of agricultural technologies has traditionally been carried out by state owned research institutions, and technological advances were then distributed through state extension services. This top-down, hierarchical system was criticised due to its apparent inability to respond to the changing demands placed on it in a more market oriented economy.

Reforms of the agricultural research system were first undertaken in the mid-1980s, with a directive to decentralise decisions on research priorities and to allow research institutes to sell the technologies they developed as a supplementary source of research funds. Since then, agricultural research and development has become more service-oriented and has been partially driven by market forces; grain-related research and development continues to be largely state funded, while research in the livestock sector is largely supported by private funding. While the reforms have increased the financial resources of research and development bodies and enhanced their responsiveness to market demand, some farmers have borne an increased share of the costs of developing new technologies. Similar reforms were also carried out with respect to the agricultural extension system (discussed in the next sub-section).

In 2001, the State Council released the Agricultural Science and Technology Development Program for 2001-2010. The programme is to focus on developing high-quality, high-efficiency, and low-cost sustainable agriculture, in order to improve farm incomes and ensure food self-sufficiency (*People's Daily* on-line, 24 May 2001).

Statistics on government funding for agricultural research and development are not reported in a consistent or systematic way by all the agencies involved. In general, however, the budgetary funding of agricultural research and development tended to increase until 1998. As shown in Table 2.8 below, the actual level of budgetary expenditure on research and development in 2001 was still lower than in 1995, although funding levels had recovered from a low point in 1999.

Agricultural extension services

The disbanding of the People's communes between 1978-1983 resulted in profound changes in agricultural extension activities. The system under which agricultural extension services could address a collective and obtain a collective decision to adopt new techniques or technologies became obsolete with the introduction of the household responsibility system. With this change, technical extension services had to target individual households and extension service activities became much more labour intensive than previously. As farmers had limited means of reducing risk, they tended to be conservative in the adoption of new techniques and technologies. These factors resulted in paralysis of the extension system in the early 1980s.

In response, the government committed to increasing extension service funding. The Agricultural Law and Agricultural Technological Extension Law were promulgated in 1993, setting up a legal framework for agro-technical services. Extension service personnel were assigned as township government officials and funding for extension service activities was made available in township budgets. Development of fee-for-service activities in the non-staple food sectors was also encouraged, although farmer up-take of these services was relatively slow. In addition, agricultural extension services were exempted from paying business tax and VAT in an effort to improve their commercial viability.

To some extent, provision of agricultural extension services has been fragmented. Many national programmes administered by many different ministries have contained an element of extending the use of improved agricultural techniques. One example of this is the "dragon-head company" approach to provision of extension services. Under this approach, a particular company (the "dragon-head") which receives preferential government treatment provides extension services. Initially developed in the 1980s in the coastal areas of eastern China, under this system the agricultural product company contracts with farmers for provision of raw materials. In order to ensure an on-going supply of quality products at fixed prices, the companies support their suppliers by providing relevant technologies, training and information.

Farmer education

Beginning in 1990, the government implemented a project of agricultural production skill enhancement and certification for farmers. This project was intended to improve farmer's skills and so, improve agricultural production. By mid-2000, over 10 million farmers had been trained under the programme, of which about 4.6 million received certificates (Sun et al., 2002). Other agricultural training initiatives include the operation of specialised technical schools, other rural vocational schools, radio and television agricultural educational programming, rural cadre schools, and remote area educational services. The government provides financial support to all these activities.

The National Bureau of Statistics of China (NBSC) collects volume statistics on formal education but does not collect data on educational expenditure. This data suggests that the distribution of educational facilities is skewed towards towns and cities for educational levels above elementary schooling. An estimate of compulsory education expenditures in 1998 suggests that the ratio of budgetary to non-budgetary expenditures for compulsory education is similar between urban and rural areas (Su, 2004).

According to the Ministry of Finance, budgetary expenditure on agricultural formal education and on extension services has tended to rise in the period 1993-2001 (Table 2.8).

Inspection services[14]

In recent years, China's exported agricultural products have frequently been banned by importing countries on the grounds of failing to meet relevant product safety standards. Food safety incidents have also occurred frequently in China's domestic food markets in the period 1990–2004. In response to this problem, the government began to improve both food safety standards and the food standards inspection system. Initiatives such as production of "hazard-free" agro-food products, "green food" certification, and the establishment of disease-free areas have been implemented with financial assistance from the government. While increased stringency of food safety standards usually increases the costs of production, and may increase product rejection rates, it is also the case that a reduction in consumer risk will work to support higher prices and may improve export opportunities and returns.

Food safety, particularly since the latter part of the 1990s, has become a high profile concern of urban consumers. Specific government responses to these concerns have included improvements in regulations on inspection, scientific developments to improve testing of foods, and improvements in the food safety standards framework.

Since 2003, nineteen research institutes in China have been engaged in developing a standardised framework of food safety standards. The framework is being developed to align Chinese domestic standards with the international food safety standards of the Codex Alimentarius Commission. The framework will be wide-ranging, specifying maximum residue limits for hazardous substances, detailing technical benchmarks for testing procedures, and setting limits on hazardous materials in agricultural inputs such as irrigation water. Technical regulations governing food storage and transport are being developed to incorporate the Hazard Assessment Critical Control Point (HACCP) system. The framework is scheduled to be completed in 2005.

According to the Ministry of Finance, in 2001 (the latest year for which data is available) inspection services received an increase in budgetary expenditure, 53% higher than the preceding year.

Production infrastructure

Investment in agricultural infrastructure is by far the largest component in the government's budgetary support for agriculture (Table 2.8). Financial inputs in the construction of the rural infrastructure have come mainly from the State, while farmers have been mobilised to make labour contributions to infrastructure developments. Traditionally, the state investment focus has been on water management projects, which benefit the industrial and urban domestic sectors, as well as providing some benefits to the agricultural sector.

Land reclamation and improvement of low yield land areas have also received funding from the state, mainly under the Comprehensive Agricultural Development Programme. Commencing in 1988 this programme was intended to transform existing medium and low-yield cultivated lands, reclaim all exploitable wasteland resources for farming purposes, and establish new agricultural production bases. Financial support for projects under the programme came from agricultural development funds in central and sub-national government budgets (including funds raised through state bonds issue – see Box 2.3), special loans from state banks, funds raised by collectives and farmers, and international funding from organisations such as the World Bank, the Asian Development Bank, and the United Nations Development Programme.

In the period 1988-2002, Wang and Li (2004b) report that total domestic inputs to the programme were CNY 171.1 billion, of which CNY 48.35 billion were contributed by the central government, CNY 43.46 billion were contributed by sub-national governments, CNY 57.84 billion was contributed by collectives and farmers, and CNY 21.36 billion was supplied through bank loans. The projects are distributed throughout the country and under which 21.5 million hectares of low yield land areas have been improved and over 2 million hectares have been reclaimed for afforestation.

> Box 2.3. **Public debt funds in agricultural development**
>
> In the period 1998-2002, the state arranged almost CNY 190 billion of state bond funded investment in agriculture, fisheries and forestry (MOA, *China Agricultural Yearbook*, 2003). According to the MOA, CNY 126.3 billion of bond funds were to be invested in water conservancy works in that period, accounting for around two thirds of central government investment in agricultural infrastructure; CNY 44.3 billion was planned to be invested in forestry and ecological works (including the so-called Grain for Green Programme discussed in the environmental policy section); while CNY 6.7 billion was arranged to be invested in other agricultural projects such as construction of cotton production bases in Xinjiang and funding to support seed and livestock improvement programmes.

In the last five years, the government has increased investment in rural environmental protection projects, such as the establishment of forest belts in northern China and in the upstream catchments of the Yangtze River, and the extension of water-saving technologies in areas with poor water resource endowments. Beginning in 2001, the national government funded the "six rural small projects", which are defined as: increasing the use of water saving irrigation, improving the supply of drinking water to people and livestock, developing energy sources from agricultural by-products, construction of small hydro-power stations, fencing of grassland, and paving country roads. Total investment reached CNY 28 billion in 2003 (Jing, 2004).

In line with the 2004 "No. 1 Document", the government planned to increase its financial support for agricultural infrastructure from CNY 120 billion in 2003 to CNY 150 billion in 2004. The main projects include spending on improved irrigation facilities, rural roads, methane production facilities, rural hydroelectric plants, pasture enclosures, and construction of agricultural high technology parks (Gale *et al.* 2005).

Overall, state investments in agricultural infrastructure have tended to increase, but agriculture's share of total capital investments declined notably. Moreover, according to the report by the Auditor General Li Jinhua to the NPC in 2004, state funds provided for agricultural infrastructure projects are not always used properly (*People's Daily* online, 24 June 2004).

Information infrastructure

The government has developed a number of initiatives to improve information dissemination in the agricultural sector. These initiatives are aimed at improving market efficiency, improving food quality, and improving marketing opportunities in the agricultural sector.

In September 2001, the Ministry of Agriculture issued the "Action plan of rural market information services during the tenth five-year plan". It is proposed to establish an information system covering all counties and the majority of townships, agriculture-related enterprises and wholesale markets. The purpose of the network is to provide market information to all agricultural stakeholders, and improve the delivery system for technical agricultural information.

The Internet is also used to provide economic and market information.[15] Many specialised agricultural web sites have been established to provide market and policy information to the agricultural sector. For example, the Ministry of Agriculture opened a vegetable products wholesale information network in January 1995, which linked vegetable and other non-staple food wholesale markets in major cities. In the same year, the China agricultural information network was opened. The China Central TV (CCTV) agricultural channel[16] started in November 1995. In the tenth five-year plan (2001-2005), the central government began to accelerate the construction of the rural economic information system.

While the various governments in China have plans to improve the flow of both technical and market information to farmers, it appears that any current benefits from electronic information delivery rely largely on non-electronic dissemination of information to the end-users. Television coverage in China is high, with satellite, cable and free-to-air broadcasting, and reports of the possible television audience vary from 84% (Thomas, 2003) to 95% (*South China Morning Post*, 31 August 2004). However, media reports *e.g.* on internet access in China highlight that many people in rural areas have either no electricity supply, or an inadequate and unreliable supply; yet, electricity is a pre-requisite for both internet and television access. The same reports point out that personal computer ownership in rural areas is likely to be considered as an extravagance.

Promotion of China's agricultural products in international markets has mainly been carried out by several specialised commercial chambers. These include the China Council for the Promotion of International Trade, and the China Chamber of Commerce for Import and Export of Foodstuffs, Native Produce and Animal By-products. While organisations such as these are assisted by the national government, major contributions to market promotion activities come from relevant member enterprises.

Public stockholding

China has engaged in large scale public stockholding of food grains, and continues to do so. China's grain reserve policies have been dominated by concern for national food security in the event of supply shocks, such as widespread harvest failures or crop destruction due to floods and other natural occurrences. The quantities of grains held in Chinese state grain reserves is a state secret, and traditionally, grain reserve policy detail has not been easily available to the general public. However, with the increasing transparency in the process of governing, an improved analysis of grain reserve policies is becoming possible.

A more tangible reserve policy than that previously in place started to emerge in the early 1990s, following the good harvests in 1989 and 1990. With grain reserves already comprising grains obtained by levying the Agricultural Tax and through the state procurement quota, the government introduced a special reserve scheme to absorb the extra grains bought under support prices (State Council, 1990). The purchasing and storage operations were carried out by the State Grain Reserve Bureau, which was established in 1990. However, due to limited funding and limited storage capacity, implementation was largely ineffective.

In early 1993, the government attempted to establish "grain risk funds" to be used for policy operations associated with grain procurement and reserves. Grain risk funds were created under a circular from the State Council and the Central Committee of the Communist Party of China, "On policy measures for agricultural and rural development". The funds are budgetary allocations at the central and provincial levels which are intended to ensure that grain policy is implemented. The funds are principally intended to ensure that funds are immediately available to undertake policy or government directive action pertinent to maintaining grain reserves and grain price stability. Provincial contributions to the funds are a set proportion of national contributions and funds contributed by the provinces are to be used at the discretion of the provincial government.

The government began to improve storage facilities and establish a system of separate central and local grain reserves. The central reserves are used for ensuring national food security and stabilisation of the national grain market. Local reserves are used for ensuring local food security under certain unusual situations, such as coping with natural disasters. SGEs are often used as agents by local governments for managing local reserves. SGEs also have their own stocks based on the requirements of their commercial operations.

It should be noted that with significant decentralisation of administrative power during the 1990s, under the GGBRS provincial governments were made responsible for implementing regional grain policies. Regional policies are generally to be funded from provincial budgets. Sub-national governments are also required to establish local grain reserves. There appears to be notable variation in the implementation of state procurement and the associated state prices, and approaches to implement national policies also vary between regions. The regional variation seems to be driven mainly by variations in resource endowments and public financial capacity.

Closely related to China's grain reserve policy, is its grain storage capacity, although only a proportion of total storage capacity is utilised to store government grain reserves. There was a severe shortage of grain storage capacity in the 1980s and the existing grain storage facilities were in poor repair. Commencing in the early 1990s, China's grain storage capacity was improved. Investment in infrastructure improvements came from both the Chinese government and international sources. For example, in 1993, China launched the "China Grain Project" to improve the grain distribution and marketing system, with external funding of USD 490 million provided by a World Bank loan. During the 1990s, over 55 million tonnes of storage facilities were constructed specifically for national reserves. The Chinese government also allocated CNY 33.7 billion between 1998 and 2001 for the construction of a further 51.5 million tonnes of storage facilities (State Grain Bureau 2002). According to Song et al. (2002), the national government grain reserve is planned to reach a capacity of 75 million tonnes.

On 9 June 2000, with the approval of the State Council, SINOGRAIN (the China Grain Reserve Corporation) was established. SINOGRAIN has a duty to manage national stocks of grains and vegetable oil products and covering procurement, stock maintenance, interregional shipments, domestic sales, and import and exports. It is also responsible for constructing and maintaining stock facilities. While the capital investment comes from the national government, the corporation is required to take full responsibility for maintaining and increasing its assets.

Budgetary expenditure on public stock holding fluctuated significantly in the period 1996–2000. In 2000, the establishment of SINOGRAIN corresponded with a 26% annual increase in expenditure on public stockholding, while the increase between 2000 and 2001 was marginal (Table 2.8).

Consumer measures

Until 1992, the supply of grains, vegetable oils and some other important food products in the urban sector was subject to rationing. In contrast, rural consumption was characterised by self-supply of most food products.

In the early 1990s, the government placed a particularly high priority on securing price and supply stability in urban food markets. As a result, in the event of a rise in state retail prices of major food products, a lump-sum subsidy was paid to wage-earners. The subsidy was designed to ameliorate the impact of the food price increases, rather than calculated to fully offset the additional food costs faced by individual urban consumers. Following the initiation of the subsidy, each subsequent food price rise entailed an associated subsidy level increase. The subsidised food prices could be expected to increase the quantity of food demanded, thus increasing prices further, and potentially contributing to an inflationary spiral. Additional administrative intervention to dampen food price fluctuations was also used in "emergency" situations, such as when inflation accelerated in mid-1994.[17] Some of these subsidies remain in place, although currently they are of marginal importance (Table 2.8).

Environmental measures

The Chinese government began to place a high priority on environmental protection only after food became relatively abundant in the late 1990s. The major environmental initiatives include large scale forestation projects in areas suffering from soil erosion, measures to prevent and control desertification and sand encroachment in arid northern regions, protection of wetland areas, wildlife and wild flora conservation, soil and water conservation, and control of non-point pollution associated with agricultural activities.

In addressing these environmental problems, both as remedial and as preventative measures, the main focus has been on command and control instruments, involving the application of technical improvements, regulations, dissemination of information together with, to a lesser extent, the use of economic instruments such as taxes, charges and various forms of financial support. Since the late 1990s, the national government has provided financial assistance to these activities mainly using funds raised by issuing government bonds (Box 2.3). The government has also attempted to formulate an effective incentive framework to encourage private participation in environmental protection measures.

Several big afforestation projects were launched in the 1990s to control soil erosion and desertification. During 1998-2000, the national government allocated CNY 22.26 billion to natural forest protection projects. While such projects have indirect benefits to agriculture, they focus primarily on protecting the natural environment. The State Forestry Administration (2004) reports that during the 1990s the total forested land area in China increased.

Grain for green project

A major environmental policy initiative is the so-called "grain for green project", launched in 1999, and officially titled the Returning Farmland to Forests Programme. Under this programme, cultivated lands in environmentally fragile areas are "retired" from crop production, and converted to pasture or forest. Initially, this project covered only 14 trial counties in Hunan, Hebei, Jilin and Heilongjiang provinces. It has recently been extended to other provinces. Under this project, participating farmers are provided with grains and cash subsidies according to the area of damage-susceptible land they "retire". According to the criteria set by the national government, for each *mu* (one-fifteenth of a hectare) retired, farmers in the upstream regions of the Yellow River basin in northern China receive yearly 100 kg of grains and CNY 20 in cash; and in the upstream regions of the Yangtze River basin they receive 150 kg of grains and CNY 20 in cash. The period for which "retired" land is subsidised is set at two years for land returned to pasture, five years for land converted to "economic" forests and eight years for land converted to "ecological" forests. Free seedlings are also made available for afforestation. The programme's costs are born mainly by the central government (Table 2.8).

Under this programme, trees were planted on about 8 million hectares of cultivated land between 1999 and 2003 (State Forestry Administration, 2004). However, the future of this project is in doubt. Farmers can be expected to keep land in "retirement" provided the economic benefits of doing so, out-weigh the benefits of farming that land. In a study addressing this question, Uchida, Xu and Rozelle (2004) found that, although farmers generally were better off when receiving payments under the programme, there was a significant likelihood that "retired" lands would be returned to cultivation when payments ceased. This is partly due to the programme stipulation that 80% of trees planted should be ecological trees (not providing a direct cash return). Thus, unless the resources freed from farming can be more productively employed not farming the set-aside plot, it is likely the plot will be returned to farming when the subsidy incentive ceases.

China is also actively addressing water usage issues. In April 1999, the Ministry of Water Resources and the Ministry of Finance jointly launched an ecological demonstration programme. Under the project, 100 counties and 1 000 basins of small rivers were selected to carry out ecological agriculture experiments and demonstrations. The second phase of this programme, beginning in 2001, covered additional 46 counties and 374 small river basins. The programme is aimed at improving both water and soil conservation.

Non-point pollution associated with agriculture has become a widespread problem in China due to inappropriate use of chemical fertilisers and pesticides. Its effects include not only contamination of products and waterways, but also poisoning of farmers. Education of farmers in improved application of chemical inputs, including use of chemicals with higher efficiency/pollution ratios, is an identified measure to address this pollution problem.

China's efforts to improve the rural environment are also supported and supplemented by several international organisations. For instance, the World Bank and other donors have engaged in a number of projects that support sustainable agriculture and use of land, water, grassland, forest, or coastal resources. The Chinese authorities have also established eco-protection zones (such as national natural parks) and agricultural ecological demonstration projects, and a genetic resources bank.

Environmental issues are also addressed in the Chinese legal framework. Laws are in place relating to soil and water conservation, wildlife conservation, control of sand encroachment, and wild flora conservation. China is a signatory to several multilateral agreements on environmental protection, including those to combat desertification and protect biological diversity and has set up a number of monitoring networks on desertification, soil and water retention, and biodiversity.

Overall budgetary outlays on agro-food policies

Agricultural capital investments by government declined in the 1980s, under an over-optimistic expectation of agricultural growth. However, it was recognised that this would have adverse impacts on long-term productivity. State investment in agriculture was increased from the early 1990s, with a special focus on improving irrigation and other rural infrastructure. In the second half of the 1990s, as previously discussed, the Chinese government funded some agriculture-related investment projects with money raised from issuing government bonds, rather than making these investments with tax revenue, allowing additional investment without increasing taxes or reducing tax based government expenditures in other areas.

The national government has declared that the central financial authorities will continue to increase the budgetary funds for the construction of rural infrastructure facilities, ecological improvements, rural anti-poverty endeavours, dissemination of improved strains of crops, establishment of a quality and safety standard system and an inspection system for agricultural products, and the strategic restructuring of agriculture.

Data on overall budgetary revenues from agriculture and expenditures on agriculture-related activities is provided in Table 2.8. Clearly, in nominal terms, the levels of budgetary support to agriculture have tended to increase, with a trend increase of around 16% each year in the period 1993-2001. However, total government budgetary expenditure in the period grew at around 17% each year, resulting in agriculture's share of budgetary expenditure declining.

Aggregate budgetary expenditure to support agriculture in 2001 was around 13% of total government budgetary expenditure, having fallen sharply in terms of budget share from 1998 (15.7%). The trend in the period 1993-2001 was for agriculture's share of government budgetary expenditure to decline, losing around one quarter of one per cent share each year.

Within the expenditure on agriculture, the major outlays include rural infrastructure (47% of outlays in 2001), public stockholding (24%), operating expenses of government administration and services related to agriculture (8%), and agricultural extension services (8%). Expenditure on the grain for green programme in 2001 amounted to around 2% of agriculture-related outlays, while pest and disease control, agricultural research and development, agricultural schools, inspection services and consumer food price subsidies were each around 1% or less of the total agriculture-related outlay.

While the share of outlays within agricultural budgetary support has been reasonably stable between 1999 and 2001, there are some significant differences between this latest period and the distribution of expenditure immediately prior to 1999. Notably, the share of expenditure on infrastructure gained almost 7 percentage points, and agricultural extension gained around 4 percentage points, while public stockholding lost budget share of 6 percentage points. In the minor expenditure categories, the share of expenditure on consumer subsidies declined by 75% and the share for research and development was halved.

The changes in funding levels and shares indicate that, at least in budgetary terms, agricultural infrastructure development and on-farm education and technical extension have become more important. Public stockholding remains important but has not grown in importance since 1998, food subsidies are becoming a very minor expenditure, and research and development has received large expenditure cuts but the most recent funding levels are improving. These adjustments in the structure of budgetary expenditure, and the initiation of a decline in government revenue from agriculture, correspond to the broad policy change introduced in 1998 in which farm incomes became a policy priority. The budgetary changes also indicate a recognition that agricultural infrastructure has been inadequate.

Poverty alleviation and natural disaster relief are social policies which, while not specifically targeted at the agricultural sector, provide substantial benefits to farmers. The strong growth in poverty alleviation expenditure reflects China's ongoing effort to provide adequate food and clothing to those in poverty and government provision of funds to support preferential loans to poor regions.

Despite the fact that funds are allocated from the central budget, as previously discussed in this chapter, responsibility for implementing policies is often devolved to sub-national elements of government. However, budgetary funds are not always used in line with the stated objectives. In 2003, for example, the state auditors found that disaster relief funds in some provinces in eastern and southern China had been diverted to balance

Table 2.8. **Total national aggregate budgetary support to agriculture**
Unit: CNY million – current prices

	1993	1994	1995	1996	1997	1998	1999	2000	2001
Agricultural extension	3 631	4 856	6 047	5 635	6 111	7 008	14 706	16 459	19 184
Pest and disease control	1 262	1 687	2 100	1 957	2 145	2 294	1 275	1 190	1 439
Return farmland to forest programme	0	0	0	0	0	0		230	4 232
Relief from natural disasters	1 743	2 042	2 042	3 791	4 042	5 453	5 023	5 317	5 956
Poverty alleviation	1 417	1 950	2 000	6 000	10 000	11 000	12 869	12 493	13 487
Research and development	2 288	3 060	3 810	3 551	4 078	4 545	2 561	2 743	3 227
Agricultural schools	125	167	208	194	211	220	440	478	525
Inspection services	655	876	1 090	1 016	1 436	1 500	1 598	1 578	2 418
Infrastructure	26 012	34 789	43 315	45 321	50 367	65 828	80 663	94 549	115 163
Public stockholding	24 142	22 085	24 043	28 773	35 335	51 027	47 416	59 551	59 685
Administration operating expenses	8 975	12 003	14 945	13 928	15 654	18 435	14 185	15 872	19 594
Consumer food price subsidies	2 986	2 541	2 417	2 012	1 875	1 766	2 670	2 340	683
Budget total	73 235	86 055	102 018	112 178	131 254	169 076	183 406	212 800	245 593

Source: Ministry of Finance (2004).

county budgets and embezzled for the personal use of some officials (*People's Daily* on line, 24 June 2004). Misdirection of poverty alleviation funds has also occurred (Zhang, 2003), with funds intended to be used as preferential loans to poor farmers being lent to well-to-do business people for investment in commercial enterprises. A survey by the Development Research Center of China's State Council found that only 30% of funds destined for agriculture are actually spent on agricultural production. In the past, a large part of these funds was used for operating expenses by various levels of government to purchase cars, to pay for banquets, or otherwise misappropriated (Gale *et al.* 2005). Thus, under some programmes at least, the level of support actually received by farmers is much lower than the amounts allocated from the budget would indicate.

2.3. Trade policies

Since 1978 China has been guided by the philosophy of increasing openness and engagement with the rest of the world, as outlined in Chapter 1. This broad policy direction has resulted in numerous adjustments to economic policy, including changes to the foreign trade system in China. In this section, the major reforms of the foreign trade system will be outlined, along with the major objectives of China's agro-food trade policy. Trade measures applied by China and its major trading partners will be examined, trade relations will be outlined, and China's trade flows will be detailed.

Overall reforms of the trade system

In the early 1990s, the Chinese leaders adopted the notion of a socialist market economy and began to switch from direct control over all economic activities to more indirect economic policy interventions. With respect to the foreign trade system, the major reform measures have included:

- Developing a legal framework to govern trade activities.
- Significantly reducing import tariffs.
- Removing quantitative controls on imports and exports.
- Extending rights to engage in foreign trade to a larger number of firms.

- Transforming the STEs into commercial firms.

Foreign trade results have also been influenced by other macroeconomic policies, such as exchange rate policy (discussed in Chapter 1).

The government enacted the Foreign Trade Law in May 1994, which was revised in 2003 to accommodate China's WTO obligations. As a consequence of WTO membership, the national government has recently adopted a series of regulations dealing with antidumping and countervailing measures, quarantine and quality inspection, and safeguard measures. Many of these measures where put in place between 1999 and 2001. The general objective of these efforts has been to establish an administrative system consistent with WTO rules.

Merchandise tariff rates have been significantly reduced in the latter part of the 1990s and following WTO accession in 2001. For example, in its 1999 bilateral agreement with the United States, China committed to reduce industrial tariffs from 1997 average level of 24.6%, to an average of 9.0% by 2005, and committed to remove all tariffs on computers, semi-conductors and internet-related equipment by 2005.

China has progressively reduced the level of quantitative control it applies to international trade. This has included removing licensing requirements for some products, abolishing some quotas and converting other quotas to TRQs. These reforms have been applied to both imports and exports.

In the central planning era, all foreign trade was conducted by specific government firms, the STEs. During the period 1990-2004, the STE monopoly on foreign trade activity was abolished. Licensing procedures for the approval to engage in foreign trade have been progressively liberalised. Initially, STEs existed to carry out the plans of the central government. During the 1990s, most STEs became commercialised firms, responsible for their own business results.

In 2004, the requirement for approval was replaced with the requirement to register as a business engaged in foreign trade.

While these are broad economy wide reforms, they all apply to trade in agro-food products and will be discussed in greater detail in the sub-section on agro-food trade measures applied by China.

Main objectives of agro-food trade policy

China's agricultural trade strategy has consistently had multiple objectives: earning foreign currency income, and ensuring food security and food market stability. Under the central planning regime, earning foreign currency income to support industrialisation of the national economy was a primary objective. Political considerations also influenced trade decisions in the central planning era.

As discussed previously, the national government places a high priority on achieving food security and stability in food markets. Agricultural trade policy has contributed to these broad policy objectives. As discussed below, trade decisions regarding staple food commodities reflect a desire to maintain a balance in the domestic food market by maintaining an adequate supply of food and preventing the deterioration of farmers' incomes. Since around 1998, market competition and other distributional efficiency considerations, and supporting rural incomes, have been major concerns in discussions of agricultural trade strategy, leading to the implementation of measures to increase agricultural exports as a means of improving farmers' incomes.

Agro-food trade measures applied by China

Under China's central planning regime, all foreign trade was controlled with import and export volumes being set in annual trade plans. China has progressively liberalised its foreign trade system since 1978, with major reforms occurring since 1990. This section provides an outline of the evolution of measures applied to imports and exports.

Import measures

State trading enterprise regime. Prior to and through-out the period 1990–2004, STEs (Chapter 1) have played an important role in agro-food imports.

Under the central planning system, decisions on agricultural trade were made by the central government and carried out mainly by China National Cereals, Oils and Foodstuffs Import & Export Corporation (CEREOILS, now COFCO). At that time, CEREOILS had a state-sponsored position in agricultural trade which was largely monopolistic. The national government determined quantities of imports for different commodities, and also set the prices at which CEREOILS could sell imported product onto the Chinese market. During the central planning era, STEs operated as policy instruments, balancing supply with demand for staple foods and contributing to the achievement of the government's price objectives. STEs were not fully accountable for their financial performance, with the government subsidising trading losses, and profits being returned to the central government.

Since the reforms began, "minor" agricultural products have been deregulated, with competition introduced by extending foreign trading rights to provincial based STEs as well as non-STEs. In addition to the increase in the number of STEs eligible to engage in international trading activities, the focus and scope of activities for STEs was broadened to include purely commercial trading activities.

With the advent of the agent system, introduced in 1994 under the Foreign Trade Law of the People's Republic of China, STEs lost their monopoly on exports of many controlled commodities, although they remained an important element of government control of "strategic" agricultural commodities.[18] Under the agent system, two levels of approval were required. One level was to approve the enterprise which actually interfaced with overseas businesses to arrange trade contracts, and the other level was to give approval or a licence to trade for a specific volume of a specific commodity within a specific timeframe (these will be discussed in the next sub-section). STEs and other "designated" firms could import on their own behalf, on behalf of the government, or on behalf of other firms provided appropriate import approval had been obtained.

In China's WTO accession agreements, the remaining STEs lost their monopoly position, although allocations were reserved for STEs for imports of chemical fertiliser, wheat, corn, rice, sugar, cotton, tobacco and vegetable oils. The current state trading arrangement operates through the TRQ system (see sub-section on TRQs).

Clearly, STEs remain an important part of China's agro-food import arrangements. While a greater array of firms are permitted to engage in foreign trade, the STEs remain deeply involved as importers with a role to play in implementing government policy, especially policies designed to stabilise domestic prices. The gradual liberalisation of China's foreign trade regime may have diminished the influence of STEs, but the government retains some influence on import volumes through the state trading system. Government influence through this system is less direct under WTO rules than was previously the case.

Import licences. Licensing has been an important measure employed by the central government to regulate agro-food imports. There are two elements of the licensing structure: approval or licence to engage in foreign trade and approval or licensing to import specific quantities of specific commodities. Both elements have evolved from the state planned trading system. The second element will be discussed in detail in the following sub-section on TRQs.

Under the agent system in place since the trade reforms of 1993-1994, in addition to the STEs, certain other firms were granted approval to engage directly with foreign entities for the purposes of trading. This was a licence to trade, but specific approval was still required to import many commodities. In the agro-food sector, these commodities were grains, cotton, wool, sugar, tobacco, and vegetable oils.

In July 2001, immediately preceding China's WTO accession, the central government adopted a system whereby any domestic firm would be permitted to engage in foreign trading activities, provided it could meet certain criteria. Provincial governments were authorised to examine applications and register qualified firms. The criteria for enterprises under this system included minimum registered capital, expected import volumes above a set threshold level (initially CNY 5 million), a suitable bank credit rating and a satisfactory level of business profitability. Enterprises must also have recorded no breaches of regulations covering trade, finance, taxation or business activities, and, in some cases, the import volume of products subject to designated trading in the previous year was also taken into account. Specific approval to import commodities subject to quota control was still required. China reserved the right to continue this "designated trading regime" for three years after accession to the WTO until the end of 2004.

In April 2004, the Foreign Trade Law was amended to further liberalise foreign trade. Commencing in July 2004, all Chinese domestic enterprises have the right to engage in foreign trade provided they:

- Are corporate entities incorporated for at least one year.
- Have registered capital of no less than CNY 5 million (CNY 3 million in central and western China).
- Have completed tax registration and paid tax according to the law.
- Their legal representative must not have served in the same capacity with a firm which lost its foreign trade rights in the last three years.

Firms need only to register to gain trading rights, although some agricultural commodities remain subject to quota licence approval and other trade restrictions (see sub-section on TRQs). This reform also paves the way for individuals to engage in trade.

Tariffs. China introduced tariffs for many products in the early 1990s, despite many of those products being subject to state planning control.

During the period 1992–2004, the government substantially reduced tariff protection for agricultural products, through a series of tariff cuts. Figure 2.3 depicts the changes in average Most Favoured Nation (MFN) tariff rates of agricultural products since 1992, along with China's tariff rate commitments for 2004 and 2005. Clearly, the average level of tariff protection provided to Chinese agricultural products has tended

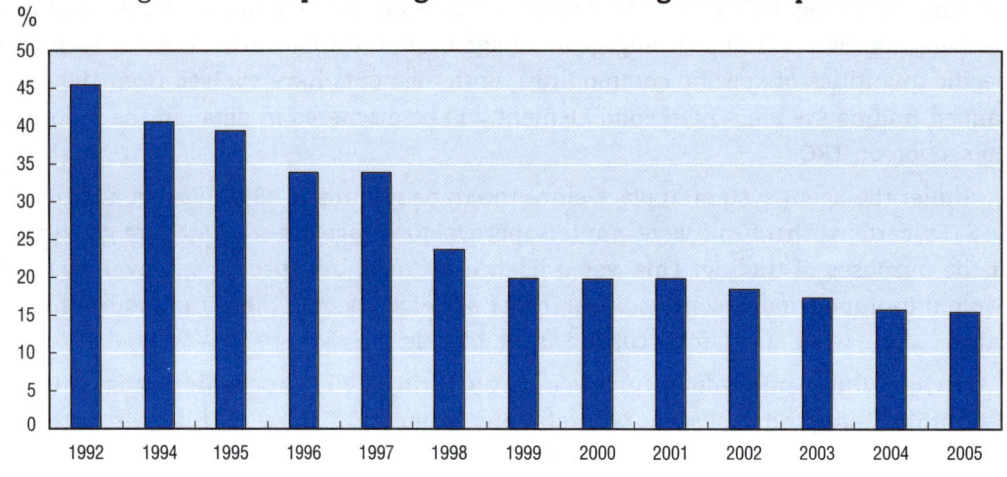

Figure 2.3. **Simple average MFN tariffs on agricultural products**

Source: China Customs Office (2005).

to decline, and was scheduled to drop in 2005 to around one third the average tariff level of 1992.

Tariffs on selected agricultural commodities are listed in Table 2.9. It is evident that, although major reductions of tariffs occurred after the mid-1990s, the introduction of the tariff-rate quota system from 1996 introduced the potential for out-of-quota imports of some commodities to incur large tariffs. Currently, the highest out-of-quota rates, at 65%, can be charged on imports of wheat, maize and rice. It is, however, notable that the in-quota tariff rates are much lower than that and even much lower than the average MFN rates. For example, in 2002 the average in-quota rate for ten TRQ products (46 tariff lines) was 6% while the average out-of-quota rate was 55%.

Since gaining WTO membership, China has continued to lower tariffs according to the schedule stipulated in the WTO accession protocol. Under this schedule, most tariff lines will be at their final bound rate by 2004; around 6% of China's agricultural tariff lines will reach their final bound rate between 2005 and 2010. The final three tariff lines – for fresh strawberries, preserved fruit and nuts other than cherries and strawberries; and other fermented beverages not elsewhere specified (such as cider, perry, mead) – are scheduled to reach their final bound rate at the beginning of 2010.

The reductions are shown in Figure 2.4 Clearly there is a shift to a greater proportion of lower tariffs. The simple average tariff rate for agriculture was 18.5% in 2002, dropping to 15.6% in 2004. The figure also demonstrates that China's tariffs are not very disperse; in 2004, 69% of tariff lines were in the 10-39% tariff range, and around 67% of tariffs lines had tariffs of less than 20%.

China has implemented tariff reduction and exemption measures in order to realise a range of policy objectives. The coverage of specific tariff reduction or exemption is determined by the State Council and all tariff reductions and exemptions are applied on an MFN basis. Goods that are exempted include those imported under inward processing programmes; domestic or foreign-funded projects encouraged by the government; articles for scientific research, for educational purposes, and for assisting people with disabilities.

Table 2.9. **Changes in MFN tariffs for basic commodities**

HS code	Description	Tariff rate applying from month									Final rate[2]
		Dec. 1992	Dec. 1993	July 1995	April 1996	Jan. 1997	Jan. 1999	Jan. 2000	Jan. 2001	Jan. 2002	
01029	Bovine animals	20	20	20	12	10	10	10	10	10	10
01039	Swine, live	40	35	35	12	10	10	10	10	10	10
0104109	Sheep, live	40	35	35	12	10	10	10	10	10	10
0104209	Goat, live	40	35	35	12	10	10	10	10	10	10
0105119	Fowls, live	40	35	35	12	10	10	10	10	10	10
02011	Bovine meat	50	50	50	50	45	45	45	40	30	20
02031	Swine meat	50	45	45	45	20	20	20	20	20	20
02041	Lamb/Sheepmeat	50	45	45	45	23	23	23	22	18.2	15
02045	Goat meat	50	45	45	45	23	23	23	23	21.2	20
020711	Poultrymeat	50	45	45	45	20	20	20	20	20	20
0401	Milk and cream	30	30	30	30	25	25	25	23	19	15
0407	Eggs	60	55	55	55	25	25	25	24	22	20
08051	Oranges	80	52	52	52	40	40	40	35	22.6	11
08081	Apples	80	40	40	40	30	30	30	30	18	10
0902	Tea	80	70	70	70	30	30	30	27	21	15
1001	Wheat[1]	0	0	0	1/114	1/114	1/114	1/114	1/114	1/71	1/65
1003	Barley[1]	3	3	3	3/91.2	3/91.2	3/91.2	3/91.2	3/91.2	3	3
1005	Maize[1]	0	0	0	1/114	1/114	1/114	1/114	1/114	1/71	1/65
1006	Rice[1]	0	0	0	1/114	1/114	1/114	1/114	1/114	1/71	1/65
1201	Soybean[1]	3	3	3	3/114	3/114	3/114	3/114	3/114	3	3
1202	Ground-nut	50	45	45	20	15	15	15	15	15	15
1205	Rapeseed[1]	50	45	45	12/40	12/40	12/40	12/40	12/40	9	9
17011	Cane and beet sugar[1]	30	30	30	30	30	30	20/90	20/90	20/65.9	15/50
2401	Tobacco	50	45	45	45	40	40	40	34	22	10
5201	Cotton[1]	3	3	3	3	3	3	3	3/90	1/54.4	1/40

1. Commodities subject to tariff-rate quota management: format is in-quota tariff/out-of-quota tariff.
2. The reduction of tariffs is phased in within the specified periods for individual goods. For most products, the final date is 2004. The latest date is 2010.

Source: Customs General Administration (2003); WTO (2001).

There are numerous criteria for the preferential tariff treatments, such as types of enterprises (STEs and foreign joint ventures), types of goods for specific purposes as mentioned above, types of specific regions (such as special economic zones). Due to these preferences, the actual level of tariff protection is much lower than the nominal rates would suggest. According to Chinese customs statistics, the ratio of customs duties collected to the total value of imports declined from 16.5% in 1985 to 2.9% in 2002. This ratio is consistently and significantly lower than the average nominal tariff rate, which Chinese Customs calculated as 18.5% for 2002.

TRQs and quantitative restrictions. Prior to 1993, the State Planning Commission determined the quantities of products to be imported and allocated quotas to provinces, according to perceived needs. In 1993, the agent system was introduced, under which STEs acted as importing agents. Businesses wishing to import had to bid for quota allocations.

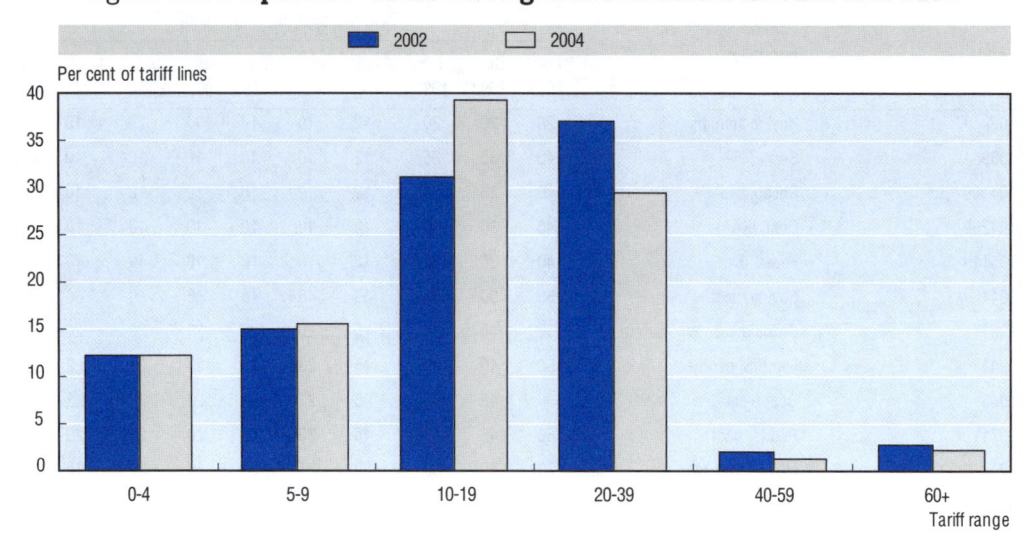

Figure 2.4. **Dispersion of China's agricultural tariffs in 2002 and 2004**

Quantitative restrictions applied to cotton, grains, and oilseeds, with quotas determined annually by the State Planning Commission.

China introduced TRQs for major grain and oilseed commodities in 1996. The establishment of TRQs and the mechanism by which the TRQs were to be managed, were significant issues in the bilateral negotiations which formed the basis of the Protocol of Accession of the People's Republic of China. It was agreed that China could apply TRQs to wheat, rice, corn, cotton, sugar, and wool. TRQs were also agreed for some vegetable oils, but the oilseeds themselves were subject only to a tariff. Unlike the TRQ systems many countries have under WTO rules, it has been stipulated that China's system includes criteria for allocating the import quotas to STEs and to non-STEs.

The process of allocation and re-allocation of quotas is managed by the National Development and Reforms Commission (NDRC) and the Ministry of Commerce (MOFCOM). The NDRC is in charge of the management of TRQs for grains and cotton, and MOFCOM is responsible for vegetable oils and sugar TRQs.

Each October, the NDRC and MOFCOM announce the quotas for the next year and the criteria which govern applicant eligibility for a quota allocation. To be eligible to receive a quota in 2005, for example, businesses must meet two sets of criteria. The first general set applies to eligibility to receive any commodity quota:

- Have registered with the State Administration of Industry and Commerce before 1st October, and passed the latest annual examination by the Authority.
- Have no record of violating import regulations in the areas of customs, foreign exchange, industry and commerce, taxation, and quality inspection.
- Have no record of violating the Temporary Measures on Management of Import Tariff Quota of Agricultural Products (a regulation issued jointly by the NDRC and MOFCOM).

Each commodity governed by a TRQ also has a set of specific criteria. In 2005, to be eligible for a sugar quota, for example, the applicant firm must also satisfy one of the following requirements:

- be a state-owned trade enterprise; or a
- central enterprise with a national stock function; or an
- enterprise which held a sugar import quota in 2004; or a
- sugar production enterprise which processes raw sugar, with a daily production capacity of at least 600 tonnes, with registered capital of at least CNY 10 million, and with annual sales of at least CNY 200 million; or an
- enterprise which uses sugar as a raw material and is engaged in processing for export trade.

MOFCOM and the NDRC have undertaken to notify successful applicants of their quota allocations before 1st January of the year in which their quota is valid. If a quota-holder has not contracted for the full quantity of their quota by 15th September in the year for which the quota is valid, then the unused portion of the quota is to be returned to the NDRC or MOFCOM to be re-allocated. Applicants wishing to obtain a share of the re-allocated quota can apply between the 1st and the 15th of September, and re-allocations will be issued on the 1st of October.

In 2004, TRQs applied to wheat, corn, rice, sugar, cotton, and vegetable oils, 90% of the wheat quota reserved for STEs, for corn the STE reserved portion of the quota is 60%, for rice 50%, for sugar 70%, for cotton 33% and for vegetable oils 10%. Following the so-called WTO transition period during which China's quotas were expanded, the 2004 TRQ settings will remain in place (unless re-negotiated) indefinitely for all commodities listed above, with the exception of vegetable oils. China has committed to eliminate the TRQ on vegetable oils as of 2006, implementing a tariff-only arrangement instead. Although tobacco has no TRQ, tobacco trade is controlled by a state monopoly. A TRQ is also in place for chemical fertilisers with the STE reserved portion ranging from 70% to 90%, depending on the type of fertiliser. China's TRQ performance on agro-food imports is discussed below (Table 2.10).

These quantitative restrictions permit the government some control over the quantity of these particular products which enter the Chinese domestic market.

VAT. The national government introduced a Value Added Tax (VAT) with effect from January 1994. The VAT was generally applied at a rate of 13% for agricultural products and agricultural inputs compared to the regular rate of 17%. However, VAT has generally not been collected on essential farm inputs such as fertilisers, seeds and water; and VAT is not levied on purchases of farm products until the primary receiver offers them for sale, at which point the tax is calculated on the value of the trader's margin, not on the value of the product. The different methods of calculating VAT obligations for domestic and imported goods may provide an additional level of protection to some domestic producers (Box 2.4) provided that importers are not exempted from VAT payments.

VAT exemptions have been applied to a wide range of agriculture-related imports, such as imports of seeds, breeding animals, certain chemical fertiliser and pesticides, raw materials for manufacturing feed products, and cotton. Application of these exemptions has been changeable, with application of exemptions fluctuating throughout the period 1990-2004.

> **Box 2.4. VAT assessment on imported agricultural products**
>
> While the same VAT rate is applied to imported and domestically produced agricultural products, differences in the method of calculating the VAT to be levied on imports and on domestic agricultural products, may provide additional protection to China's agriculture sector.
>
> Imports of agricultural products are assessed a VAT levy based on the Customs clearance price of the goods. The clearance price consists of the CIF price of the goods plus any tariff and consumption taxes which apply. Agricultural imports are assessed a VAT of 13% of the Customs clearance price.
>
> Domestically produced agricultural products are assessed a VAT levy according to the marketing margins at each point in the marketing chain, except for the primary handler. The primary producer of agricultural products is not liable for VAT on their products. The primary handler can assume an input VAT of 10%. Assuming a wholesale or primary handler margin of 15% for grains and soybeans, VAT collected on domestic products is only around 4% of the value of the goods.
>
> To demonstrate, assume a primary handler purchases CNY 1 000 worth of soybeans from a farmer, and on-sells to a vegetable oil mill for CNY 1 150. The VAT obligation faced by the primary handler is:
>
> (13% of CNY 1 150) – (10% of CNY 1 000) = CNY 49.5
>
> The VAT collected is 4.3% of the value of the shipment. If the same primary handler imported soybeans to sell to the mill at the same price, assuming the same margin, their VAT obligation would be the sum of the VAT on their margin and the VAT on the customs clearing price:
>
> (13% of CNY 1 150) = CNY 149.5
>
> The VAT collected is 13% of the value of the shipment. Considering this issue from the perspective of the primary handler's profit, in this simple example, the profit from the domestic purchase is total primary handler revenue less their total costs:
>
> CNY 1 150 – CNY 1 000 – CNY 49.5 = CNY 100.5
>
> To achieve the same level of profit handling imported soybeans, the Customs clearance price must be CNY 900. VAT on the imports will be (13% of CNY 900) = CNY 117. Assuming the primary handler pays (CNY 900 + CNY 117) = CNY 1 017, their VAT obligation would be (13% of CNY 1 150 less VAT already paid on the shipment) = CNY 32.5. Primary handler's profit in this case is:
>
> CNY 1 150 – CNY 1 017 – CNY 32.5 = CNY 100.5
>
> Thus, for the same level of profit to be achieved handling both the imported and domestic commodity, the price received by the foreign primary producer must be 10% lower than that received by the domestic primary producer. This method of VAT assessment may provide additional protection to domestic producers – additional nominal border protection of around 9% for grains and soybean farmers in this example.
>
> Source: Huang and Rozelle (2002b); Wu (2002); Huang, Rozelle and Chang (2004); Liu and Liu (2004); Wade, Branson and Qing (2002).

During the 1990s and early 2000s, selective reduction of VAT rates (to zero per cent), was an important means by which the government sought to guide import decisions. Application or exemption of the 13% VAT on agricultural imports can significantly alter the price competitiveness of those imports. For instance, in 1999 when China had a domestic oversupply of grains, the national government removed the VAT exemption on grain imports, effectively raising the price of imported grains and increasing the relative competitiveness of domestic grains.

In November 1999, prior to China's WTO accession, China and the United States reached an agreement under which China committed to apply all taxes and tariffs uniformly to both domestic and foreign businesses. This agreement was intended to alleviate the uncertainty associated with China's inconsistent application, refunds and waivers of VAT.

VAT applied to imports has been, and in 2005 remains, an effective component of Chinese border protection for domestic producers.

SPS and TBT.[19] Chinese consumers, particularly since the latter part of the 1990s, have become more concerned with food safety issues. While the public concern is relatively recent, the Chinese central government had been developing a food safety framework and infrastructure prior to that emerging concern.

The government has devoted efforts to establishing a system of quality certification for agricultural products since the early 1990s. Under the system, certificates are issued for "hazard free food", "green food" and "organic food".[20] Enterprises are encouraged to adopt the Hazard Analysis and Critical Control Point (HACCP) system for quality control. Food products for direct consumption are subject to compulsory certification and inspection.

Given that production is carried out by numerous smallholders, it is technically very complex and economically very costly to exercise effective control over food safety; a very large number of producers need to have their goods tested to ensure compliance with safety guidelines. In an attempt to rectify this situation, in 2002 Zhejiang provincial government introduced regulations which stipulate that sub-provincial governments allocate funds for vegetable pesticide residue monitoring. The regulation also requires that wholesale vegetable markets and supermarkets establish residue monitoring capacity.

China's accession to the WTO meant that it became bound to WTO rules on SPS and TBT measures. Under these rules, members must notify the SPS and TBT measures they apply. In April 2004, China advised the WTO Committee on Technical Barriers to Trade that work had begun on revising the 21 000 national technical standards. China advised that each standard would be checked to confirm its current relevance and alignment with international standards, and those that did not comply with these criteria would be annulled. It was expected that the revision would be completed by mid-2005.

Similarly with SPS measures, China has made numerous notifications to the WTO advising new SPS standards to be applied. An examination of WTO notifications reveals that China made 20 notifications in 2004, with most measures being aligned to relevant international standards.

China has also banned some products from entry, due to concerns over consumer safety. One example occurred in 2002 – China blocked imports of US soybeans on the grounds that they contained genetically modified material. The ban applied for 3 months, after which temporary import permits were issued while China completed a new safety evaluation of the product. In mid-2004, China banned imports of soybeans from Brazil on the basis that some shipments were contaminated with a dangerous fungicide. The contamination occurred as a result of mingling food and seed soybeans (the latter had been treated with the fungicide carboxin) in the commodity transport system. Brazil satisfactorily addressed China's concerns and trade resumed some two months after the ban was initiated.

SPS and TBT standards are clearly an emerging issue in China. It is likely that food safety and other human, animal and plant health issues will continue to be a priority for consumers, producers and government. The government is continuing to take positive steps to ensure that human, animal and plant health is safeguarded.

Export measures

During the period 1990–2004, China utilised a range of export interventions, including: quantitative restrictions, licensing, tax rebates, and export subsidies.

State trading enterprise regime. Throughout the period 1990-2004, China used a system of state trading for exports of certain commodities. In the beginning of the 1990s, rice, soybeans, corn, tea, cotton, and peanuts, along with some other agricultural products such as raw silk and soybean meal, were subject to planned export quota control. Export plans took into account the expected level of domestic production and demand, and sought to earn foreign exchange, maintain stable prices of "strategic" agricultural commodities, and ensure adequate supplies of inputs to state-run processing industries.

In China's WTO accession agreement, state trading was permitted to be retained for rice, corn, soybeans, tea, cotton and silk. As a consequence of the continued use of state trading for exports of certain agro-food products, the Chinese central government is able to influence the domestic prices of these products, to meet policy objectives including maintaining and improving farmer incomes.

Export licences. In a set-up similar to the arrangement for import licences, China had licensed exports of certain agro-food commodities for the period 1990-2004. The main criteria used in determining whether a product was subject to export licensing were: maintenance of national security or public interests; protection against shortage of supply in the domestic market or exhaustion of natural resources; limited market capacity of importing countries or regions; or obligations stipulated in international treaties. In addition to providing a means of regulating supply on domestic markets, export licensing was also used for statistical purposes.

In the early part of the 1990s, rice, corn, soybeans, tea, cotton, raw silk, soybean meal, and peanuts were subject to export licensing and quota control. In the latter part of the 1990s, livestock and poultry were added to the list of goods subject to licence control.

Under the terms of China's accession to the WTO, export licensing was phased out with controlled goods remaining subject to the state trading regime, under which the Chinese government can achieve the same outcomes as the licensing system achieved.

VAT rebates. China introduced a tax rebate system for exports in 1985. Since 1994, VAT has been rebated for exports. The VAT rebate has been used as a measure to influence export volumes. The national government has adjusted the rates of rebate and the coverage of goods frequently, based on policy considerations. When domestic supply is abundant, the government often adjusts the rate of VAT rebate upward to encourage exports. Otherwise, the rate is adjusted downward or even removed. For instance, the VAT rebate on sugar ceased in 1995 when there was a shortage in the domestic market and a rebate of 9% (less than the 13% VAT applying to agricultural products) was restored in late 1998 to encourage exports.

In principle, VAT should be reimbursed once goods are exported. However, implementation of this scheme has met some problems. In response to budgetary constraints, the central government stipulates a budgetary allocation from which VAT rebates will be paid. As a result of the growth in exports, the allocation has often been insufficient, causing rebate payments to be delayed.

Since WTO accession, China has continued to pay VAT rebates as permitted under WTO rules. Following the issue of VAT rebate arrears which developed in 2000, the central government has reportedly paid most outstanding rebate obligations. The official Chinese news agency reports that between October 2003 and June 2004, 96% of outstanding rebates were paid (Xinhuanet, 29 June 2004).

Clearly, rebates of VAT have the potential to influence exports, making products more competitive on international markets. In China's case, the degree of assistance provided is difficult to determine as rebate rates have been variable.

Export subsidies. Prior to becoming a WTO member, China provided export subsidies for corn and rice. These subsidies were available from the late 1990s as a means of easing the downward pressure on domestic prices which was brought about by large domestic production surpluses. According to Gale (2002b), subsidies for corn exports were of the order of CNY 368 per tonne in 1999 and CNY 378 per tonne in 2001. China was obliged to cease all export subsidies in 2002 in line with its WTO accession commitments.

Although not an export subsidy as such, in April 2002 the national government did implement a subsidy which may benefit exporters. Rail shipments of grains, cotton and soybeans were exempted from payment of the railway construction fee from April 2002 until December 2005. While this measure applied to products destined for export and the domestic market, it did provide an incentive to export from the north-eastern provinces, with a reduction of shipment cost of CNY 33 per tonne per thousand kilometres. This compares favourably with the CNY 38.9 per tonne standard freight costs to ship grains, cotton and soybeans a distance of 1 000 kilometres. The effect of the railway construction fee reduction was to achieve a substantial reduction in domestic freight costs for these commodities. The government ended this measure in May 2004 in order to ensure domestic supply.

Trade relations

WTO

During the period 1990-2004, a major influence on China's trade relations was the WTO accession package of negotiations, and subsequent membership of the WTO.

China submitted a request to restore its GATT membership just after the Uruguay Round started in 1986. Subsequently, China was involved in the Uruguay Round negotiations in an observer status. Following 15 years of accession negotiations, China became a WTO member in December 2001. During the period in which accession negotiations were in progress, China carried out a wide range of reforms on its economic system and policies, partly to meet the requirements of major parties in the negotiations and partly to extend the market-oriented reforms which began in 1978. As a result, China has increased its market openness and trade competitiveness significantly.

In the protocols of accession (WTO, 2001), China made comprehensive commitments regarding import measures for agricultural products (Box 2.5). Under the protocols, China is permitted to maintain a state trading system for the import of grains (wheat, rice and corn), vegetable oils, sugar, tobacco, cotton and chemical fertilisers. Meanwhile, China committed to use tariff-rate quotas (TRQs) to replace the previous planned quotas for products considered to be "sensitive". In order to ensure that the state trading system does not constitute a barrier to imports, a special arrangement of allocating part of the quotas to non-STEs and reallocating STEs' unused quotas within each year is included in the protocols (discussed previously). It is expected that implementation of the commitments will result in significant improvement in market access for imported farm products.

In addition to the above commitments, China also agreed to phase out licence controls on traded commodities by the end of 2004 and agreed to abide by the WTO rules governing technical barriers to trade and the use of export subsidies.

Since WTO accession, China has experienced an increase in both imports and exports of agro-food products, with the value of imports growing from USD 12 billion in 2001 to USD 30 billion in 2004. Export growth has been less spectacular, with agro-food exports being valued at USD 16 billion in 2001, growing to USD 23 billion in 2004.

Despite the large increase in the value of imports, the quotas in China's TRQ system have generally not been filled and in the case of cereals, the fill rates were very low until 2003 (Table 2.10). However, it is to early to properly assess China's TRQ performance on the basis of two years results, particularly as the performance in 2002 suffered from administrative problems in implementing the system. Particularly in 2002, the allocation of import quotas was delayed, leaving importers only about eight months to import the annual quota (Lohmar and Skully, 2003). In 2004 quota fill-rates for wheat increased to 75%, but for corn remained at 0%. Notwithstanding the fact that China is under no obligation to import any specific quantities of agricultural commodities, it does appear that the grain quotas may be under-utilised; the price differential between international prices and Chinese domestic prices for corn, for example, did provide an economic incentive for China to import corn in 2002 and 2003.

As previously mentioned, China has been gradually engaging with the major world economies since the late 1970s. The WTO accession negotiations and subsequent WTO membership have been consistent with China's longer term trend towards a market-oriented economy.

Regional trading arrangements

In addition to engaging in multilateral trading arrangements, the Chinese government has recently begun to develop regional and bilateral trade arrangements as an integral component of its trade strategy. A China-ASEAN[21] free trade area was proposed in 2000. Significant progress has been made so far. China and the ASEAN nations agreed in 2001 to form a free trade area and in November 2002 signed a framework agreement to progress closer economic relations between ASEAN and the People's Republic of China (ASEAN, 2002), and agreements on tariff reduction and dispute settlement were subsequently signed in 2004. While a China–ASEAN free trade agreement covering agricultural products would almost certainly lead to some benefits for consumers, as a result of the parties to the agreement having similar resource endowments (being relatively short of arable land and being relatively abundant in labour

Box 2.5. **Major WTO accession commitments by China – agricultural trade**

With respect to commodities

Cereals	A tariff-rate-quota system is applied to wheat, rice and corn. The quotas are to be increased annually at different rates till 2004. The final quotas are 9.636 million tonnes for wheat, 7.2 million tonnes for corn and 5.32 million tonnes for rice. Part of the quotas is retained for non-state trading enterprises according to pre-specified proportions. With respect to corn, this share will increase from 25% in 2002 to 40% in 2004. The in-quota tariff rate remains at 1% while above-quota rate declines from 71% in 2002 to 65% in 2004. Unless re-negotiated, the 2004 tariff-rate-quota settings will be maintained thereafter. A non-quota "tariff-only" system will be applied to other minor cereals, including barley (3%). Ban on TCK wheat from the US is lifted.
Cotton, wool and wool top	A tariff-rate-quota system is applied. The quotas are to be increased to 894 000 tonnes for cotton, 287 000 tonnes for wool and 80 000 tonnes for wool tops by 2004. The in-quota tariffs are 1% for cotton, 1% for wool and 3% for wool tops, while the above-quota tariffs are reduced to 40% for cotton and 38% for wool and wool top by 2004. Two-thirds of the cotton import quota is retained for non-state trading enterprises. Unless re-negotiated, the 2004 tariff-rate-quota settings will be maintained thereafter.
Vegetable oils	A tariff-rate-quota system is applied to soybean oil, rapeseed oil and palm oil. By 2005, the import quotas will rise to 3.5871 million tonnes for soybean oil, 1.243 million tonnes for rapeseed oil and 3.168 million tonnes for palm oil. The in-quota tariff rates are uniformly 9% during the implementing period. The above-quota tariff rates will be reduced from 52.4% in 2002 to 9% in 2005. Two-thirds of the import quota is initially with-held for non-state trading enterprises, rising to 90% in 2005. China commits to eliminate vegetable oil quotas after 2005, committing to the adoption of a tariff-only system starting from 2006.
Oilseeds and meals	A tariff-only system is applied to soybeans with a tariff rate of 3%. The tariff of soybean meal remains at 5%.
Animal products	A tariff-only system is applied. Import tariffs are to be reduced gradually by varying extents. Sharp reduction of tariff is committed in frozen beef, frozen pork, edible offal of bovine animals, frozen cuts and offal of poultry, and dairy products. China commits to remove scientifically unjustified restrictions on importation of meat products.
Sugar	A tariff-quota system is applied to sugar. Import quota is raised from 1.764 million tonnes in 2002 to 1.945 million tonnes in 2005. The in-quota tariff rate is reduced from 20% in 2002 to 15% in 2005; the above-quota tariff rate will be reduced from 69.4% in 2002 to 65% in 2004. Throughout the three year WTO transitional period, 30% of the import quota is allocated to non-state trading enterprises. Unless re-negotiated, the 2004 tariff-rate-quota settings will be maintained thereafter.
Fishery products*	A tariff-only system is applied. Import tariffs are to be reduced gradually by varying extents. Sharp reduction of tariff is committed in many frozen products of sea fish species and lobster.

> **Box 2.5. Major WTO accession commitments by China – agricultural trade** (cont.)
>
> | | |
> |---|---|
> | Horticultural products | A tariff-only system is applied. Import tariffs are to be reduced gradually by varying extents. Sharp reduction of tariff is committed in a range of fruits and nuts produced in the temperate zone. China commits to remove restrictions on the importation of horticultural products, unless the restrictions are scientifically justified. |
> | Processed food and beverages | A tariff-only system is applied. Import tariffs are to be reduced gradually by varying extents. Sharp reduction of tariff is committed in malt, wines, beer, tobacco, etc. |
>
> **With respect to services**
>
> | | |
> |---|---|
> | General | China will permit all approved enterprises, including non-state domestic enterprises and foreign enterprises, to engage in a full range of trading and distribution activities. |
> | Specific to agriculture | China will permit all approved enterprises to engage in all services affiliated to agriculture, forestry and livestock, including wholesale, retail, and warehousing. |
>
> * Fishery products are not classified as agricultural commodities in the Uruguay Round.
> Source: Compiled from WTO (2001).

Table 2.10. **China's TRQ performance**

Commodity	2002		2003	
	Quota (tonnes)	Fill-rate (%)	Quota (tonnes)	Fill-rate (%)
Wheat	8 468 000	7	9 052 000	5
Corn	5 850 000	0	6 525 000	0
Rice	3 990 000	6	4 655 000	6
Soybean oil	2 518 000	35	2 818 000	67
Palm oil	2 400 000	71	2 600 000	90
Rapeseed oil	878 900	9	1 018 600	15
Sugar	1 764 000	67	1 852 000	42
Wool	264 500	72	275 750	62
Cotton	818 500	22	856 250	102

Source: AMAD, OECD (2004).

resources), the benefits to the agricultural sectors of China and the ASEAN nations are likely to be muted. Although a free trade agreement between the ASEAN nations and China would be expected to generate increased trade activity, the similarity of the agricultural sectors of the economies involved is likely to result in only modest increases in trade of agricultural products.

As well as negotiating a free trade agreement with ASEAN, China is actively pursuing bilateral trade deals with New Zealand, Australia and Chile. It is not, however, officially pursuing bilateral trade arrangements with the Republic of Korea or Japan, the two most important economies in terms of China's agricultural trade.

Trade measures applied by partners

Since the late 1970s, China has been opening its economy to engage more closely with other economies active in international trade. While China has progressively dismantled and lowered its levels of border protection for agricultural products, many of its trade partners persist with relatively high border protection measures.

Japan and the European Union are important trading partners who apply significantly higher tariff protection to agriculture than does China. Japan's average agricultural tariff in 2002 was 43.6%, more than double China's 18.5%, while the European Union's average agricultural tariff was 32%. Both Japan and the European Union have more dispersed tariff schedules than China, with a larger proportion of low tariff levels and a larger proportion of high tariff levels. Both partners have a much higher maximum tariff than China (AMAD, 2004).

In terms of Chinese exports, Japan, the Republic of Korea, the European Union, and the United States are the most important markets.[22] The import measures applied by these countries are examined in more detail in the following sub-sections.

Japan

Japan is currently the largest market for Chinese agro-food exports, with 28% of China's exports valued at USD 6 billion in 2003. It has been noted that Japan tends to provide relatively high levels of tariff protection; for example, 310% for wheat, 821% for rice and 268% for sugar. In products such as corn, the tariff is still relatively high at 30%, while the tariffs on apples (11%), citrus (12%), tea (10%) and poultrymeat (7.5%) are lower. Clearly, in many of the agricultural products in which China has production advantages, the Japanese tariffs are lower than the average agricultural tariff (AMAD, 2004).

China is a low cost producer of many agricultural products and Japanese tariffs are not prohibitively high for many Chinese producers. Japan has, however, instituted safeguard measures against Chinese exports in the past. In 2001, Japan applied safeguard measures to imports of Chinese welsh onions, shiitake mushrooms and irises (Park, 2002). The safeguard comprised a TRQ for a 200-day period (from 23 April to 8 November 2001), with above-quota tariff rates of 256% (onions), 266% (mushrooms) and 106% (irises) compared with the normal rates of 3, 4.3 and 6% respectively.

As Park (2002) notes, Chinese farm products are not perceived in Japan as being of a high quality and there are concerns regarding the safety of Chinese agricultural products. The issue of food safety has proven to be a significant barrier to meat exports from China. Japan does not accept Chinese pigmeat as China is not free of a number of diseases affecting pigs, including Foot and Mouth Disease, and exports of poultry have been disrupted in recent years as a result of outbreaks of Avian Influenza.

Japan has relatively stringent SPS regulations. Enhanced maximum residue limit regulations and increased testing have reportedly led to increases in the costs of testing and in the level of export rejection, as well as a reduction in the level of Japanese orders for certain agricultural products (CFNA, 2002b).

Republic of Korea

China's exports to the Republic of Korea face significant tariff barriers. The Republic of Korea protects its corn and rice farmers with a high tariff (334%) for corn and a true quota for rice. For labour intensive agricultural products exported by China, the Republic of Korea also tends to have high tariff protection; for example, citrus faces a tariff of 99%, apples are also well protected with a tariff of 48%, tea faces a tariff of 294%, while poultry and pigmeat of 22% and 26% respectively (AMAD, 2004). China supplies around two thirds of the Republic of Korea's vegetable imports. Potatoes, sweet potatoes, onions, garlic and peppers are the major vegetable types exported, with tariffs of around 50% in TRQs which have prohibitively high above-quota tariffs. Clearly, Chinese exports face significant tariff barriers in the Republic of Korea.

The Republic of Korea has also instigated safeguard action against China in the past. In 2000, the Republic of Korea imposed a safeguard tariff totalling 315% on frozen and processed garlic, and total safeguard tariffs of 436% on peeled garlic. These measures were in response to a surge in Chinese garlic exports to the Republic of Korea. This particular dispute has since been resolved (Park, 2002).

While the Republic of Korea has a food safety monitoring system and food safety regulations, this has not proven to be a major NTB for Chinese exports in the period 1990-2004.

The European Union

In 2003, around 10% of China's agro-food exports were destined for the European Union, making the European Union a major trading partner. As discussed above, the European Union has relatively high average agricultural product tariff levels. In terms of the average tariff levels faced by China's major export categories, these range from low to high. Tea has free entry, fruit (13%) and vegetables (16%) have moderate average tariffs but are covered by an entry-price regime, preparations of fruit and vegetables (24%) have moderately high tariffs, while cereals (50%) and sugar (81%) have quite high average tariffs (AMAD, 2004).

In addition to quite high average tariffs for meat (53%), China's exports of livestock products and fishery products to the European Union have encountered SPS and TBT barriers since the mid-1990s. In 1995, the European Union banned the import of poultry products from China on the basis that China was unable to establish an effective veterinary system and to control inappropriate use of feed additives.

In January 2002, the European Union suspended import of all products of animal origin from China. The ban was triggered by a customs inspection that reportedly found 24 batches of Chinese frozen shrimps contaminated with the antibiotic chloramphenicol. In July 2004, the European Union agreed to lift the ban. The move will allow China to resume exports of shrimps, farmed fish, honey, royal jelly, rabbit meat; and a number of other products of animal origin in recognition of China's "significant improvements" in meeting veterinary standards. It is required that all products must be checked by the Chinese food safety authorities and each consignment will be certified as meeting the relevant European Union food safety standards. However, the ban on the import of poultry products still remains in place because of safety concerns, particularly given the recent re-emergence of avian influenza in East Asia.

China's exports of tea products have also encountered NTBs. Enhanced testing measures and smaller maximum residue limits in recent years have caused a decline in tea exports to major markets, including the European Union. The major safety problems with China's tea products are reported to be: residues of pesticides and heavy metals, and contamination by harmful micro-organisms, foreign matter and dust.[23] Many pesticides used in tea production are included in the prohibited list. For instance, the list issued by the European Union in 2003 includes 10 popular pesticides used in Chinese tea production.

United States

China's 2003 exports of products of animal origin, vegetables, preparations of meat and fish, and preparations of vegetables and fruits (Harmonised System Chapters 05, 07, 16, and 20) accounted for 72% of Chinese agro-food exports to the United States. Typically, these products face very moderate or low tariffs. Animal products (not elsewhere specified) have an average tariff of less than 1%, edible fruit and nuts attract a tariff of around 9%, while food preparations of meat have an average tariff of 4% and preparations of vegetables fruit and nuts attract an average tariff of 11% (AMAD, 2004). These tariffs are relatively low and are unlikely to be prohibitive to low cost producers in China.

China does face barriers in accessing US markets. The major barriers faced by Chinese producers have been anti-dumping tariff measures. For example, the US Department of Commerce (DOC) conducted an investigation into potential dumping of apple juice in 1999, determining in June 2000 that dumping was occurring. The United States International Trade Commission voted in May 2000 to impose anti-dumping tariffs of up to 51.74%, on apple juice imports from China. Nine Chinese fruit juice firms (representing 70% of China's juice exports) appealed to the US-based Court of International Trade against the US Department of Commerce's determination of July 2000. The Court decision, issued in June 2002, overturned some key aspects of the anti-dumping order released by DOC, ruling that an anti-dumping tariff rate of 4% should apply to four of the firms involved and that no anti-dumping tariff was warranted for the other five companies which appealed.

The United States also announced anti-dumping measures against imports of shrimp from China in 2004, with tariffs tentatively set to range between 8% and 113% (*People's Daily Online* 7 July 2004).

China has also encountered SPS barriers in exporting to the United States, similar to those encountered in exporting to Japan and the European Union.

Trade flows

China has become more engaged in world markets since the reforms began in 1978 and increasing trade has been matched by increases in domestic agricultural output. Between 1992 and 2003, the value of Chinese agro-food exports increased by 88% in nominal terms, but the ratio of exports to the gross value of agricultural output declined marginally, from 7.2% in 1992, to 6.4% in 2003. The nominal value of imports increased four-fold while the ratio of value of imports to gross value of agricultural output increased from 3.4% to 5.8%. These results suggest China has remained largely self-sufficient in agro-food products, with foreign trade playing a relatively minor part in the agro-food sector.

Overall agricultural trade performance

Throughout the period 1992-2003, China was a net exporter of agro-food products, but according to preliminary data it became a large net importer in 2004 (Figure 2.5). Clearly, exports have been more stable in terms of aggregate value than imports, with two dramatic changes in net trade largely due to shifts in the aggregate value of imports.

First, net exports dropped significantly in 1995 compared to the previous three years. This was a result of a significant increase in imports of cereals. In the same year, policy makers in China believed that a shortage of food grains was imminent and planned imports of cereals surged in 1995 and remained historically high in 1996.

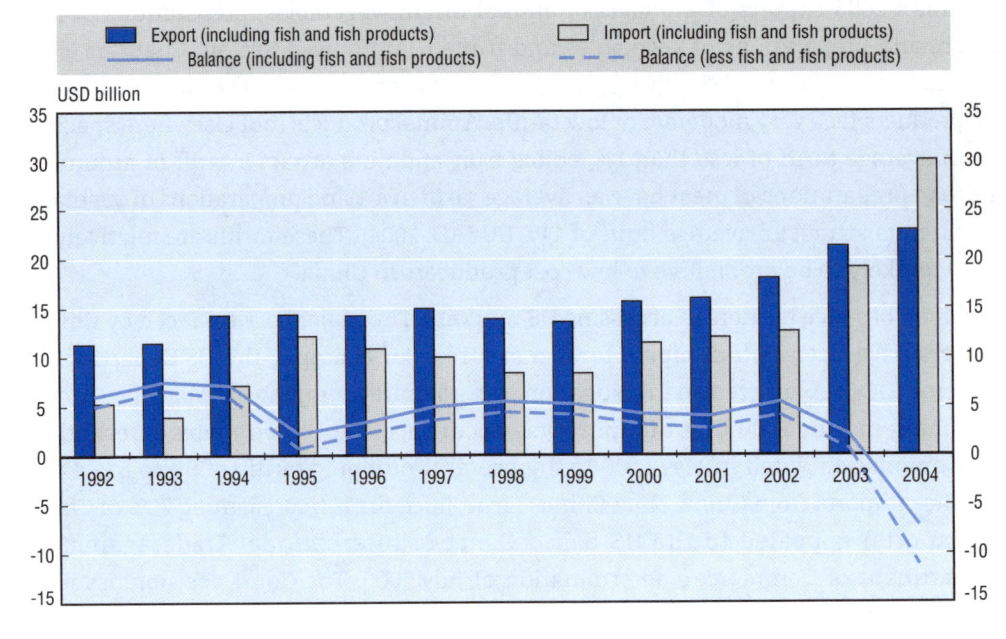

Figure 2.5. **China's agricultural trade, 1992-2004**

Source: Comtrade database; China Customs Statistics for 2004.

The second feature evident in Figure 2.5 is the sharp drop in net exports which occurred in 2003, followed by a large deficit in 2004. China's accession to the WTO in December 2001 ushered in a new set of trade rules for China. China's access to foreign markets improved and foreign traders' access to Chinese markets was also scheduled to improve. Delays in finalising the administration of quotas under the TRQ system contributed to a delay in the expected increase in agro-food imports. In 2002, China increased exports of cereals and vegetable, fruit and nut food preparations; in 2003, these sustained increases were more than offset by large increases in imports of cotton, soybeans and vegetable oils as problems implementing the new trade rules were overcome. In turn, growing grain prices on domestic markets in the last quarter of 2003 and in the first quarter of 2004 urged the government to buy 7 million tonnes of soft and durum wheat to replenish the strategic stocks. Sharply growing wheat imports combined with a continued fast growth in imports of soybeans and cotton contributed to net imports of agro-food products at about USD 11 billion in 2004, even if agro-food exports (excluding fish and fish products) reached a record value of USD 16.5 billion.[24]

Composition of agro-food trade

During the period 1992-2003, China's agricultural trade shows notable structural change. The structural change in China's agricultural product trade is graphically illustrated in Figure 2.6. Clearly, net exports (exports less imports) in land intensive commodities have been trending down in the period 1992-2003, and since 1995 there has generally been a trade deficit in these commodities. This is mainly due to a large increase in cereal imports in 1995 and 1996, and an increasingly high level of oilseed imports since 1997. Labour intensive commodities have had a trade surplus throughout the period 1992-2003 with a virtually flat trend. The apparent shift in the broad composition of agro-food trade is in-line with China's perceived advantage in producing labour intensive agricultural products.

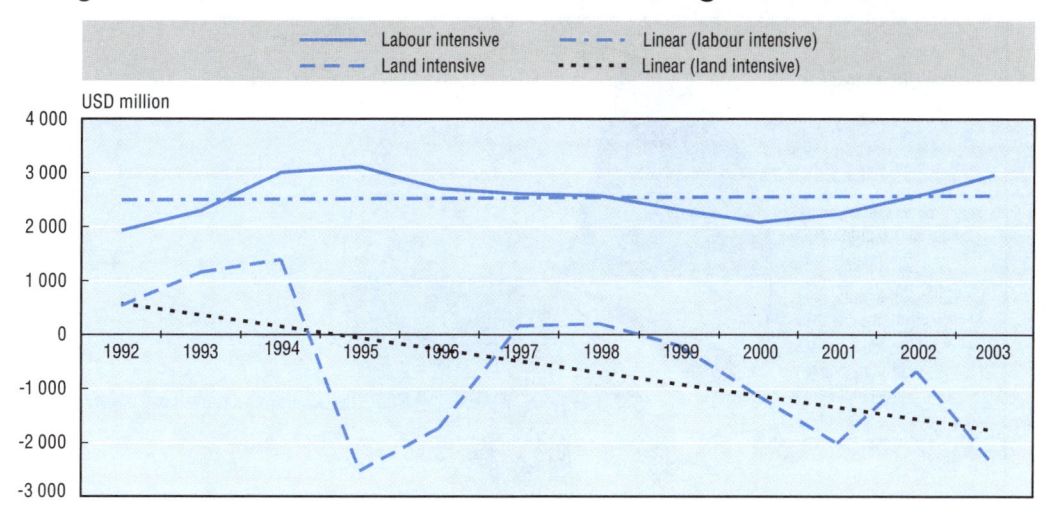

Figure 2.6. **Net trade in land and labour intensive agricultural commodities**

Notes: Land intensive defined as HS10 – cereals and HS12 – oilseeds. Labour intensive defined as HS07 – vegetables, HS08 – fruits and nuts, HS09 – tea, and HS24 – tobacco.
Sources: Adapted from Huang and Chen (1999); derived from Comtrade data.

An examination of China's import performance during the period 1992-2003 reveals that total imports of agro-food products (including fish and fish products) averaged growth of around 12% per year from 1992 to 2003, with oilseeds imports achieving the highest rate of 35% per year. Import of meat and edible offal, edible fruit and nuts and preparations also increased at high rates, reflecting income growth and lifestyle changes, mainly in the eastern provinces. Only two groupings of agro-food commodities experienced declining levels of nominal import value, with cereals imports declining by around 12% per year, and sugar and sugar confectionary imports declining by 2% per year during the period.

In 2003, as can be seen in Figure 2.7, oilseeds constituted almost one third of the value of Chinese agro-food imports, with animal and vegetable fats and oils contributing 17% of the value of agro-food imports, meaning that almost half the value of Chinese imports of agro-food products is contributed by edible oils or the primary product from which the oils are extracted. This reflects the relatively poor oil yield from Chinese soybeans, the existence of a strong domestic crushing industry, increasing demand for edible oils in China, and low barriers to trade in these commodities. Fish, cotton, and wool are also important imports, particularly as they are inputs to value-adding industries in China. For example, in 2003, cotton and wool imports were valued at around USD 2 billion, while wool and cotton textile product exports were valued at around USD 12 billion.

The nominal value of agro-food product exports in the period 1992-2003 grew at a rate of around 6% a year. Almost all the important product groupings exhibited export growth, with food preparations of meat and fish achieving annual growth of almost 17% to become the second most valuable export grouping at the HS 2 digit level in 2003. Fish and crustaceans (the most valuable group) achieved growth of around 8%. In those categories which experienced declining trends in real values of exports, sugar exhibited the largest annual decline at around 11%.

Changes in China's export performance reflect the ongoing structural adjustment in China's agro-food sector, as well as an increased level of engagement with international agro-food markets. The liberalisation of foreign trade policy in China, particularly with regard to products other than cereals and cotton, has contributed to the increase in exports

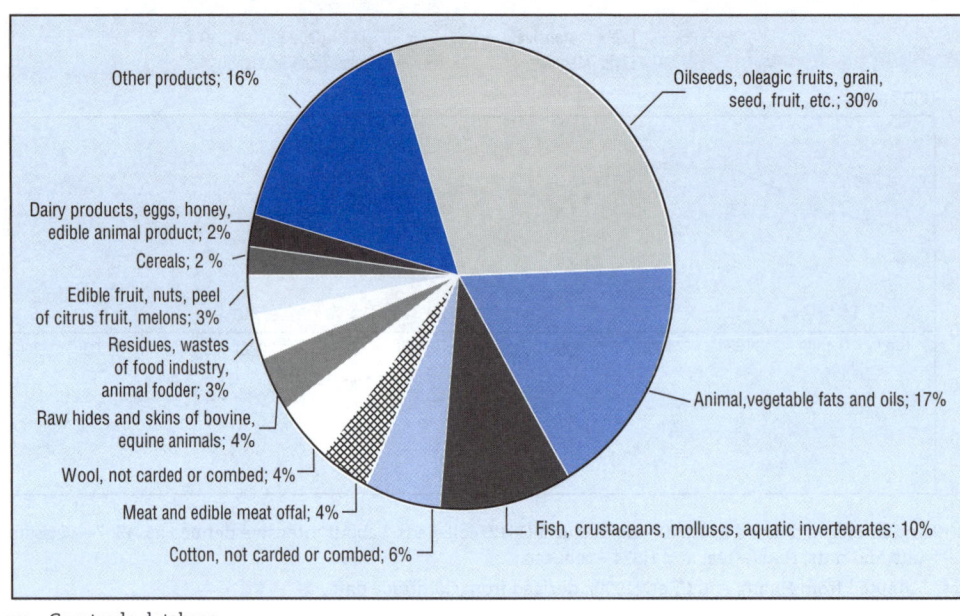

Figure 2.7. China's main agro-food imports, 2003

- Other products; 16%
- Oilseeds, oleagic fruits, grain, seed, fruit, etc.; 30%
- Dairy products, eggs, honey, edible animal product; 2%
- Cereals; 2%
- Edible fruit, nuts, peel of citrus fruit, melons; 3%
- Residues, wastes of food industry, animal fodder; 3%
- Raw hides and skins of bovine, equine animals; 4%
- Wool, not carded or combed; 4%
- Meat and edible meat offal; 4%
- Cotton, not carded or combed; 6%
- Fish, crustaceans, molluscs, aquatic invertebrates; 10%
- Animal, vegetable fats and oils; 17%

Source: Comtrade database.

of those product groupings. Domestic policies which have contributed to a significant increase in domestic production of agro-food products, coupled with a focus on increasing farm incomes, have also created a situation in which agro-food exports have grown.

In 2003, as is shown in Figure 2.8, meat, fish, and meat and fish food preparations comprised a third of the value of agro-food exports, while vegetables, fruits, nuts, and vegetable, fruit and nut food preparations comprised a quarter of the value of exports. Cereals and oilseeds comprised 17% of the value of exports, down from 21% in 1992 despite government policies that focussed on increasing production of cereals.

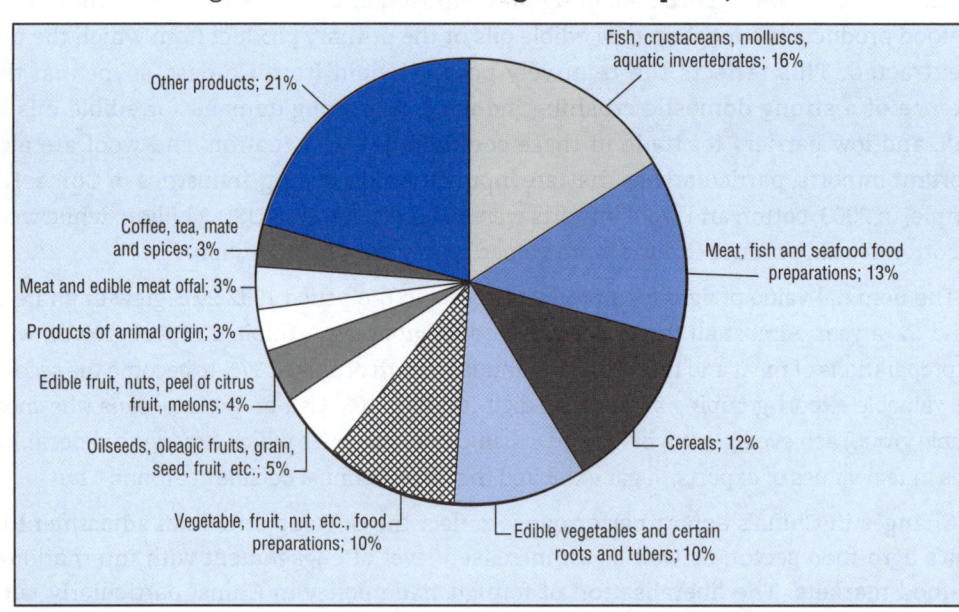

Figure 2.8. China's main agro-food exports, 2003

- Other products; 21%
- Fish, crustaceans, molluscs, aquatic invertebrates; 16%
- Coffee, tea, mate and spices; 3%
- Meat and edible meat offal; 3%
- Products of animal origin; 3%
- Edible fruit, nuts, peel of citrus fruit, melons; 4%
- Oilseeds, oleagic fruits, grain, seed, fruit, etc.; 5%
- Vegetable, fruit, nut, etc., food preparations; 10%
- Edible vegetables and certain roots and tubers; 10%
- Cereals; 12%
- Meat, fish and seafood food preparations; 13%

Source: Comtrade database.

Destination and origin of agro-food trade

In addition to changes in the aggregate levels of China's imports and exports, between 1992 and 2003 there have been significant changes in the make-up of China's trade flows.

China's exports have grown significantly more to some regions than others, as is evident from Figure 2.9. While exports to Asia have increased in absolute value terms, exports to Asia in 2003 constituted 70% of China's exports, down from 78% in 1995. Japan is the biggest single export market for China. As a close neighbour, there are well established trade links between China and Japan, with more than one fourth of China's agro-food exports in 2003 destined for Japan (Figure 2.10). However, Japan is a mature agro-food market and the growth of exports into that market has been slower than in some markets in which China is a relative new-comer. It is also notable that China's exports of meat and edible meat offal have declined by around USD 400 million between 1995 and 2003. Most of this decline was a result of large declines in poultrymeat exports in 2002 and 2003 due to disease out-breaks in China's poultry industry.

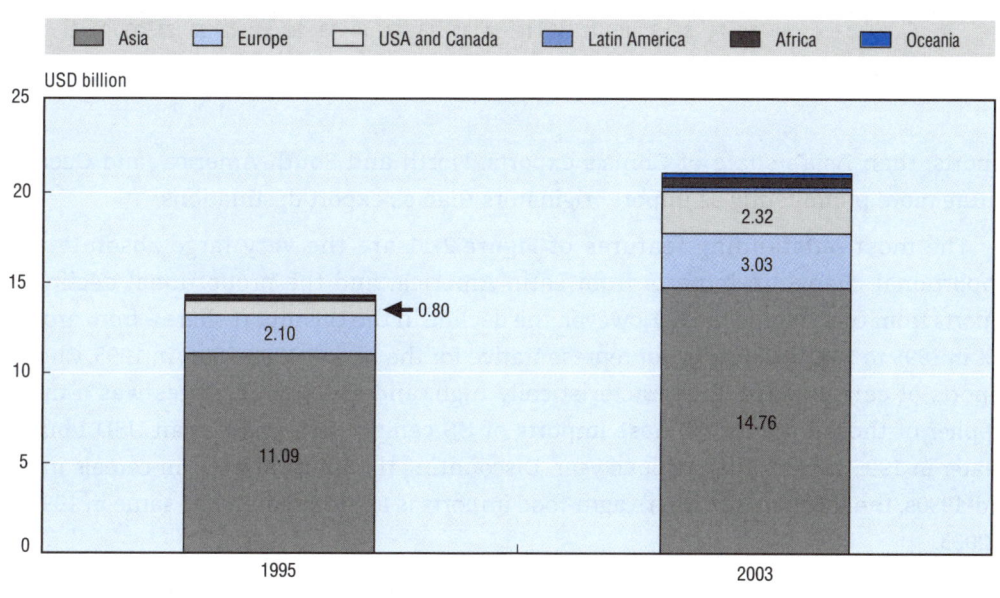

Figure 2.9. **China's agro-food exports (including fish and fish products) by region**

Region	1995	2003
Asia	11.09	14.76
Europe	2.10	3.03
USA and Canada	0.80	2.32

Source: Comtrade database.

China's exports to the United States, for example, have grown from being around 5% of China's agro-food exports in 1995 to 10% in 2003. This growth has been a result of large increases in exports of fruit, vegetable and nuts, and particularly a consequence of increases in exports of food preparations of fish, meat, fruit, nuts and vegetables. This result reflects China's comparative advantage in labour and Chinese policies supporting the development of vertically integrated food processing companies.

While Latin America, Africa and Oceania have approximately doubled their respective shares of China's exports, their shares remain very small at 1% in aggregate.

The situation for imports is quite different. As can be seen in Figure 2.11 and Figure 2.12, China's imports from Asia constitute a much lower proportion of total Chinese

2. POLICY TRENDS

Figure 2.10. **Main export markets for Chinese agro-food products (including fish and fish products), 2003**
As per cent of total agro-food exports

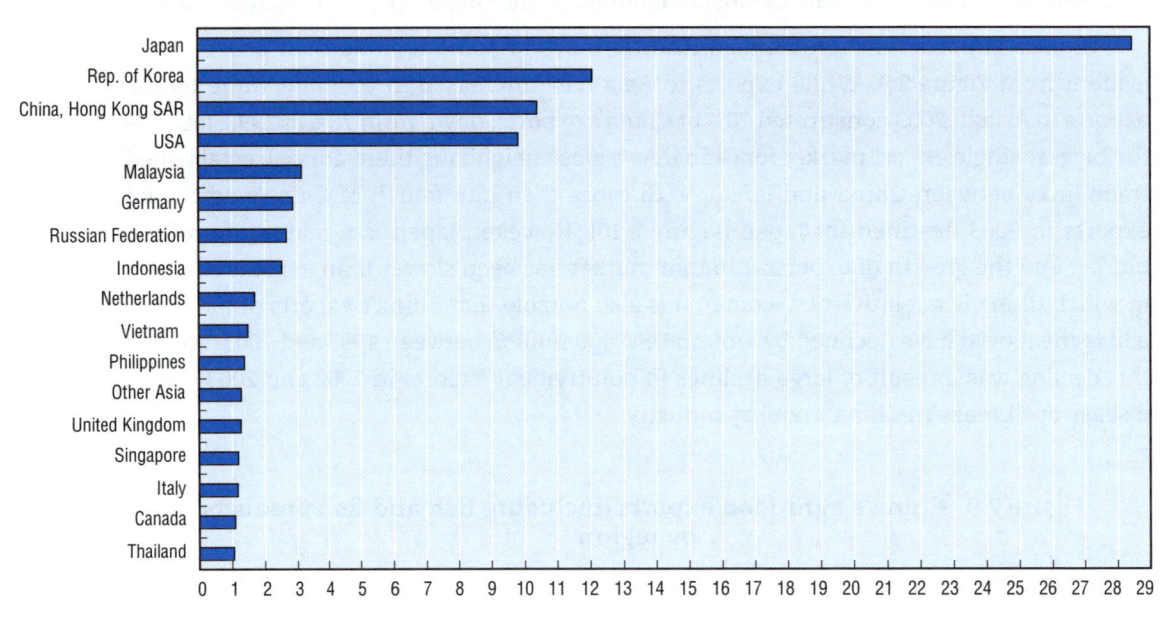

Country	%
Japan	~29
Rep. of Korea	~12
China, Hong Kong SAR	~10
USA	~10
Malaysia	~3
Germany	~3
Russian Federation	~3
Indonesia	~2
Netherlands	~2
Vietnam	~1
Philippines	~1
Other Asia	~1
United Kingdom	~1
Singapore	~1
Italy	~1
Canada	~1
Thailand	~1

Source: Comtrade database.

imports, than Asia's share of China's exports. North and South America, and Oceania feature more prominently as import originators than as export destinations.

The most outstanding features of Figure 2.11 are the very large absolute and proportional change in imports from Latin America, and the proportional decline in imports from the United States. However, the decline in the US import share – from around 35% in 1995 to 26% in 2003 – is not representative for the period 1990-2004. In 1995, China's imports of cereals were uncharacteristically high, and the United States was a major supplier of those imports. Chinese imports of US cereals were more than USD 1 billion greater in 1995 than in the previous year. Discounting the unusual trade in cereals in the mid-1990s, the US share of China's agro-food imports is approximately the same in 1995 as in 2003.

The increase in imports from Latin America is very significant. The increase is as a result of a major increase in Chinese imports of oilseeds from Brazil and Argentina – imports worth USD 3.2 billion in 2003 compared to around USD 25 million in 1995. This change in import levels is a result of Chinese policies which supported the development of a domestic oilseed crushing industry and trade policies since 2002 which have allowed oilseed imports (particularly soybeans) at relatively low tariffs with no quantity restrictions.

Clearly, foreign trade is becoming more diversified both in terms of the balance of products being traded and in terms of the proportion of trade taking place in the various regions of the world. Overall, the picture is one of gradually increasing trade volumes, the maintenance of established trading relationships and the development of new trading partners. There is evidence of China's advantage in labour contributing to improved levels of trade with strong growth in exports of food preparations, and in the growth of oilseed

Figure 2.11. **China's agro-food imports (including fish and fish products) by region**

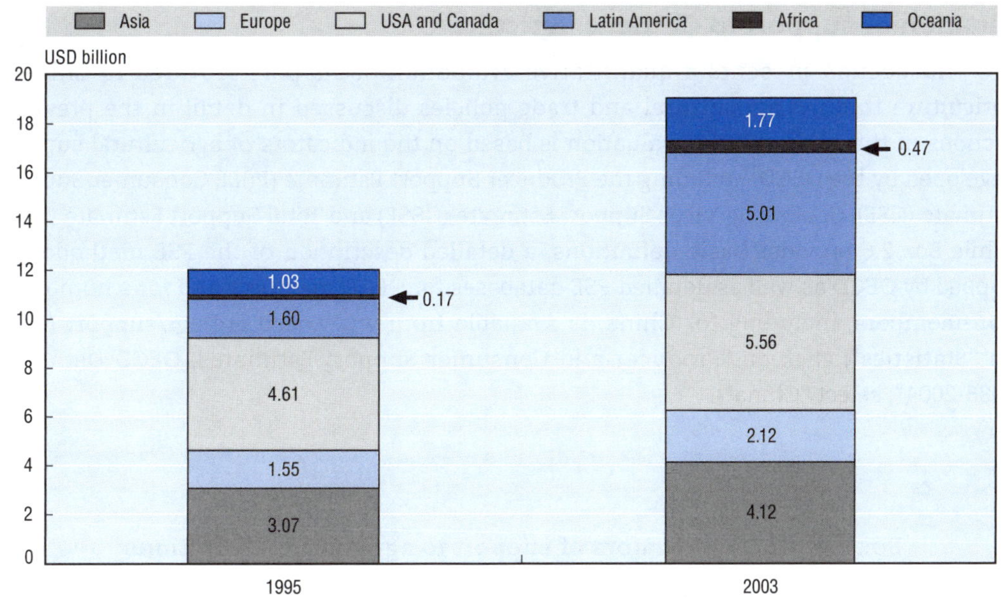

Legend: Asia, Europe, USA and Canada, Latin America, Africa, Oceania

USD billion

1995:
- Asia: 3.07
- Europe: 1.55
- USA and Canada: 4.61
- Latin America: 1.60
- Africa: 0.17
- Oceania: 1.03

2003:
- Asia: 4.12
- Europe: 2.12
- USA and Canada: 5.56
- Latin America: 5.01
- Africa: 0.47
- Oceania: 1.77

Source: Comtrade database.

imports there is a suggestion that China's relative scarcity of land is also influencing the composition and direction of trade flows.

These trends accord with Chinese trade policy for the agro-food sector, with China progressively becoming more open and engaged in international markets, especially in the more labour intensive farm products sector. The absence of significant increases in the

Figure 2.12. **Main suppliers of agro-food products (including fish and fish products) to China, 2003**

As per cent of total agro-food imports

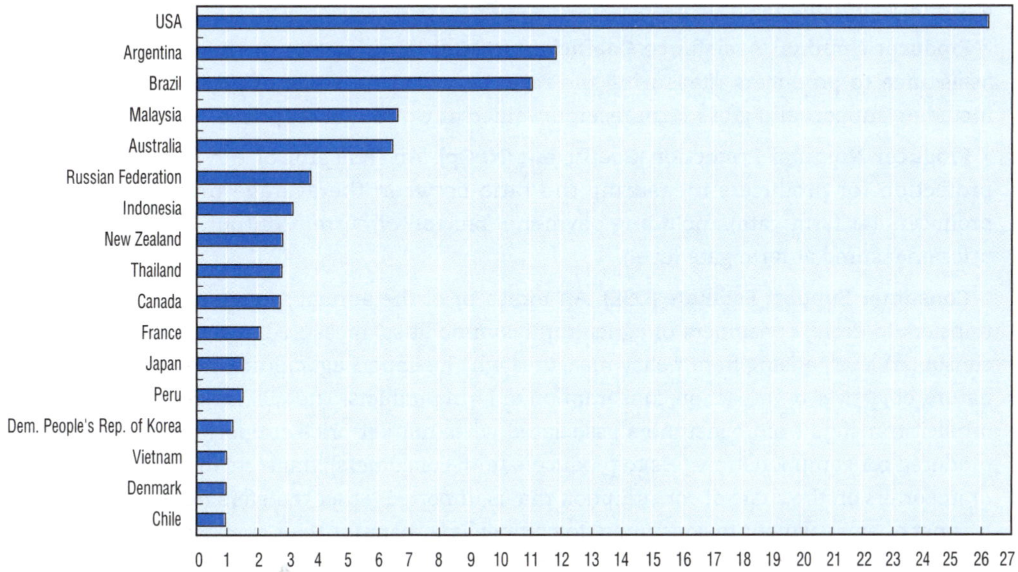

Countries listed top to bottom: USA, Argentina, Brazil, Malaysia, Australia, Russian Federation, Indonesia, New Zealand, Thailand, Canada, France, Japan, Peru, Dem. People's Rep. of Korea, Vietnam, Denmark, Chile.

Source: Comtrade database.

level of food grain imports also accords with the policy objective of maintaining a high level of grain self-sufficiency.

2.4. Evaluation of support to Chinese agriculture

This section presents a quantitative evaluation of support provided to Chinese agriculture through agricultural and trade policies discussed in detail in the previous sections of this chapter. The evaluation is based on the indicators of agricultural support developed by the OECD, including the Producer Support Estimate (PSE), Consumer Support Estimate (CSE), General Services Support Estimate (GSSE) and Total Support Estimate (TSE). While Box 2.6 provides basic definitions, a detailed description of the PSE methodology applied by OECD as well as detailed PSE databases for OECD members and for a number of non-members, including for China, is available from www.oecd.org/agr/support (click on "Statistics"; click on "Producer and Consumer Support Estimates, OECD Database 1986-2004"; select "China").

Box 2.6. OECD indicators of support to agriculture: definitions

Producer Support Estimate (PSE): An indicator of the annual monetary value of gross transfers from consumers and taxpayers to support agricultural producers, measured at farm gate level, arising from policy measures, regardless of their nature, objectives or impacts on farm production or income. The PSE measures support arising from policies targeted to agriculture relative to a situation without such policies – i.e. when producers are subject only to general policies (including economic, social, environmental and tax policies) of the country. The PSE is a **gross** notion implying that any costs associated with those policies and incurred by individual producers are not deducted. It is also a **nominal assistance** notion meaning that increased costs associated with import duties on inputs are not deducted. But it is an indicator **net** of producer contributions to help finance the policy measure (e.g. producer levies) providing a given transfer to producers. The PSE includes implicit and explicit transfers. The **%PSE** is the ratio of the PSE to the value of total gross farm receipts, measured by the value of total production (at farm gate prices), plus budgetary support.

Producer Nominal Assistance Coefficient (NACp): An indicator of the nominal rate of assistance to producers measuring the ratio between the value of gross farm receipts including support and gross farm receipts valued at world market prices without support.

Producer Nominal Protection Coefficient (NPCp): An indicator of the nominal rate of protection for producers measuring the ratio between the average price received by producers (at farm gate), including payments per ton of current output, and the border price (measured at farm gate level).

Consumer Support Estimate (CSE): An indicator of the annual monetary value of gross transfers to (from) consumers of agricultural commodities, measured at the farm gate (first consumer) level, arising from policy measures which support agriculture, regardless of their nature, objectives or impact on consumption of farm products. The CSE includes explicit and implicit transfers from consumers associated with: market price support on domestically produced consumption (transfers to producers from consumers); transfers to the budget and/or importers on the share of consumption that is imported (other transfers from consumers). It is **net** of any payment to consumers to compensate them for their contribution to market price support of a specific commodity (consumer subsidy from taxpayers); and the producer

Box 2.6. **OECD indicators of support to agriculture: definitions** (cont.)

contribution (as consumers of domestically produced crops) to the market price support on crops used in animal feed (*excess feed cost*). When negative, transfers from consumers measure the implicit tax on consumption associated with policies to the agricultural sector. Although consumption expenditure is increased/reduced by the amount of the implicit tax/subsidy, this indicator is not in itself an estimate of the impacts on consumption expenditure. The **%CSE** is the ratio of the CSE to the total value of consumption expenditure on commodities domestically produced, measured by the value of total consumption (at farm gate prices) minus budgetary support to consumers (consumer subsidies).

Consumer Nominal Assistance Coefficient (NACc): an indicator of the nominal rate of assistance to consumers measuring the ratio between the value of consumption expenditure on agricultural commodities domestically produced including support to producers and that valued at world market prices without support to consumers.

Consumer Nominal Protection Coefficient (NPCc): an indicator of the nominal rate of protection for consumers measuring the ratio between the average price paid by consumers (at farm gate) and the border price (measured at farm gate level).

General Services Support Estimate (GSSE): An indicator of the annual monetary value of gross transfers to services provided collectively to agriculture and arising from policy measures which support agriculture, regardless of their nature, objectives and impacts on farm production, income, or consumption of farm products. It includes taxpayer transfers to: improve agricultural production (research and development); agricultural training and education (agricultural schools); control of quality and safety of food, agricultural inputs, and the environment (inspection services); improving off-farm collective infrastructures, including downstream and upstream industry (infrastructures); assist marketing and promotion (marketing and promotion); meet the costs of depreciation and disposal of public storage of agricultural products (public stockholding); and other general services that cannot be disaggregated and allocated to the above categories due, for example, to a lack of information (miscellaneous). Unlike the *PSE* and *CSE* transfers, these transfers are not received by producers or consumers individually and do not affect farm receipts (revenue) or consumption expenditure by their amount, although they may affect production and consumption of agricultural commodities. The **%GSSE** is the ratio of the GSSE to the *Total Support Estimate*.

Total Support Estimate (TSE): An indicator of the annual monetary value of all gross transfers from taxpayers and consumers arising from policy measures which support agriculture, net of the associated budgetary receipts, regardless of their objectives and impact on farm production and income, or consumption of farm products. The TSE is the sum of the explicit and implicit gross transfers from consumers of agricultural commodities to agricultural producers net of producer financial contributions (in *MPS* and *CSE*); the gross transfers from taxpayers to agricultural producers (in *PSE*); the gross transfers from taxpayers to general services provided to agriculture (*GSSE*); and the gross transfers from taxpayers to consumers of agricultural commodities (in *CSE*). As the transfers from consumers to producers are included in the *MPS*, the TSE is also the sum of the *PSE*, the *GSSE*, and the transfers from taxpayers to consumers (in *CSE*). The TSE measures the overall transfers associated with agricultural support, financed by consumers (transfers from consumers) and taxpayers (transfers from taxpayers) net of import receipts (budget revenues). The **%TSE** is the ratio of the TSE to the GDP.

The methodology applied in this study is fully consistent with that applied for OECD and other non-member countries. Box 2.7 provides basic information on how this has been done. It also discusses some data limitations which should be seen in the context of more general problems with China's agricultural statistics, discussed in Box 1.4.

> ### Box 2.7. **China's PSEs: what and how?**
>
> ***Period covered:*** 1993-2003
>
> ***Products covered:*** wheat, maize, rice, rapeseed, soybeans, peanuts, apples, sugar, cotton, milk, beef and veal, pigmeat, sheepmeat, poultry, eggs. These 15 commodities accounted for about 80% of the total value of gross agricultural output (GAO) in China in 1994-1995, but this share fell to 53% in 2002 and then increased to 61% in 2003. The share of the 9 crop products in the total crop production fell from 60-69% between 1993 and 1995 to just 37% in 2002 and then increased to 43% in 2003. The share of 6 livestock products in the total livestock production remained high during the whole period at above 80%.
>
> Changes in the shares reflect restructuring in China's agriculture (switch from grains and other traditional PSE products to fruits and vegetables; see Chapter 1), changes in relative prices, and, most likely, an overestimation of fruit and vegetable production in China in more recent years. An attempt has been made to include such products as tea, tobacco and oranges (0.7-1.0% of GAO each), but insufficient price information and quality gaps made it impossible to assess the level of support for these commodities.
>
> *Market Price Support*
>
> **Exchange rate CNY/USD:** weighted average for 1993 to reflect two exchange rates then applied for trade transactions: official and secondary (Chapter 1). As exporters were obliged to sell 20% of foreign currencies earned at official rate and 80% they could sell at secondary markets, the weighted average was calculated as follows: $(0.8 \times 8.28) + (0.2 \times 5.76) = 7.776$. Following the devaluation of the Yuan at the beginning of 1994, official rate was used for trade transactions. Therefore, for the period 1994-2003 official exchange rates were applied for the price gap calculations.
>
> **Producer prices:** unit values of above mentioned agricultural commodities sold by rural households through different marketing channels. Data originate from the yearly rural household surveys conducted by the NDRC in various regions.
>
> **External reference prices:** FOB prices for exportables and CIF prices for importables registered at the Chinese border.
>
> **Marketing margins:** estimated on the basis of price gaps between domestic wholesale and farm gate prices for a given commodity. Available technical coefficients were used when needed (*e.g.* to convert paddy to milled rice; sugar cane to sugar or live weight to slaughter weight). As data on wholesale prices were not sufficient to assess the level of margin, this source was supplemented for almost all products by phone interviews with various traders in China. A marketing margin for a given commodity was expressed as a percentage of a farm gate price. While it was assumed that the percentage margin remained at the same level over the whole period, its equivalent in absolute terms varied depending on the level of farm gate price in a given year. The absolute value of the margin in a given year was subtracted from the border reference price.
>
> **Transportation costs** (between China's border and domestic wholesale markets): assessed on the basis of phone interviews with traders and expressed as a percentage of the border reference price. These percentages have been converted into absolute values and added to the CIF price for importables and subtracted from the FOB price for exportables.

Box 2.7. **China's PSEs: what and how?** *(cont.)*

Quality adjustments: all efforts have been made to select such products traded by China whose quality corresponds best to products produced domestically (like with like comparisons). In most cases identifiable quality gaps reflected in price differences were small, 1-5% of the reference price. Therefore, quality adjustments have not been made with the exception of wheat. In the case of wheat, there have been two tendencies: a share of higher quality wheat in the overall wheat production has been growing, but at the same time the share of high quality durum wheat in total wheat imports has also been increasing (until 2003). Therefore, the CIF import price of wheat on the Chinese border has been adjusted by the same coefficient of 0.85 for the whole period under analysis. The coefficient has been calculated on the basis of detailed price survey conducted by Jikun Huang and Scott Rozelle in October 2001.

Price gap estimates: for all the above mentioned products relevant data have been collected and price gaps calculated. But, as for selected exportable products such peanuts, apples, beef and veal, pigmeat, poultry and eggs; no export subsidies and no other market price policy supporting or taxing producers have been identified, in line with the OECD methodology applied for other countries; the price gaps for these products have been set at zero.

Budgetary support

Budgetary information for the period 1993-2001 originates from the Ministry of Finance. Expenditures for different programmes in 2002 and 2003 have been estimated on the basis of partial information from *China Statistical Yearbooks* 2003 and 2004.

While all budgetary expenditures from various government bodies and at various levels of government administration should be accounted in the Ministry of Finance reporting, it is difficult to verify if this is the case. A general problem is that publicly available budgetary data, including on expenditures related to agricultural policy, tend to be strongly aggregated and do not allow precise assessment of the amounts actually spent on various policy measures and thereby evaluate their effectiveness. Moreover, more detailed information is not available for free and is released with long delays.

Even if information received from the Ministry of Finance is more detailed than that available in the Statistical Yearbooks, it remains very aggregated and for many programmes the coverage of payments within a given programme is so large that it is impossible to separate:

i) PSE-type payments from those which could be classified as General Services; and

ii) support to agriculture from support to rural areas in general, including for non-agricultural activities in rural areas.

In particular, it concerns support for agricultural infrastructure which is by far the largest component of transfers (about USD 13.9 billion in 2001). One of the examples is the so-called Comprehensive Development Plan for Agriculture being handled by a special office within the Ministry of Finance. It mainly gives support to (original wording): "improvement of low and medium-yielding fields; building of small-scale reservoirs; building of irrigation and drainage systems; building of electrical pumping wells; improvement of soil; purchase of agro-facilities for dry farming; building of roads; building of shelter-forests, building of agro-technical service stations and facilities for farmers' training". While part of the above mentioned (and similar) budgetary expenditures which have the objective of supporting rural infrastructure could be treated as input subsidies (*e.g.* "purchase of agro-facilities for dry farming"), other expenditures (*e.g.* on water supply or flood prevention included in other programmes under the general label of agricultural infrastructure) provide benefits to urban and industrial centres (*e.g.* township and village enterprises) in the vicinity. For now, due to a lack of accurate information, these expenditures are allocated to General Services.

As for other transition or developing economies, the results have to be interpreted carefully bearing in mind recognised limitations with respect to policy and commodity coverage, and data availability. In addition, the macroeconomic and institutional framework within which agricultural policy measures have been applied may have an impact on the results. Thus, the market price support (MPS) element may capture the effects not only of agricultural policies as such, but also macroeconomic policies (in particular through the exchange rate) and of imperfect price transmission from the border to the farm gate level. In the case of China, with very stable exchange rates, the impact of macroeconomic factors is weak. However, other factors such as a continuing inefficient downstream sector, a large share of agricultural production consumed on farms (Tian et al., 2002), weak price transmission compared to mature market economies, and data collection systems lagging behind the changes in the economy, may distort the measured level of support.

Aggregate results

Producer Support Estimate

As measured by the aggregate percentage PSE, producer support in China fluctuated within a range of minus 14% to plus 6% between 1993 and 1998. After falling to minus 3% in 1999, it increased each year and was plus 8% in 2003. A comparison of producer support for China and selected OECD and non-OECD countries, including principal world players, indicates that China has a low level of producer support. The percentage PSE in China, at 6% on average in 2000-2003, is above that in countries with the lowest support (New Zealand, Brazil, Australia), but much lower than the OECD average (31%) and far below that in Japan and Korea (58% and 64%, respectively), the closest OECD neighbours and main export markets for agro-food products (Table 2.11 and Figure 2.13).

Table 2.11. **Evolution of producer support (% PSE) and consumer support (% CSE) in China and selected countries, 1993-2004**

	1993	1994	1995	1996	1997	1998	1999	2000	2001	2002	2003	2004
Producer support estimates (% PSE)												
China	−14	1	6	1	1	1	−3	3	5	7	8	n.c.
Brazil	n.c.	n.c.	−1	1	1	6	1	4	3	3	4	3
Japan	57	62	61	57	53	57	59	60	57	58	59	56
Korea	73	73	72	64	63	57	65	67	62	65	61	63
Mexico	30	23	−5	5	15	18	18	24	19	26	19	17
United States	17	14	10	13	13	21	26	24	22	18	15	18
Turkey	23	14	12	15	25	26	23	21	4	20	29	27
EU[1]	38	36	36	33	34	37	39	33	32	34	36	33
OECD	35	34	31	29	29	33	35	32	29	31	30	30
Consumer support estimates (% CSE)												
China	15	0	−6	1	0	2	5	−2	−3	−5	7	n.c.
Brazil	n.c.	n.c.	1	2	3	−3	2	−2	0	−1	−2	−1
Japan	−51	−53	−53	−49	−47	−52	−54	−50	−49	−52	−52	−50
Korea	−71	−69	−71	−63	−61	−53	−63	−63	−59	−64	−58	−58
Mexico	−25	−11	18	6	−8	−12	−15	−19	−14	−22	−14	−10
United States	2	4	7	4	4	−2	−2	0	0	4	7	6
Turkey	−23	−8	−8	−11	−22	−27	−23	−23	−2	−17	−26	−22
EU[1]	−27	−25	−23	−20	−20	−24	−28	−20	−18	−21	−22	−19
OECD	−28	−27	−24	−21	−21	−24	−27	−23	−20	−22	−21	−20

n.c.: not calculated.
1. 1990-1994: EU12; 1995-2003: EU15; 2004: EU25.

Source: OECD PSE/CSE databases 2005.

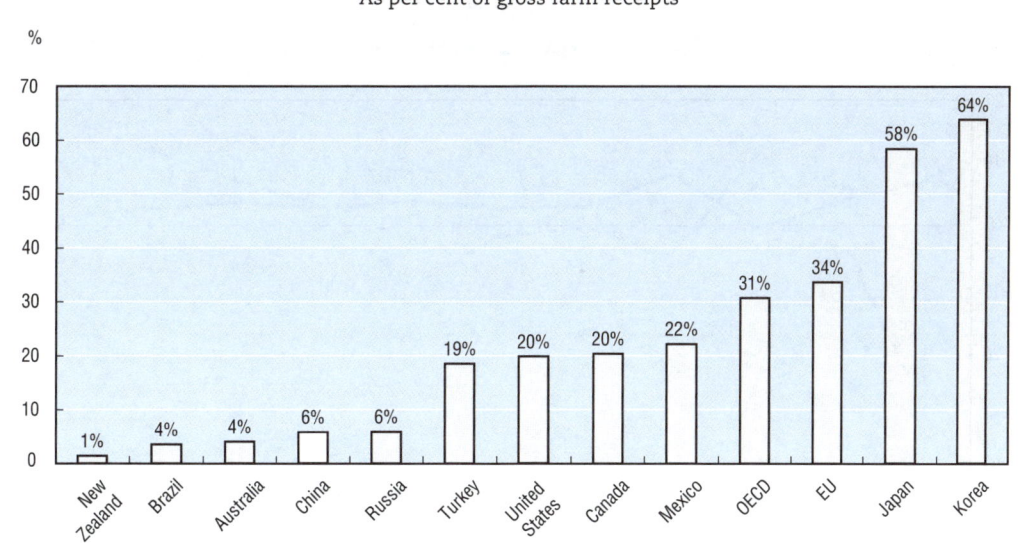

Figure 2.13. **Percentage PSEs for China and selected countries, average 2000-2003**
As per cent of gross farm receipts

Country	%
New Zealand	1%
Brazil	4%
Australia	4%
China	6%
Russia	6%
Turkey	19%
United States	20%
Canada	20%
Mexico	22%
OECD	31%
EU	34%
Japan	58%
Korea	64%

Note: EU15.
Source: OECD PSE/CSE databases 2005.

Changes in the level of support for China are driven mainly by the evolution of support for crop products, in particular for grains (Figure 2.14). As explained in Section 2.1 and Section 2.2, the Chinese government pays special attention to grain policies and the adequate supply of grains is a major driver, not only of grain policies specifically, but also of broader agricultural policies in general. Therefore, some volatility in the level of support, in particular in the 1990s, was to a large extent created by relative changes in the domestic and international prices for grains. For example, China's government raised state procurement prices sharply between 1994 and 1996 (Figure 2.2), but an increase in the world market prices represented by prices on the Chinese border was even stronger, which resulted in a fall in the support for the Chinese producers to 1% in 1996. In turn, a fall in world market prices for grains in 1997 and 1998 was fully transmitted on China's domestic markets. As a result, the level of support stabilised as measured by the PSE. Partial grain market reforms in mid-1998 combined with the pressure of huge grain stocks accumulated in the previous years contributed to the fall in grain prices on China's domestic markets in 1999, in particular for wheat and rice. As a result, the level of support fell again to minus 3%. Since then, the level of support has increased each year, which may seem paradoxical taking into account China's accession to WTO in 2001 and a continued fall in the level of import tariffs.

It should be noted, however, that up to the end of the 1990s prices for basic crops (cereals, soybeans and cotton) were fixed by the government, that state trading played a key role in foreign trade transactions, and that domestic grain supplies were secured by the grain quota system. Therefore, the level of tariffs, even if much higher than in the 2000s, had very limited impact on trade flows and on the level of domestic prices in China, including for importables. Tariffs were at most a source of budgetary revenues, but their regulatory impact on trade flows and prices was outweighed by the other more direct regulation instruments. The situation started to change at the end of the 1990s, when grain surplus encouraged the government to discontinue grain quotas and to engage in the

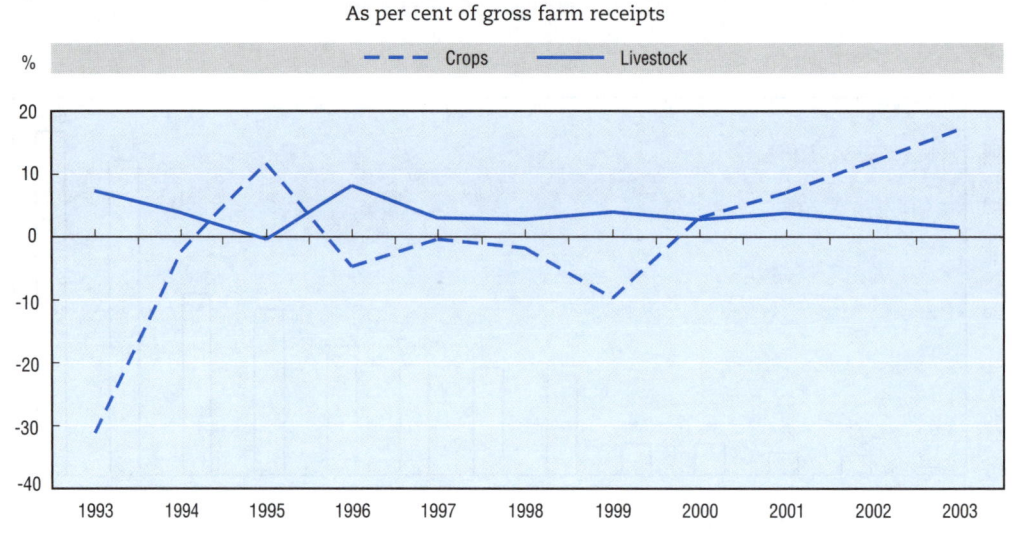

Figure 2.14. **Percentage PSE for crops and livestock products in China, 1993-2003**
As per cent of gross farm receipts

Source: OECD PSE/CSE databases 2005.

process of continued liberalisation of domestic grain markets. China's WTO commitments allowed private enterprises to participate to a growing extent in foreign trade transactions (even if for the most sensitive grains the share of private traders remains small) and registration procedures for enterprises active in foreign trade transactions have been substantially simplified.

Within such a framework, tariffs, even if falling, started to play a more active role in the determination of domestic prices, in particular for importables. The same applies for other measures such as VAT paid on imports at a higher rate than that charged on domestically produced agricultural commodities (Box 2.4). This could be one reason that, within the context of falling grain production in China between 1999 and 2003 and the growing expectation that China will become a net import of grains, the level of support for China's producers has tended to increase since 1999. Incidentally, the potential of China becoming a net importer coincided with the declared switch in policy objectives from the maximisation of agricultural production to policies supporting rural incomes (Section 2.1).

Consumer Support Estimate

The Consumer Support Estimate (CSE) is a PSE-related indicator measuring the cost of producer support to consumers of agricultural products. In the OECD methodology; the consumer is understood as the first buyer of these products. In the absence of consumer support policies, CSE generally mirrors the developments in the market price support (Table 2.11). An overall low degree of producer support in China means that agricultural support puts a relatively small burden on consumers, but overall taxation of Chinese consumers through agricultural policy measures has been growing in more recent years from 2% in 2000 to 7% in 2003.

2. POLICY TRENDS

Composition of the PSE

As is seen from Figure 2.15, the level of producer support in China is determined predominantly by the Market Price Support (MPS). However, during the period covered by the study, the contributions of MPS to the PSE varied, in particular in the 1990s, reflecting fluctuations in the levels of domestic prices relative to world prices.

It is worth noting that the share of MPS in the PSE has substantially increased between 2000 and 2003. While budgetary support has almost constantly been growing in absolute terms, its share in the aggregate has been falling in the 2000s. However, within budgetary support, a large part is provided through input subsidies (Chapter 2). Both MPS and input subsidies are known to be most trade distorting and least efficient channels of providing agricultural assistance. In particular, low transfer efficiency means that only a small part of support is effectively received by producers (Box 2.8).

Total Support Estimate

The Total Support Estimate (TSE) is the broadest indicator of support, representing the sum of transfers to agricultural producers (the PSE), expenditure for general services (the GSSE), and direct budgetary transfers to consumers.

The aggregate TSE in China reached USD 43 billion per year in 2000-2003. The TSE expressed as a percentage of GDP, indicates the cost that the support to the agricultural sector places on the overall economy. Between 1993 and 1998, the Chinese percentage TSE fluctuated between minus 1.9% and 3.4% and then, after falling to 1.1% in 1999, it increased each year and was 3.7% in 2003 (Table 2.12). This suggests a relatively high burden of the agricultural support on the Chinese economy. China's average percentage TSE at 3.3%

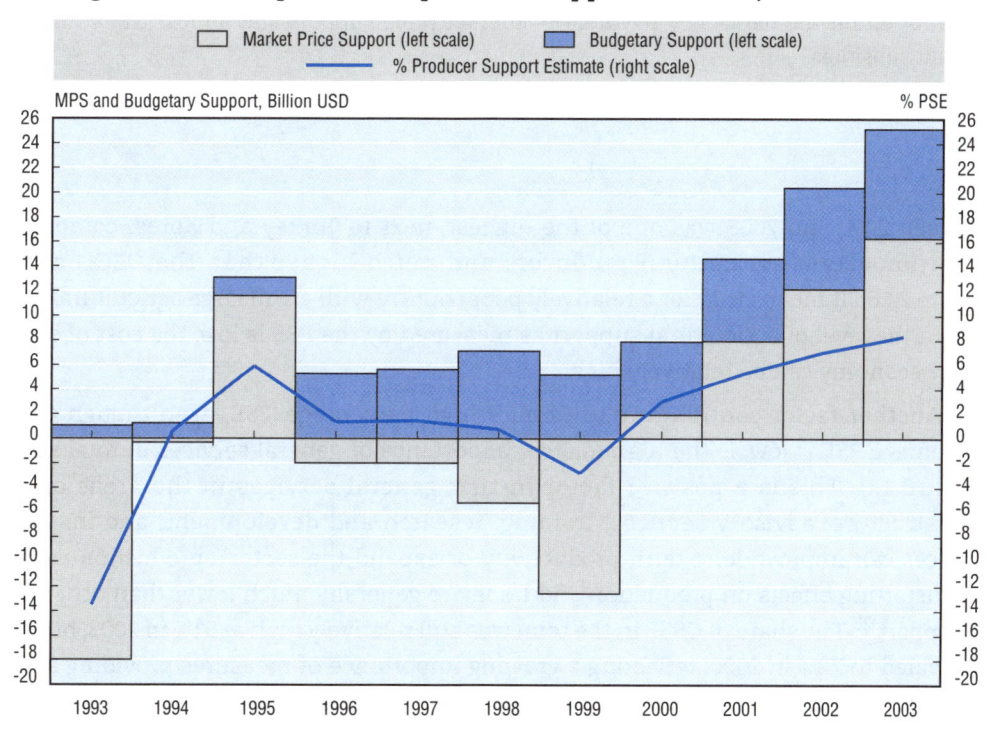

Figure 2.15. **Composition of producer support estimate, 1993-2003**

Source: OECD PSE/CSE databases 2005.

> **Box 2.8. Transfer efficiency in agricultural support policies**
>
> PSE/CSE methodology estimates the support aimed at agricultural producers, not the support effectively received by producers. It is important to note that a part of the support aimed at agricultural producers is captured at other stages of the food chain, such as upstream and downstream sectors, and part of it is a dead-weight loss. The higher the support effectively received by producers out of total costs incurred by consumers and taxpayers to provide such support, the higher the transfer efficiency of agricultural support policies is.
>
> The Policy Evaluation Matrix (PEM), which is an OECD model showing the effects of "small" changes in support on production, trade and economic welfare, indicates that the effects of a given amount of support may differ substantially depending on the type of support measures used. The results show that the estimated effects on farm household income of support in the form of payments based on area are systematically higher than for the other support measures (market price support, payments based on output, payment based on input use). The model also shows that the estimated effects on farm household income of support provided in the form of payments based on the use of purchased inputs are always lower than when the same amount of support is provided through other measures. It means that transfer efficiency of payments based on area is relatively high and that transfer efficiency of input subsidies is relatively low. The results confirm that input subsidies constitute the least efficient way of supporting producers, as most of the support is captured by input suppliers and part of it is a dead-weight loss.
>
> Quantitative estimations of transfer efficiency of various policy measures have been made for several OECD countries. The results show that on average only 17% of the additional support given in the form of input subsidy is transferred to farm households. In the case of market price support and deficiency payments, the percentage is higher at 24-25%. The highest transfer efficiency is for area payments at 47%, which is about double that of either the deficiency payment or market price support and almost triple that of input subsidies.
>
> *Source:* OECD (2003).

between 2000 and 2003 was one of the highest, next to Turkey and Korea, compared to other important agricultural producers and much higher than the OECD average (Figure 2.16). It means that for a relatively poor country with a still large agricultural sector, even if the level of agricultural support as measured by the PSE is low, the cost of support to the economy can be relatively high.

Another factor contributing to China's high percentage TSE, even though China's percentage PSE is low, is the high relative importance of general services in total support (Table 2.12). This is a positive factor in that general services in the areas of rural infrastructure, advisory services, training, research and development, and inspection services can improve long-term productivity or expand the sector's production capacity, the distorting effects on production and trade are generally much lower than other forms of support.[25] The share of GSSE in the total was still relatively high at 51% in 2003, but lower compared to 72% in 2000, reflecting a growing importance of measures providing support to producers (the PSE component). However, even the 2003 share compares favourably to

Table 2.12. **Total support to Chinese agriculture**

	1993	1994	1995	1996	1997	1998	1999	2000	2001	2002	2003p
Total Support Estimate (TSE), million CNY	−66 168	83 642	199 256	122 330	139 446	159 201	88 605	242 522	322 437	379 315	428 072
of which:											
Producer Support Estimate (PSE)	−131 351	8 122	109 426	27 535	30 490	15 880	−60 928	65 411	121 142	168 965	208 392
General Services (GSSE)	62 197	72 979	87 413	92 783	107 081	141 555	146 863	174 771	200 612	209 750	219 080
Transfers to consumers from taxpayers	2 986	2 541	2 417	2 012	1 875	1 766	2 670	2 340	683	600	600
Total Support Estimate in:											
Million USD	−8 509	9 705	23 860	14 713	16 821	19 229	10 703	29 296	38 956	45 828	51 718
Million EUR	−9 796	8 165	18 240	11 604	14 833	17 169	10 046	31 798	43 533	48 694	45 824
TSE as share of GDP, %	−1.9	1.8	3.4	1.8	1.9	2.0	1.1	2.7	3.3	3.6	3.7

p: provisional.
Source: OECD PSE/CSE databases 2005.

Figure 2.16. **Total support estimate in China and selected countries, average 2000-2003 – as per cent of GDP**

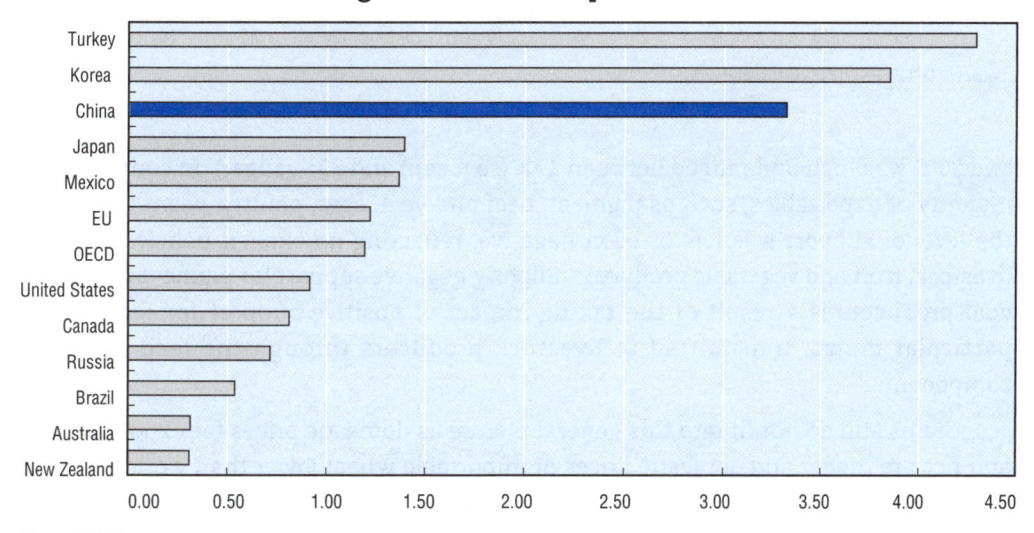

Country	% of GDP
Turkey	~4.35
Korea	~3.90
China	~3.35
Japan	~1.40
Mexico	~1.35
EU	~1.20
OECD	~1.15
United States	~0.90
Canada	~0.80
Russia	~0.70
Brazil	~0.50
Australia	~0.25
New Zealand	~0.25

Note: EU15.
Source: OECD PSE/CSE databases 2005.

the OECD average at 18% in 2001-03. Only in countries with the least distorting policies, such as Australia and New Zealand, the average shares were at or above 40% (OECD, 2004a).

Transfers to consumers from taxpayers are of marginal importance (Table 2.12).

Commodity profile of producer support[26]

Level of producer support by commodity

While China's aggregate producer support is low, the level of support varies significantly across commodities. The spread in support levels across commodities is a potential source of distortion. There is a clear distinction between the levels of support for importable and exportable products (Figure 2.17). For the majority of importables, such as soybeans, sugar, milk, sheepmeat and cotton, the average level of support between 2000

Figure 2.17. **Chinese % PSE by commodity, average 2000-2003**
As per cent of gross farm receipts

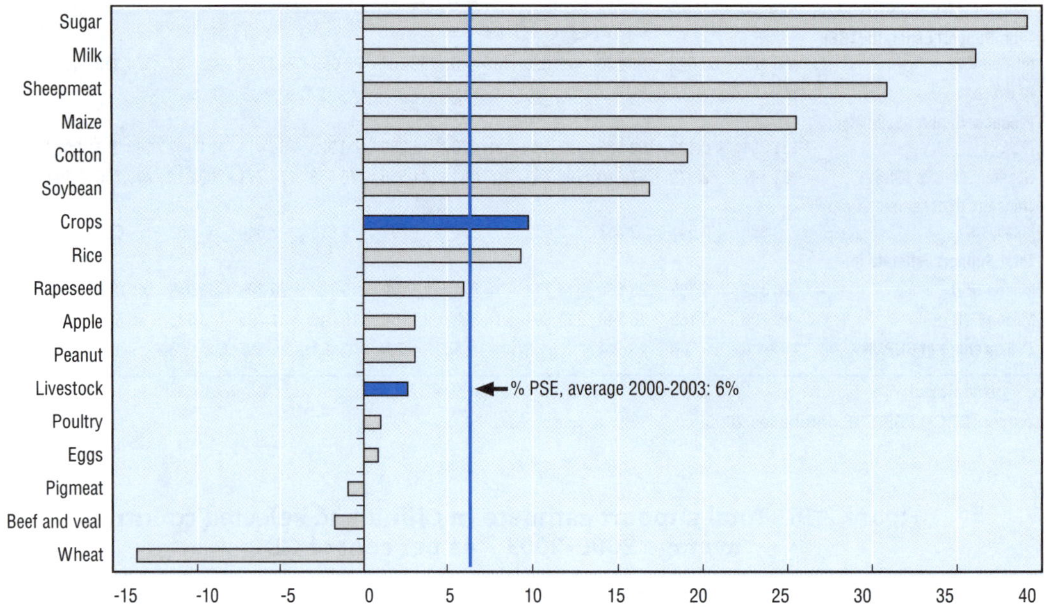

← % PSE, average 2000-2003: 6%

Source: OECD PSE/CSE databases 2005.

and 2003 was high and ranged between 17% (soybean) and 39% (sugar). In contrast, for the majority of exportables, such as pigmeat, beef and veal, eggs, poultry, peanuts and apple, the level of support was low or even negative, reflecting no explicit policies supporting livestock, fruit and vegetable producers. Slightly negative support for pigmeat and beef and veal producers is a result of the taxing impact of positive support for feed crops, in particular maize, transmitted to livestock producers through the feed adjustment component.

Grains still do not fit into this general picture as domestic prices for exportable maize and rice are higher and domestic prices of importable wheat lower than world prices. One of possible explanations is a dominant role of state trading in grain transactions, even if the role of private traders increased in line with China's WTO commitments. Trade flows in grains are still not driven by profits and relative price levels but rather by the government decisions reflecting concerns over food security and the level of grain stocks. For example, in 2004 (not yet covered by the PSE calculations) the government decided to import more than 7 million tonnes of wheat to replenish government stocks at prices reportedly higher than domestic ones. On the other hand, between 1999 and 2003, China exported large amounts of maize at lower prices than domestic prices to get rid of huge stocks accumulated in the second half of the 1990s.

Distribution of producer support across commodities

The distribution of overall producer support across commodities reflects relative price changes on domestic and world markets, the scale of budgetary assistance to specific commodities, and the relative importance of these commodities to overall agricultural production. In line with China's policy focus on key crop commodities, about 75% of overall PSE transfers (on average about USD 7 billion) between 2000 and 2003 was provided to crop

2. POLICY TRENDS

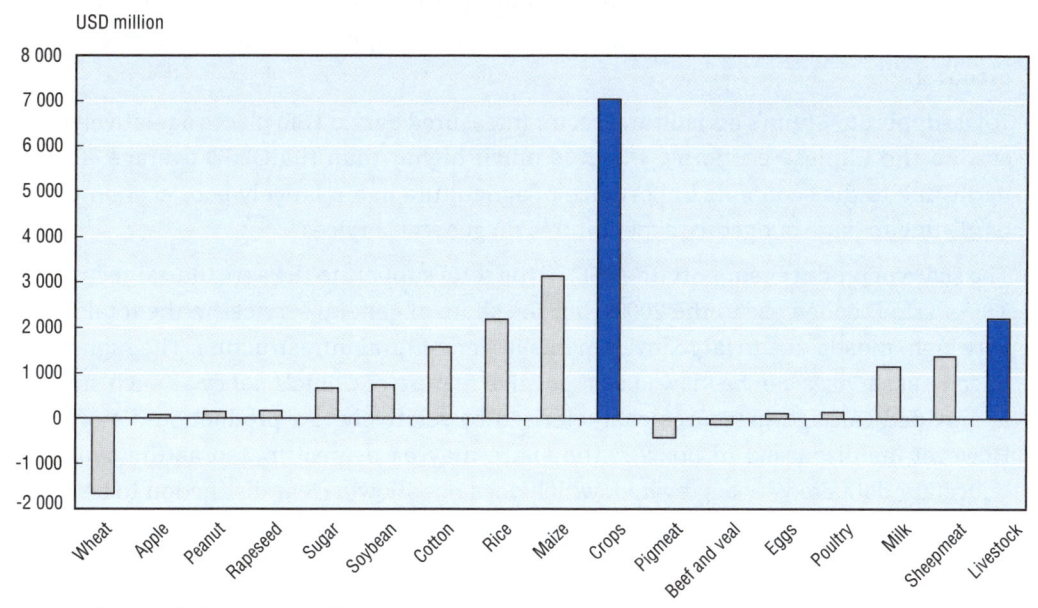

Figure 2.18. **Distribution of producer support by commodity, 2000-2003 average**

Source: OECD PSE/CSE databases 2005.

products, in particular to maize and rice producers. In the livestock sector, only milk and sheepmeat producers were the recipients of any significant support (Figure 2.18).

Conclusions

Bearing in mind data problems discussed at the beginning of this section, from the above analysis of agricultural support in China, the following general conclusions can be drawn.

- In the 1990s China's government was still applying a large number of distortive policies such as grain quotas, government fixed prices for selected crops and state trading. But domestic prices, including those set by the government, were usually fixed at levels close to world prices. Budgetary support for producers was low. As a result, the level of support, as measured by the percentage PSE, although fluctuating was generally low.

- The level of support in the 2000s increased, but remained far below the OECD average. The increase in support may mean that, within the context of largely liberalised domestic commodity markets, producer prices started to adjust to reflect market conditions as well as border protection, in particular for imported commodities. Therefore, even as border protection declined, tariffs and other border measures started to have a stronger impact on domestic prices compared to the 1990s. At the same time, budgetary support tended to increase, which contributed to a rise in the level of support to 8% in 2003.

- While China's producer support is low, the level of support varies significantly across commodities, which is an indication of distortive policies. The highest levels of support are for import-competing commodities, such as sugar, milk, sheepmeat, soybeans and cotton, as well as some export commodities such as maize and rice. The distortions on grain markets are still high, mostly due to state trading which continues to drive a wedge between domestic and world prices.

- The mix of measures used to support China's farmers is dominated by market price support and input subsidies, categories known to be amongst the least efficient and most trade distorting ways of providing agricultural assistance. In particular, low transfer efficiency means that only a small part of support is effectively received by producers.

- Total support to China's agricultural sector (measured by the TSE) places a relatively high cost on the Chinese economy, which is much higher than the OECD average. This is partly due to the economic importance of agriculture in a relatively poor economy, and partly due to large budgetary expenditures on general services.

- The share of producer support (the PSE) in the total support to the agricultural sector (the TSE) started to increase in the 2000s, but the share of general services in the total is still very high, mostly due to large investments in agricultural infrastructure. The high share of general services can be viewed as a positive feature of China's policy as such support is provided through measures characterised by relatively low production-distortions. However, as discussed in Box 2.7, the share may be overestimated as the available budgetary data are very aggregated, which does not allow a clear distinction to be made between payments supporting producers and those improving the performance of the agricultural sector as a whole.

Notes

1. In fact, according to the Chinese definition of grains (see Chapter 1), the self-sufficiency rate was much lower at 82% in 2003, mostly due to massive imports of oilseeds (Han Jun, Development Research Center of the State Council, personal communication, April 2005).
2. The "No. 1 Document" is the highest priority document of the Chinese authorities identifying the top priority issues for a given year and addressing attention of the party organs and governments (ministries, departments) at various levels.
3. Additional information on the agricultural policy development process and on institutions designing and implementing agricultural policies in China can be found in OECD (2005a).
4. The State Grain Administration (SGA) was supervised by the Ministry of Internal Trade in the early 1990s and is currently supervised by the National Development and Reform Commission.
5. Also known as Foreign Trade Corporations (FTCs).
6. The Ministry of Foreign Trade and Economic Co-operation (MOFTEC) was merged with Ministry of Internal Trade to form the Ministry of Commerce in 2002.
7. For instance, while the national government imposes quantitative restriction on local governments for approving transfers of agricultural lands to non-agricultural purposes, local governments may sometimes take the approach of piecemeal approval in order to avoid control by the central government – large parcels of land are broken into smaller parcels which are then approved for transfer by the local government. Also, some local governments are reported to have either over- or under-reported rural incomes for either personal gain from exaggerating their performances or for local benefit from obtaining special assistance such as poverty alleviation funds.
8. A detailed description of policies by commodity is provided in Annex B.
9. Price, determined according to both state-set and market prices, at which government would purchase above quota quantities offered for sale to the state.
10. Unlike a normal commercial transaction, the terms of the loan are not always enforced.
11. Under the previous arrangement, the prices were calculated as the state-set final user prices minus handling costs.
12. Village electricity infrastructure and user charges are the responsibility of the village administration. The village usually trains an "electrician" who is responsible for maintaining the

local distribution grid and collecting user charges from end-users. The village has payment responsibility for electricity delivered to the village transformer and must raise this payment from end-users. Especially prior to the Electricity Network Reform Program instituted in 1997-98, loss of energy in obsolete local distribution networks resulted in end-users effectively paying higher prices for electric power than was indicated by nominal electricity fees at the provincial level.

13. Tibet has been exempt from paying agricultural taxes since the Autonomous Region was established.

14. A comprehensive overview of China's food safety system is provided in Annex C. In this sub-section some basic issues are discussed briefly.

15. Information is sent electronically to local Agricultural Bureaux and extension services, which disseminate information to farmers by word-of-mouth, leaflets, and public notice boards.

16. CCTV-7 carries shows principally for children, military affairs programming, and agriculture-related programming. The agricultural programmes mainly target rural life and the agricultural economy, featuring news, entertainment, agricultural science and information services.

17. In March 1994, the government issued an instruction to supervise the prices of 20 essential goods and services; including wheat flour, rice, vegetable oils, pork, eggs, milk and sugar. To protect urban consumers, the national government authorised sub-national governments to take measures to prevent price escalation, such as announcing reference prices, controlling retail mark-ups, and instituting price ceilings.

18. As defined in Chapter 1, these are: cereals, oilseeds and vegetable oils, sugar, cotton, tobacco, tea and silk.

19. See Annex C for detailed information on China's food safety regulations.

20. These categories are classified based on a series of criteria and safety standards. The standards for green food and organic food products include requirements on environmental conditions of production sites as well as stipulations on production practices.

21. ASEAN members are Brunei Darussalam, Cambodia, Indonesia, Laos, Malaysia, Myanmar, Philippines, Singapore, Thailand and Vietnam.

22. Hong Kong SAR of China received around 10% of China's agro-food exports in 2003. While China's trade flows with Hong Kong are registered separately, it is not a sovereign country.

23. Available from: *http://info.china.alibaba.com/news/detail/v2-d5294515.html.*

24. However, balance data may be to some extent misleading as China imports large quantities of raw materials such as cotton, hides and skins which are classified as agricultural imports and then exports processed products such as apparel which is not included in the agricultural export statistics.

25. Support for general services to agriculture does not depend on individual farmer's production decisions regarding output or use of factors of production, and does not directly affect farm receipts (OECD, 2004a).

26. A detailed overview of agricultural policy measures and trends in support for individual commodities is presented in Annex A.

ISBN 92-64-01260-5
OECD Review of Agricultural Policies
China
© OECD 2005

Chapter 3

Policy Impacts

This chapter reports analysis on the potential economic impact of reducing China's agricultural trade protection and domestic support and comparing the effects with those that might accompany widespread reductions in support afforded farmers in OECD countries. Section 3.1 investigates, with the use of an economy-wide general equilibrium model (GTAPEM), the source and size of the sectoral and economy-wide gains to China from multilateral trade liberalisation. The estimated changes in prices and quantities due to liberalisation are then used as inputs to simulate how welfare gains and losses are distributed across various types of rural households in China.

The last two sections of the chapter are devoted to an analysis of the potential impact on China's main agricultural commodity markets of hypothetical reforms to agricultural trade and domestic policy both within China and globally (Section 3.2), and to an investigation of the impact of China's grain policies on mid-term projections of its grain imports and the implications for grain self-sufficiency (Section 3.3).

3.1. Welfare impacts of trade and agricultural policy reforms

Simulated impacts of multilateral and multi-sectoral policy reform

Results reported here were obtained in analysis undertaken as part of a larger OECD project aimed at measuring distributional impacts of agricultural policy reforms – both among and within selected OECD and non-OECD countries. A report done for that project, (OECD, 2005b) contains a comprehensive description of the model, the data and the policy simulation experiments performed. The GTAPEM model used to estimate reform impacts is a modified version of the GTAP global general equilibrium model widely used in applied agricultural and trade policy analysis (Hertel, 1997).

Initial rates and pattern of trade protection

The estimated market effects of policy reform are determined in large measure by: a) the initial level and pattern of trade protection and b) the reform scenario implemented. (The policy reform scenarios are discussed in the following sub-section.) The data presented in Table 3.1 indicate the initial level and pattern of trade protection in China, in the OECD and in the non-OECD regions respectively. The table displays averages of tariffs levied and faced by China, and compares these with the averages of tariffs levied and faced by OECD and non-OECD countries.[1]

Generally speaking, the rates of trade protection for China's agriculture are relatively lower than what it faces in foreign markets. In primary agriculture, the tariffs levied by China are somewhat lower than those applied to processed agricultural commodities and even lower when compared to averages of tariffs levied on both primary and processed agricultural products entering OECD countries. On the other hand, China levies tariffs in non-agriculture sectors that are somewhat higher than the average for OECD countries. Finally, the tariffs faced by China in non-agriculture sectors are somewhat higher than those faced on average by either OECD or non-OECD countries.

These initial conditions for border protection imply a potential gain for China's agricultural economy from global agricultural liberalisation (higher tariffs faced than

Table 3.1. **Tariffs levied and faced (%)**

	Tariffs levied			Tariffs faced		
	China	OECD	Non-OECD	China	OECD	Non-OECD
Agriculture	8	39	12	28	12	18
Primary agriculture	5	36	7	24	19	17
Processed	12	44	17	33	26	20
Non-agriculture	9	2	10	8	3	3
Textiles and wearing apparel	14	8	18	17	7	11

Source: GTAP version 6 database.

levied).[2] However, given the relatively large economic weight of the non-agriculture sector (see related discussion in Chapter 1), important gains should also be realised here. China's abundant labour supply gives it a potentially important competitive edge in certain sectors, such as textiles and wearing apparels. Here, the tariffs faced by China's exporters are relatively high, particularly to important markets such as the US (12.3%) and the EU (9.5%). However, the success of these sectors also depends on other key factors such as appropriate investment in modern capital and management practices, and reasonably free movement of labour from surplus areas (typically rural).

Policy reform scenarios

The policy scenario comprises a multilateral 50% reduction in domestic farm support for OECD and select non-OECD countries, and a 50% reduction in tariffs (*ad valorem*, specific and TRQ) and export subsidies for all sectors. The results should be interpreted as the medium term impact (about 5 years) of the policy change after all necessary adjustments on consumption and production have taken place.

Table 3.2 contains estimates of the combined impacts of the reform scenario on incomes globally, for the OECD and non-OECD regions and for China. For each country/region, the total income gain in the first column is decomposed into the impact due to OECD and non-OECD agriculture and non-agriculture reforms respectively. The global gains are about USD 44 billion, three-quarters of which go to OECD countries. For these countries, most of the gains arise from reforms of their agricultural policies, and reform of non-OECD non-agriculture policies. In contrast, the largest gains for non-OECD countries are due to liberalisation of OECD non-agriculture sectors.

The overwhelming share of the estimated welfare gains in China are attributed to the assumed reductions in trade protection afforded non-agricultural merchandise, particularly textiles and wearing apparels both in OECD countries and in China itself. The results are illustrated in Figure 3.1.

The potential implications for China of combined global and domestic agricultural policy reforms have been the subject of many analyses in recent years. In some analyses, including the one reported here, the net welfare effects are slightly negative; in others slightly positive. To understand this outcome, it is useful to recall that the total welfare effect adds the impacts across both consumers and producers. The higher world market prices for agricultural goods that would come with global agricultural policy reform bestow

Table 3.2. **Welfare effects of multilateral policy reform, USD millions**[1]

	Total welfare	OECD agriculture	Non-OECD agriculture	OECD non-agriculture	Non-OECD non-agriculture
World	44 268	23 361	3 124	6 694	11 357
OECD	33 459	21 407	1 871	−248	10 680
Non-OECD	10 809	1 954	1 253	6 943	677
Brazil	1 730	1 178	94	367	96
China	3 739	−73	−199	3 373	635
India	1 723	72	544	378	735
South Africa	253	69	25	23	137

1. Welfare is measured in terms of the "equivalent variation" in monetary income.

Source: OECD Secretariat.

Figure 3.1. **Welfare gains (losses) by source of liberalisation**

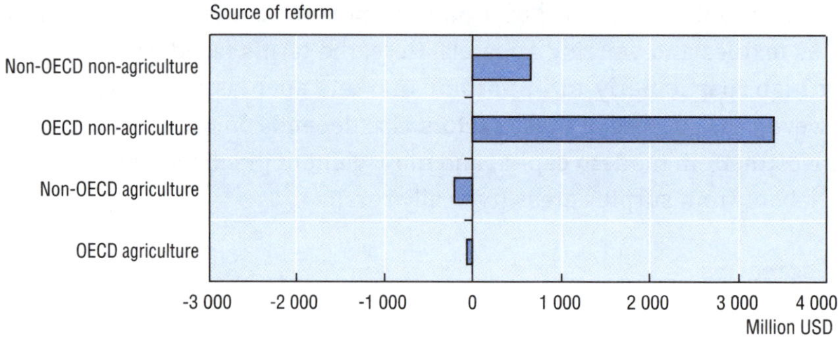

[Bar chart showing Source of reform, in Million USD:
- Non-OECD non-agriculture: ~700
- OECD non-agriculture: ~3 400
- Non-OECD agriculture: ~ -200
- OECD agriculture: ~ -50]

Source: OECD Secretariat.

economic benefits on most of China's farmers (*e.g.* those producing exportable agricultural commodities), but also higher costs of agricultural imports on some consumers. Because agricultural trade constitutes only a relatively small share of total agricultural output in China (domestic agricultural production largely equals domestic consumption), these two effects offset each other leaving small net impacts on China's overall economy from agricultural trade reforms. There is no doubt, however, that global agricultural policy reform is in the interest of China's agriculture.

As the discussion in Chapter 1 makes clear, the past ten to fifteen years of rapid economic growth in China occurred in a situation of persistent, even widening, divergence in returns to labour employed in agriculture versus non-agricultural sectors of the economy. That discussion also highlighted the necessity for structural adjustment and the pressures created by rural to urban migration arising from the higher potential earnings from off-farm jobs. A critical question therefore is whether further, multi-sectoral reductions in trade protection will ameliorate or attenuate these intersectoral adjustment pressures. Figure 3.2 contains estimates showing that the simulated policy reforms would boost returns to labour and capital employed, whether this was to occur in China's agriculture or non-agriculture sectors. Importantly however, the estimated gains are uniformly spread both across sectors and factors suggesting that this particular reform scenario might add little to ongoing pressures for inter-sectoral factor adjustment and hence for rural to urban migration.

Distributional effects of policy reform across various types of households in rural China

Introduction

Since the start of economic reforms, inequalities in income and consumption have been rising in China, while poverty has been falling. Most poverty reduction was achieved through growth in the rural economy, including in agriculture. But rising rural inequality has been shown to slow down poverty reduction (Ravallion and Chen, 2004). Given this record, the impact of ongoing trade and domestic reforms on agricultural prices, and through these on rural inequality and poverty, are important.

Various studies suggest that ongoing economic reforms may bring overall gains, but need to be accompanied by policies targeting pockets of poverty and vulnerable

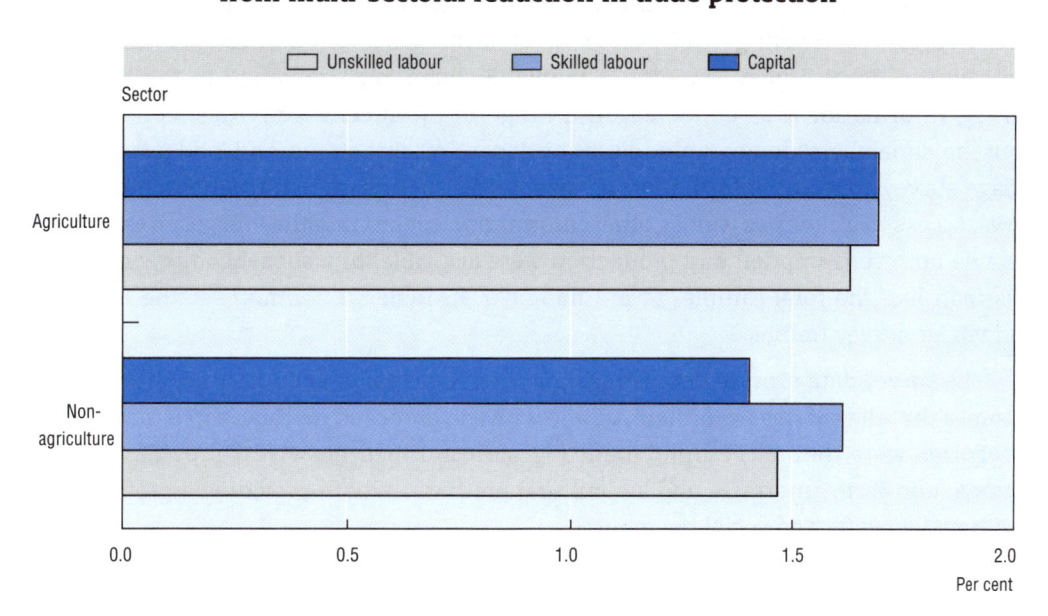

Figure 3.2. **Changes in factor returns to agriculture and non-agriculture resulting from multi-sectoral reduction in trade protection**

Source: OECD Secretariat.

households in order to minimise localised and temporal trade-offs between border price liberalisation and poverty reduction. This section draws on household and village survey data to explore distributional effects and identify the characteristics of winners and losers from reforms.

Method

In this section, a procedure modified from McCulloch (2003) is employed to assess the impact of structural reforms on household incomes, applicable to data sets with household-level information on production and consumption. The policy scenario assumed is a global 50% liberalisation as described above, and the GTAPEM results presented above are used as input into this analysis of welfare implications in a simple household model.

A change in household income in the neighbourhood of its optimum, with given endowments, is calculated. Household income changes due to the exogenously (by GTAPEM) given changes in both the prices and quantities of output (most household are agricultural producers), factors (including wages) and consumption goods. Thus, both first and second order effects of policy reform are incorporated in the analysis. Rising (falling) incomes increase (decrease) the consumption bundle that households can consume and imply a positive (negative) income effect of price policy reform. Rising (falling) prices of consumption goods decrease (increase) the consumption bundle that households can consume and imply a negative (positive) consumption effect of price policy reform. The total welfare effect, for each household, is the sum of the consumption and production effects. Then, the distributional effects of policy reform within the sample are evaluated.

Data

Data originate from a November-December 2000 rural household survey among 1 200 households, covering 4 387 people, in 60 townships in six Chinese provinces across the country: Hebei, Hubei, Liaoning, Shaanxi, Sichuan and Zhejiang.[3] In the household survey, information was collected on a range of productive activities, consumption, household members' demographic situation, labour market status, and land use. A parallel village survey collected information on size, infrastructure, agricultural potential, labour flows, access to utilities as well as other information concerning the villages in the sample. Data on both consumption and production were available for a subset comprising slightly less than half the total sample, as in Chang (2004). Table 3.3 summarises the household and village survey frames.

The survey data provide detailed insight into households' economic activities and the incomes that they derive from them. In line with the proposed method, three main income categories were defined: employment (agricultural or otherwise[4]) providing a wage income; non-farm enterprise and agricultural production, both providing a profit income; and transfers (gifts and social payments).

Wages comprise both conventional employment income (reported individually and aggregated over households) and *corvée* labour (*yi wu gong* or *ji lei gong*), which many households supply to local authorities, mostly for the construction of local infrastructure. In some cases households reported paying money in lieu of supplying *corvée* labour. This was treated as negative income.

Typical of agricultural household incomes, particularly in smaller farms in low and middle-income countries, is the large importance of produce that is consumed by the farm household. Valuing this correctly is a challenge since there is no market price for subsistence production. One solution is to value at official, nationally published prices. But regional variation of retail prices is typically large, and this is especially true in this sample of households across China, a huge country with imperfectly liberalised markets. In order

Table 3.3. **The 2000 rural household survey frame**

Province	Number of townships	Number of households	Number of people
Hebei	10	200	746
Hubei	10	200	667
Liaoning	10	200	701
Shaanxi	10	200	770
Sichuan	10	200	744
Zhejiang	10	200	759
Total Sample	60	1 200	4 387
Sub sample			
Hebei	10	98	369
Hubei	10	92	306
Liaoning	10	87	299
Shaanxi	10	75	270
Sichuan	10	110	421
Zhejiang	10	79	309
Sub Sample	60	541	1 974

Source: Survey findings.

to approximate local product prices in this study, home-consumed output was valued at unit values implied by households' reported quantities and revenues (as in Ravallion and Chen, 2004) aggregated by province.

Agricultural profit based on these prices derives from five categories of productive activities:

- Hog production, where net income is the value of pork consumed by the household plus revenues from hog or pork sales, net of all costs connected to hog production.
- Grain crop growing. Grain crops include rice, wheat, millet, sorghum, and others, depending on the household's location. Net income from crop production is the value of grains consumed or exchanged by the household plus revenues from grain sales and from sale or is of its by-products (such as chaff), net of all costs connected to crop production. In the Chinese dual marketing system, grains are sold both on the open (typically local) market and to "grain stations", typically at different prices (Chapter 2).
- Production of orchard fruits, with identical profit definition.
- Livestock rearing, other than hog production but including fowls. Revenues included sale of meat and of live animals and meat consumed by the family, valued at sale prices averaged by province.
- Income from "overhead" factors, including income from renting out land (in money or in-kind) net of land rent paid, depreciation and maintenance costs of productive assets, and agricultural tax.

Income from agricultural production[5] is the sum of the above five items. Non-agricultural enterprise income consists of both profit from production and sale of processed goods (mainly food products) closely connected to the household's farming activities, and profit from other typically micro-enterprise independent activity. Transfers included pensions and other social payments, and the balance of gifts received and given. Based on these definitions, it is possible to construct the structure of household income, represented by income shares of each of the above categories. Table 3.4 summarises results by province.

Table 3.4 demonstrates that there is large variation in the composition of net income over provinces. On average, revenues from crop production contribute more to agricultural income than revenues from livestock rearing. The importance of wage income equals that of

Table 3.4. **Structure of household income, provincial averages (% of total income)**

Province	Crops and orchards revenues	Livestock revenues	Non-farm enterprise revenues	Capital outlays	Job income	Transfers	Corvée income	Total
Hebei	27.4	28.1	–2.9	–2.5	34.4	15.3	0.0	100
Hubei	52.3	33.7	8.7	–62.9	52.8	15.4	0.0	100
Liaoning	20.1	8.9	39.6	–16.4	36.5	11.6	–0.5	100
Shaanxi	–77.3	–3.6	-13.1	140.4	48.2	6.8	–1.4	100
Sichuan	38.0	36.4	12.0	–49.7	54.5	9.0	–0.2	100
Zhejiang	237.8	72.5	9.0	–285.9	52.8	14.7	–0.9	100
Sample	48.8	29.7	9.3	–46.2	46.5	12.2	–0.4	100

1. Totals may not sum to 100 due to rounding. The sample average is the population-weighted average of provinces.
Source: Survey findings.

crop revenues. Also non-farm enterprise and non-earned income contributes considerable shares of income. The table also shows that both the revenues and the cost side of net income will matter in the determination of income effects below. The level of outlays on capital goods for both agricultural production and non-agricultural enterprise is a factor in total net household income of equal magnitude as wage income and crop revenues.

The survey provides detailed data on consumption. Consumption of food included money expenditures, home consumption of own produced food, exchanged food, and gifts. Food consumption in the last three categories (i.e. not paid for by money expenditure) was valued at unit values imputed from households' reports in the survey, averaged over provinces. Food consumption recorded in the survey included 32 separate items or categories. Non-food consumption included expenditure on 35 goods and services. Table 3.5 shows per capita consumption levels, poverty rates, and shares of food consumption in households' budgets, per province.

Table 3.5. **Consumption and poverty, provincial averages**

Province	Consumption per capita (yuan)	Poverty rate (%)	Share of food in consumption (%)
Hebei	2 028	7.0	49
Hubei	1 512	30.4	38
Liaoning	2 227	2.3	47
Shaanxi	1 510	20.7	55
Sichuan	1 261	28.8	46
Zhejiang	2 128	2.7	29
Sample	1 770	15.2	45

Note: The sample average is the population-weighed average of provinces.
Source: Survey findings.

Table 3.5 shows that the sample is roughly divided into three high-welfare provinces, one medium-welfare and two relatively poor provinces. The absolute measure for defining poverty used is the official poverty line for 2000 of CNY 625 consumption value per person (NBSC, 2004). According to this measure, 13.1% of rural Chinese households (15.2% of people) in these provinces are in poverty. This compares to a 2004 figure calculated by Ravallion and Chen (2004) of 12.5%, based on a CNY 850 per capita poverty line.[6] Table 3.5 shows that poverty varies much over provinces in line with provincial-average consumption levels, reflecting China's regional development differences. There is no clear link between provincial-average welfare levels and the provincial-average share of consumption budgets spent on food. This cautions against using the food share as a poverty proxy, as is often done. The variation in food shares also suggests that consumption effects of agricultural price reforms will vary markedly by region.

Analysis

The price and quantity changes resulting from the general equilibrium model cover 16 agricultural product categories, 4 non-agricultural product categories, and factor inputs (land, labour and capital – separately for agriculture and non-agricultural sectors). The general equilibrium model produces simulated price and quantity changes, from which changes in household income are calculated through cost and revenue changes (the income effect) as well as changes in consumption through consumption price changes (the consumption effect). To give a flavour of the changes involved, the general equilibrium model output includes an average 0.7% increase in agricultural revenues, a 1.7% increase in wages of agricultural unskilled labour, a 2.0% increase in the prices of paddy rice and wheat for consumption and a 10% decrease in the prices of other cereals for consumption. These figures show in the GTAPEM scenario that households' income components are increasing while changes in the prices of various items for consumption are mixed.

Applying these results to our sample, the calculated welfare effects are presented in Table 3.6. On average, the consumption effect is close to zero because of offsetting price changes resulting from GTAPEM. Depending on households' budget composition, consumption effects vary between plus 0.2% and minus 0.3%. The income effects are positive because the revenues from the goods and factors the households sell, and the wages they earn, mostly rise. Total welfare gains add up to an average welfare gain of 2.8%, varying between 2.2% in Shaanxi and 4.0% in Hubei province.

Only 9% of the sample households are net losers from the changes (i.e. they experience negative welfare effects). While the share of households gaining increases continuously over consumption levels, the difference between 89% among the poorest quintile and 95% among the top quintile is small. Poor households gain significantly more in relative terms than non-poor households: their welfare levels increases by 4.6% on average, compared to a 2.6% increase among non-poor households. This is due to the income effect; in contrast, the tiny consumption effect is negative for the poor but positive for most of the others, thereby slightly increasing inequality.

Welfare effects are conventionally expressed as percentages of the before-change consumption levels, as above. We also examine the distribution of welfare gains in absolute value terms in Table 3.7. It is shown that money metric welfare gains add up to an average CNY 137 extra consumption per household per year. While percentage welfare gains decreased with incomes, money metric welfare gains increase continuously over income groups. Poor households gain just over half the amount of non-poor households,

Table 3.6. **Proportional welfare effects of price changes**

	Consumption effect (%)	Income effect (%)	Welfare effect (%)	% of households gaining
Total	0.0	2.8	2.8	91
... by province				
Hebei	0.2	2.7	2.9	93
Hubei	0.1	4.0	4.0	96
Liaoning	0.1	2.4	2.5	89
Shaanxi	0.0	2.2	2.2	90
Sichuan	−0.3	2.7	2.4	87
Zhejiang	0.0	3.3	3.3	89
... by poverty status				
Non-poor	0.0	2.5	2.6	91
Poor	−0.2	4.8	4.6	89
... by quintile				
1st quintile (poorest)	−0.2	4.6	4.5	89
2nd quintile	−0.1	3.8	3.7	88
3rd quintile	0.1	2.6	2.7	90
4th quintile	0.1	2.3	2.3	90
5th quintile (richest)	0.1	1.4	1.5	95

Source: Survey data and OECD Secretariat's calculations.

Table 3.7. **Money metric welfare effects of price changes**
(yuan per household)

Welfare bracket	Consumption effect	Income effect	Welfare effect
Non-poor	7	139	146
Poor	−4	83	79
1st quintile	−4	92	88
2nd quintile	−2	122	120
3rd quintile	4	124	128
4th quintile	6	148	154
5th quintile	19	159	178
All	6	132	137

Source: Survey data and OECD Secretariat's calculations. There may be differences between the sum of the income and consumption effects and the total welfare effects due to rounding.

and the highest-income fifth quintile households have more than twice the gains of poorest-quintile households. Still, this difference in gains is smaller than the difference in consumption levels, so that the poor still gain more proportionally than the non-poor, as Table 3.6 showed. The able also shows the small absolute values of consumption effects, which rounded to zero in percentage terms.

The survey data were finally used to construct a profile of the 9% of households who did not gain from reforms. Compared to households who did gain, they were found to live in communities which on average:

- Are smaller (1 160 compared to 1 220 inhabitants).
- Have a third less arable land per 1 000 population.
- Live further from the nearest paved road (1.4 compared to 1.2 km) and from the nearest town (30 km compared to 25 km).
- Have less mobile labour forces, with fewer emigrants and immigrant per 1 000 population (19% and 25% less, respectively).

This suggests that households in communities poorly endowed with agricultural potential, infrastructure and human capital are particularly vulnerable to any negative side effects of reforms. Although sample size, as well as the small share of households who did not gain from reforms, prohibits a more detailed analysis, these and similar data could be used to guide policies targeted to support vulnerable households during reforms that bring long-run overall benefits.

Summary

In this section, results from GTAPEM estimations of changes in prices and quantities due to trade reforms were used as inputs to simulate the welfare effects in a sample of rural Chinese households. Due to increasing output and supply prices and falling consumption prices for some goods resulting from GTAPEM, absolute welfare changes are substantial at 2.8% of current welfare levels. Virtually all of the gains come from increasing incomes. The reform reduces inequality since poorer households receive larger relative welfare gains even though their absolute value welfare gains are lower. There are few losing out from the reform (9% of the sample), and the percentages of those who lose are similar among the poor and the non-poor. Additional analysis suggests that these households typically live in communities poorly endowed with agricultural potential, physical infrastructure and human capital.

3.2. The impact of liberalisation on Chinese agricultural commodity markets

As China's domestic grain and livestock markets become increasingly integrated with world markets, agricultural policy developments there can have effects that spill over to other countries. Domestically, food security issues remain important for the large population and will continue to significantly influence development of Chinese agricultural policies (Chapter 2). Internationally, the sheer size of the Chinese markets makes world markets sensitive to changes in the Chinese trade regime induced by even small relative adjustments in domestic supply and demand. Such effects may need to be considered when discussing Chinese policy alternatives even from the domestic perspective. This section discusses the implications of potential agricultural market and policy developments based on simulations undertaken with the AGLINK model.

Baseline projections

The starting point for the analysis was a ten-year baseline projection that embodied specific assumptions about future trends in a wide range of economic and policy variables applying to all the individual countries in the AGLINK model, including for China. Some of the features of the baseline projection are described in Box 3.1. It should be noted that the baseline projections generally assume average macroeconomic and weather conditions and status-quo policies, and therefore do not represent forecasts.

Using the baseline projections described in Box 3.1 as a reference, the following section describes how these market outcomes over the medium term are affected by changes in the policy assumptions that underlie the projections. To assess the impacts of reforms in different countries or groups of countries, three different scenarios are analysed: first partial liberalisation is assumed in OECD countries, leaving China's and other non-OECD countries' policies unchanged. Secondly, China is assumed to unilaterally liberalise. Finally, a third scenario assumes a multilateral liberalisation including policy changes in China, India, South Africa, and OECD countries. The main assumptions of the liberalisation scenarios are presented in Table 3.8.

It should be noted that reductions in the border measures examined in this section start from the bound levels agreed to in the URAA. In contrast, reductions in domestic supports start from actual levels. Thus, the scenarios examined here are somewhat different from those considered for the GTAPEM simulations reported in the previous section. There, tariff cuts were assumed to be made to the actual tariff, export subsidy and domestic support rates applied. Here, reductions up to 50% of tariffs and export subsidy commitments and increases in TRQs may lead to much less or even no effective reductions in actual protection rates, depending on the commodity and country. This is because in many cases there is considerable latitude to reduce binding levels without significantly altering effective protection. Furthermore, all instruments, although reduced, remain in place. Thus, in the scenario, in some cases, even though a commitment (say, on export subsidies) may be reduced and actual use of that policy instrument may increase up to the commitment level in the simulation, thus counter-acting the liberalisation effects of other instruments and leading to lower effects on world and domestic markets that may have occurred otherwise.

Box 3.1. **The baseline projections for Chinese and world agricultural markets**

China has recently seen the beginning of stagnating consumption of rice, and declining use of wheat – two major staple food commodities. Projections are for a continuation of these recent trends, with food consumption of wheat and rice declining by 8% each between 2003 and 2013 (Figure 3.3). This is partly compensated by an increase in feed use, which however is relatively small in the case of wheat.

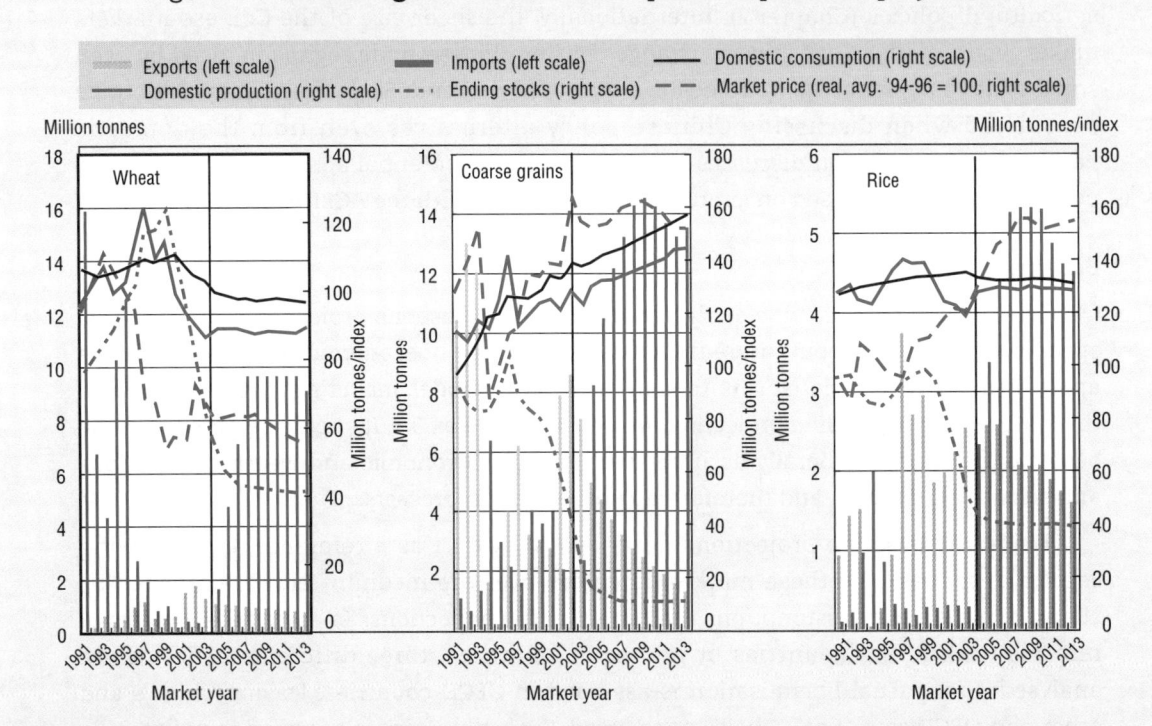

Figure 3.3. **Chinese grain market developments: past and projections**

Feed use of coarse grains is also projected to increase significantly over the decade to come. With livestock production continuing its rapid growth (see below), feed use growth of 19% is expected over the 10-year period – compared to stagnating food consumption. Total cereal use therefore is projected to increase at a moderate pace and to reach 385 million tonnes by 2013, compared to 376 million tonnes in 2003.

Even though grain production is projected to increase more strongly than domestic use, it is not expected to close the current deficit. Chinese production of wheat, coarse grains and rice is projected to reach 362 million tonnes by 2013 with stronger growth for coarse grains than for wheat and rice. While in the recent past the deficit was balanced by significant stock sales, stock declines are projected to slow down and to end by 2008. Consequently, most of the gap between consumption and production needs to be filled by increased imports and decreased exports.

Tariff rate quotas for wheat and rice are assumed to become filled by 2007 and 2008, respectively, even though they should remain binding only for a few years as domestic use slows down. Coarse grain imports are projected to reach more than 14 million tonnes per year, and hence twice the TRQ for maize, in 2008-2010, before slowing down somewhat as well. Consequently, the baseline assumes significant amounts of other feed grains to be imported, mainly barley.

> **Box 3.1. The baseline projections for Chinese and world agricultural markets**
> *(cont.)*
>
> With the increased reliance on foreign imports, Chinese market prices for grains are expected to increase beyond the inflation rate over the projection period, largely in line with developments on international grain markets. Average cereal prices are projected to increase by 4% per year in nominal terms or 1.2% per year in real terms between 2003 and 2013. Market prices are projected to strengthen particularly for rice, whereas wheat prices should decline in the medium term. All market prices develop more strongly in the first years of the projection when import rise, compared to the later years when import dependency declines. With domestic cereal prices being increasingly determined by international grain price developments, implications for world markets become more and more important when discussing Chinese market and policy changes (see below).
>
> Chinese meat markets are characterised by substantial growth in both domestic production and consumption, and this growth is projected to continue over the next decade, though at somewhat lower rates. Total meat production could reach 77 million tonnes by the end of the projection period in 2013, some 22% more than produced in 2003. The strongest growth is expected for ruminant meat: both beef and sheep production are expected to grow at more than 3% p.a., while pork and poultry are projected to increase by 2% and 1% p.a., respectively. Still, more than two thirds of the additional meat between 2003 and 2013 would be pork (Figure 3.4).
>
> **Figure 3.4. Development of Chinese meat production**
>
> *Source:* Historical data from OECD, AGLINK database, projections from OECD (2004c)
>
> While growth in Chinese meat consumption is projected slightly higher than production growth, foreign trade remains small compared to the domestic disappearance. In particular, net pork exports are expected to decline by about one third and net poultry exports to increase from almost zero to about 0.5 million tonnes over the decade to come – but even that represents less than 5% of total poultry use by Chinese consumers.

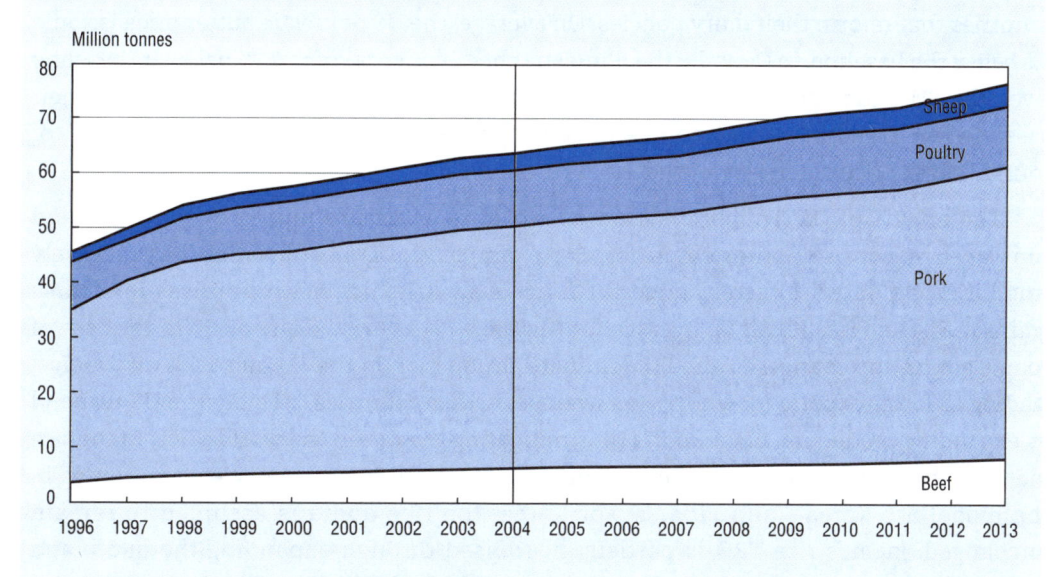

Table 3.8. **Principal assumptions of the liberalisation scenarios**

Policy	Scenario assumptions, relative to 2004 policy levels
OECD and other countries	
Tariffs (in-quota, over-quota and non-quota)	Reduction by 50%
Tariff rate quotas	Increase by 50%
Limits on subsidised exports	Reduction by 50%
Direct payments	Reduction by 50%
Support prices, target prices, loan rates	Reduction of total support benefits by 50%
China	
Tariffs (in-quota, over-quota and non-quota)	Reduction by 50%
Tariff rate quotas	Increase by 50%
Adjustment path developed countries	5 equal steps of 10%-points each, starting from 2005
Adjustment path developing countries	10 equal steps of 5%-points each, starting from 2005

Source: OECD Secretariat.

Impacts on Chinese and world crop markets[7]

Liberalising OECD markets

A partial liberalisation of OECD policies by 50% confirms the fact that the biggest changes take place in the OECD countries where reforms occur and for products with relatively high levels of protection. For example, dairy is among the most protected sectors in many OECD countries. Partial OECD liberalisation leads to declining domestic prices in countries that reform their dairy policies. On average, the US domestic butter price is some 5% below the baseline, in Canada, the domestic cheese price is some 20% below its baseline level, the EU butter price is some 15% below and coarse grains is some 6% below their baseline levels and in Japan, the rice price is some 9% lower. In each of these cases, of course, larger price changes are evident in certain years.

The effects of OECD liberalisation on world markets are muted by producer and consumer response in other countries. For example, OECD liberalisation has relatively small impacts on world cereal prices. It triggers slightly higher world prices for coarse grains (0.8%) and rice (0.8%) on average for the 2005-2013 period (Figure 3.5). In the case of coarse grains, the changes in the EU dominate. Production in the EU falls 2.7% on average, leading to lower exports (down 25% on average). Other countries, notably the US respond by expanding production and exports thus mitigating the rise in the world price. In the rice market, increased imports by Japan dominate the results. OECD rice policies represented in the model are Korea's and Japan's. For Korea, the rice quota is assumed to remain unchanged. Japan's rice TRQ is partially liberalised through expanding the quota and lowering the tariffs. These policy changes are sufficient to generate larger rice imports in Japan leading to the higher world price reported. But, the OECD countries comprise a small share of the world rice market, hence the relative modest changes reported.

At the same time, as support reduction leads to land reallocation in the EU, EU production and exports of wheat would grow (exports increase about 15% on average), mostly offsetting falling exports from other exporting countries leading to slightly declining world wheat prices of about 0.7% on average. Assuming no change in Chinese policies, the price increases for coarse grains and rice would be largely transmitted to the domestic markets in China (Figure 3.6), causing some reduced domestic (mainly feed) use

3. POLICY IMPACTS

Figure 3.5. **Impact of 50% liberalisation on world crop markets, average 2005-2013**

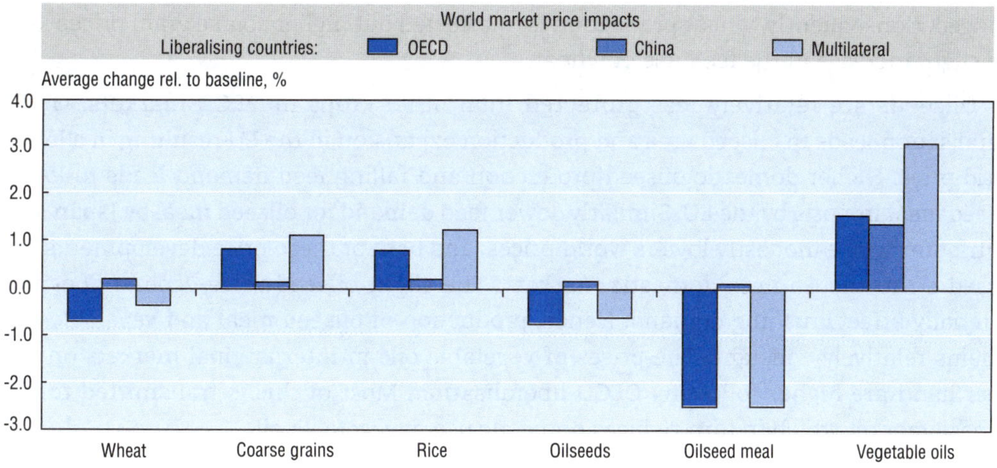

World market price impacts
Liberalising countries: ■ OECD ■ China □ Multilateral

Source: AGLINK simulation results, OECD.

Figure 3.6. **Impact of 50% liberalisation on Chinese crop markets, average 2005-2013**

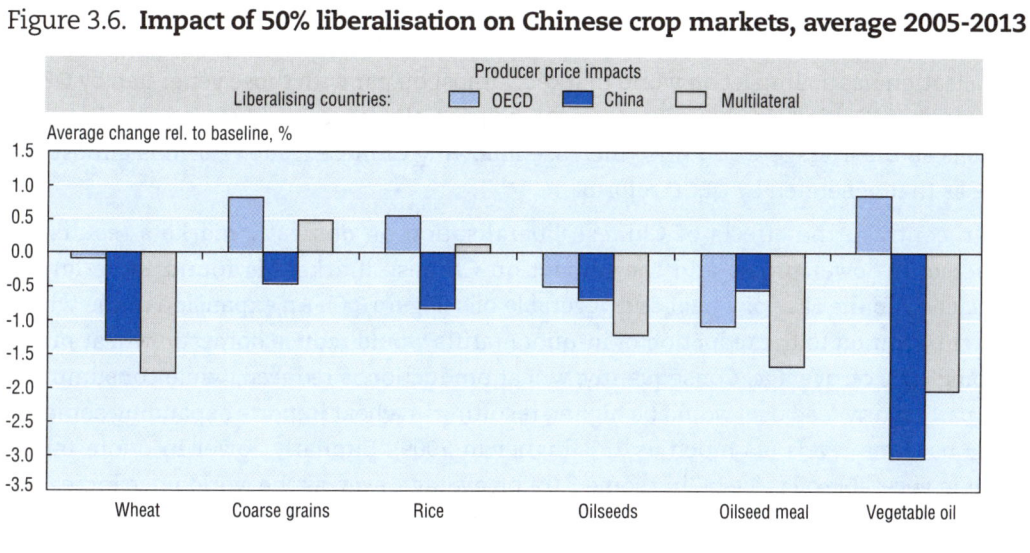

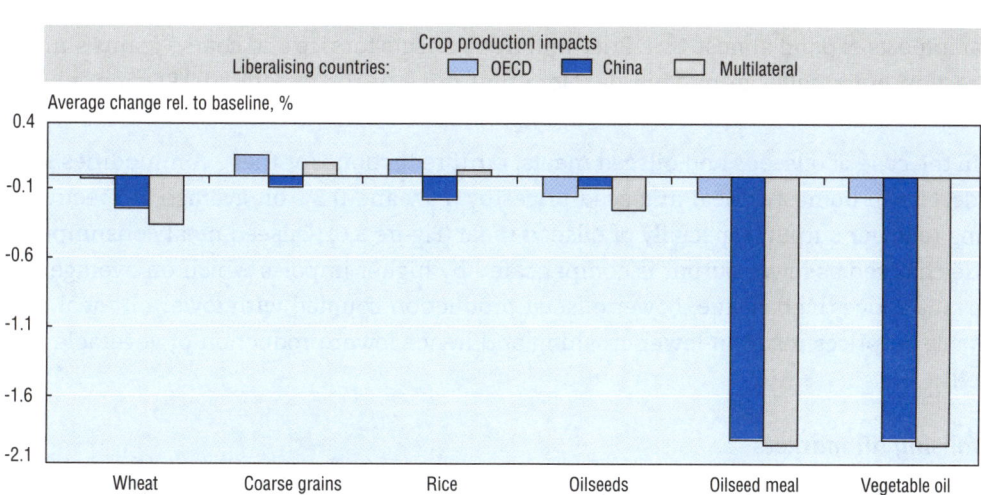

Producer price impacts
Liberalising countries: □ OECD ■ China □ Multilateral

Crop production impacts
Liberalising countries: □ OECD ■ China □ Multilateral

Source: AGLINK simulation results, OECD.

OECD REVIEW OF AGRICULTURAL POLICIES – CHINA – ISBN 92-64-01260-5 – © OECD 2005 but increased production of these commodities. Given that wheat imports to China are largely determined by the tariff rate quota (TRQ), however, domestic wheat prices are little changed. Consequently, wheat production is unaffected but higher coarse grain prices lead to a slight increase in the feed use of wheat.

Oilseeds are relatively less protected than other crops in OECD markets. OECD liberalisation leads to oilseed area and production expansion in the EU resulting in a lower world price. Higher domestic oilseed production and falling feed demand leads to lower oilseed meal imports by the EU. Similarly, lower feed demand for oilseed meal by Japan also contributes to the modestly lowers world prices, and both of these price developments are passed on to the Chinese domestic markets. The slight oilseed price decline does not materially affect crushing demand. Hence, production of oilseed meal and vegetable oils remains relatively constant. The prices of vegetable oils in international markets on the other hand are higher following OECD liberalisation. Most of this is transmitted to the Chinese market which in turn reduces domestic use of vegetable oil.

Liberalising Chinese markets

Compared to a partial liberalisation in all OECD countries, liberalising Chinese trade policies by 50% would have a positive effect on the world prices of the crops reported in Figure 3.5, albeit with much less of an effect. However, in the case of vegetable oils, Chinese liberalisation has an impact on world markets almost on par with those generated by OECD country liberalisation. In this case, Chinese policies play an equally strong role in world markets as the average world price increase following Chinese trade reform is almost the same as that generated by OECD reform.

In contrast, the effects of Chinese liberalisation on domestic markets results, as expected, in lower prices and the impact on Chinese markets is found to be larger, particularly in the case of wheat and vegetable oils (Figure 3.6). An expansion of the wheat quota in addition to the reduction of in-quota tariffs would reduce domestic wheat prices by some 1.3% on average. Consequently, wheat production is reduced, while consumption (and particularly feed use) would be higher, resulting in wheat imports expanding some 7% above baseline levels (as much as 15% higher in 2009). Similarly, lower in-quota tariffs result in vegetable oil price falling some 3.0% on average, even as the world price increases, leading to an average drop in production of 2% while consumption expands some 3% above the baseline on average (but it is more than 8% greater in the last two years). The net effect is that imports expand almost 18%. Price and trade effects for rice and coarse grains similar in direction but smaller in magnitude. Note that none of the expanded TRQs for wheat, vegetable oils, and rice would become binding.

In the case of oilseeds and oilseed meals, tariff reductions for these commodities lead to a decline in domestic meal and seed prices by 0.6% and 0.8% on average, respectively, leading to lower output, especially of oilseed meal (Figure 3.6). Oilseed meal consumption is little changed as lower output is compensated by higher imports which on average are some 40% above the baseline. Lower oilseed production coupled with lower oil meal and vegetable oil prices result in lower crushing and hence lower production of vegetable oils as well.

Liberalising all markets

A multilateral liberalisation would result in market impacts that very much represent the total of the abovementioned scenarios. While world price changes would be dominated by the effects of liberalising OECD markets, domestic markets within China would be equally affected by the Chinese policy changes. The impact of the Chinese liberalisation is particularly pronounced for wheat, where the expansion of China's TRQ, which was binding from 2008 to 2011 in the baseline, has a strong influence on the Chinese market projections. Lower international prices would put some limited pressure on domestic market prices which would be about 1.8% lower than under baseline conditions on average resulting in more demand, especially for feeding. Feed use of wheat, although relatively low in absolute terms, would increase by 1.7% on average, mostly at the expense of coarse grains.

Conclusions

To a certain extent, liberalising OECD and Chinese agricultural policies would have opposite, and therefore partly offsetting, effects on Chinese domestic markets for crops and, in particular, meat and milk. For a number of commodities, the effects of a partial multilateral liberalisation would therefore be rather small. The effects of OECD and Chinese policies individually are generally more pronounced in crop markets than in livestock. Higher meat prices would generally follow from a liberalisation of OECD policies, but decline following a Chinese policy cut. Given that feed prices generally change in the same direction but more strongly than meat prices, Chinese meat production would decline as a response to higher feed prices due to an OECD liberalisation, but increase with lower feed costs if Chinese policies were partly liberalised.

A partial liberalisation in OECD countries would generally result in higher world and domestic prices for most agricultural products and would therefore not only benefit Chinese producers, but at the same time negatively affect food consumers. This potential negative effect would be largely eliminated in a multilateral liberalisation scenario where prices for the main food commodities would be either reduced (vegetable oils and wheat) or little affected on average (particularly rice, and most meats). Unilateral Chinese liberalisation on the other hand will have minor effects on world prices but lead to larger declines in domestic prices.

The relatively small effects reported may be a reflection of the partial nature of the reform. All policies remain in place in the scenarios, only their levels differ. Complete liberalisation would be expected to have a more substantial impact. In addition, the results are conditional on the underlying baseline. A different baseline may lead to different results. Furthermore, the results are dependent upon the commodities represented in the model which does not include all agriculture. Although the commodities represented are among the most protected, a fuller set of commodities would give a more complete picture of the effects of liberalisation.

It should be noted that both the baseline projections themselves and the impacts of the partial liberalisation of Chinese policies strongly depend on the actual implementation of Chinese TRQ policies. Within WTO, large shares of the agreed grain TRQs remain under control of STEs, and hence of the government. As in the past these entities have been found to be rather restrictive with respect to grain imports, it remains unclear whether, in case of grain shortages on the Chinese markets, TRQs would become filled as assumed in the baseline projections. Another uncertainty is related to the actual levels of Chinese grain stocks; data for them have been revised several times in the past in light of the large stock sales observed since the end of last decade. An analysis of both these issues and possible implications for the Chinese grain markets can be found in the following section.

3.3. Domestic and world market implications of alternative grain stock estimates and trade policies in China

This section discusses the impact on China's and the international market from changes in two factors driving Chinese grain imports. First, an alternative set of market projections based on the recent estimates of Chinese grain stocks is compared to the projections discussed below to verify the validity of the baseline. Second, an analysis is provided of the market implications that would arise from a significant under-fill of import quotas for technical or political reasons, for example because the Chinese state traders could prevent imports within the TRQ shares that remain under their control, or from an overfill if Chinese authorities decided to expand the import quotas.

Recent developments on China's grain markets

Recently, Chinese grain markets were characterised by a significant deficit in domestic production when compared to consumption. Despite production failing to meet consumption by 34 million tonnes per year, on average, between 2000/01 and 2003/04, China remained an important net exporter of grains, exporting 10 million tonnes of wheat, coarse grains and rice per year, on average, in the same period, while at the same time reducing imports to the lowest levels since the mid-1970s. Obviously, the deficit is filled from large cereal stocks built up during the 1990s in particular when domestic production outperformed consumption by significant amounts (Figure 3.7).

Clearly, with consumption remaining larger than production, China's status as a net exporter cannot be maintained forever. Once stocks are reduced to levels considered to be minimum reserves by Chinese authorities, the deficit will have to be filled by imports from abroad. Recent developments indicate that Chinese grain prices have increased significantly over the past several months, indicating the first signs of scarcity in the domestic markets. Between February 2003 and February 2004 maize and wheat prices increased by 12% and 36%, respectively, while rice prices rose between 40% and 60%. If the domestic supply-demand ratios remain as in the past four years, the drop in stock supplies would mean a switch from 7 million tonnes net exports to 34 million tonnes net imports. This change in the net trade position by some 40 million tonnes per year compares to current global trade quantities of around 230 million tonnes of wheat, coarse grains and rice – a ratio that indicates the relative importance of the issue.

When China joined the WTO, it agreed to open its domestic market through the introduction of TRQs for wheat, maize and rice, totalling to some 22 million tonnes per year, with a certain and, in the case of maize, increasing share of import quotas allocated to private traders other than the official Chinese state trading enterprise. In addition, any STE quota shares unused by the third quarter of a given year were to be opened to non-STE traders.

Indeed, the OECD *Agricultural Outlook 2004-2013* (OECD 2004c) projects a significant rise in cereal imports. With vanishing supplies from domestic stocks, the *Outlook* expects both wheat and rice import quotas to become filled, and coarse grain imports (comprising both maize and other coarse grains) to far exceed the TRQ for maize (see next section for details). There are several unanswered questions, though. First, the exact magnitude of

3. POLICY IMPACTS

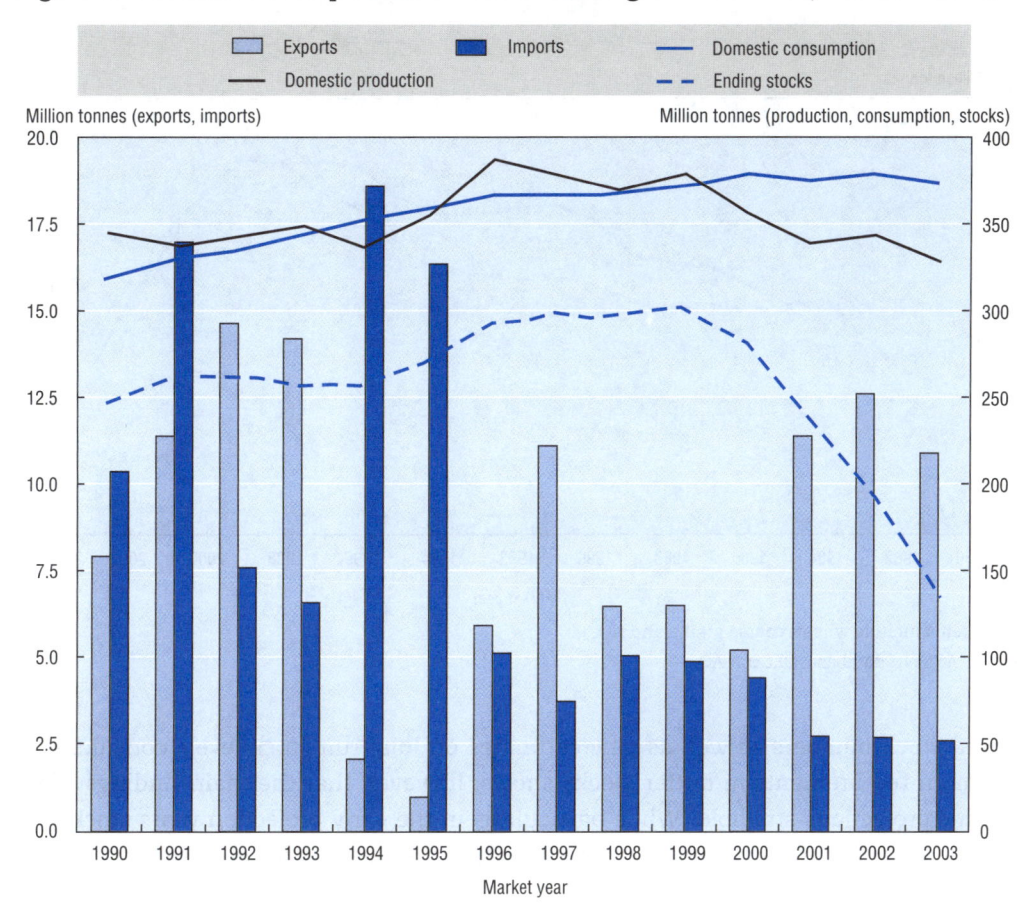

Figure 3.7. **Recent developments in the Chinese grains balance, 1990/91-2003/04**

Note: Grains include wheat, coarse grains and rice.
Source: AGLINK database, OECD.

current Chinese grain stocks is unknown (and is indeed regarded as a state secret by the Chinese authorities). Recent revisions of Chinese grain stock estimates provided by the UN Food and Agriculture Organisation (FAO) could therefore lead to a reconsideration of the Outlook projections for China, as any change in the total stock figures obviously may result in a change in the pressure for cereal imports. Second, opening up TRQs may not be sufficient to allow for the necessary grain imports; domestic policy reforms, *e.g.* the reduction of trade control power of the current Chinese STEs, as well as improvements in infrastructure are equally necessary to allow the import quotas to be actually filled.[8]

Recent grain balance estimates

More recently, FAO has provided new estimates on Chinese grain balance data. In particular, and unsurprisingly given the debate on the issue in the past, stock estimates have been revised. In 2003 and 2004, total stocks are now estimated at 163 and 148 million tonnes, respectively, some 19% and 30% above the data currently used in AGLINK (Figure 3.8). While wheat stocks are actually revised downwards, it is coarse grain stock estimates that have more than doubled for 2003 and 2004.

Such important data revisions obviously could trigger significant revisions in the baseline projections presented in Box 3.1. A scenario analysis taking into account the

Figure 3.8. **Chinese total grain stocks: recent FAO revisions**

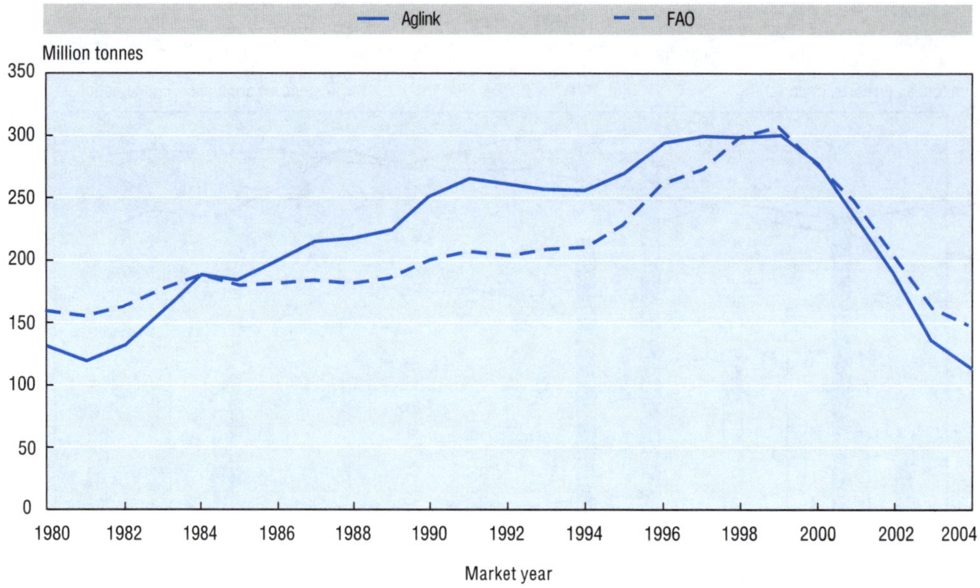

Note: Grains include wheat, coarse grains and rice.
Sources: AGLINK database, OECD; FAO.

revised stock numbers as well as new estimates on long-run stock levels considered as minimum requirements in buffer stocks shows, however, that the main findings of the baseline projections still hold. While particularly in the early projection years stock sales could be higher than assumed for coarse grains, and somewhat lower in the case of rice, projections for production, consumption and trade as well as domestic and international prices would change only modestly. Both domestic and world market prices for all grains could be slightly lower than projected in the first couple of years, with the exception of wheat in 2004 and rice in 2005 following the lower stock sales for those commodities. But changes for the baseline would remain below 2% in almost all cases and years.

Medium-term projections for Chinese grain imports in general remain valid as well. The larger availability of coarse grain stocks would, however, result in lower imports by some 10% on average. In particular, the increase in Chinese coarse grain imports could be expected to follow a less rapid path, although very similar levels would be reached by the end of this decade. In contrast, imports of wheat and rice could fill the tariff rate quotas more quickly given the lower stock availabilities for these commodities (Figure 3.9). Projections for meat markets and trade are barely affected by the grain stock revisions. Lower feed grain prices would trigger slightly higher production of beef, pork and poultry, but the impact would not exceed 0.2% on average for any of the meat types.

Market implications of alternative licensing policies for Chinese grain TRQs

As mentioned above, the baseline assumptions (Box 3.1) of completely filled TRQs for cereals may not be entirely realistic if the introduction of the TRQs is not supported by domestic reforms and necessary infrastructure development. In particular, it was said that the persistent system of STEs controlling large shares of the total TRQs might represent a *de facto* reduction of the import quotas.[9] On the other hand, in a severe deficit situation the Chinese authorities may opt to expand imports at in-quota tariffs beyond the agreed TRQ levels.

3. POLICY IMPACTS

Figure 3.9. **Impact of Chinese grain stock revisions: import projections of wheat, coarse grains and rice**

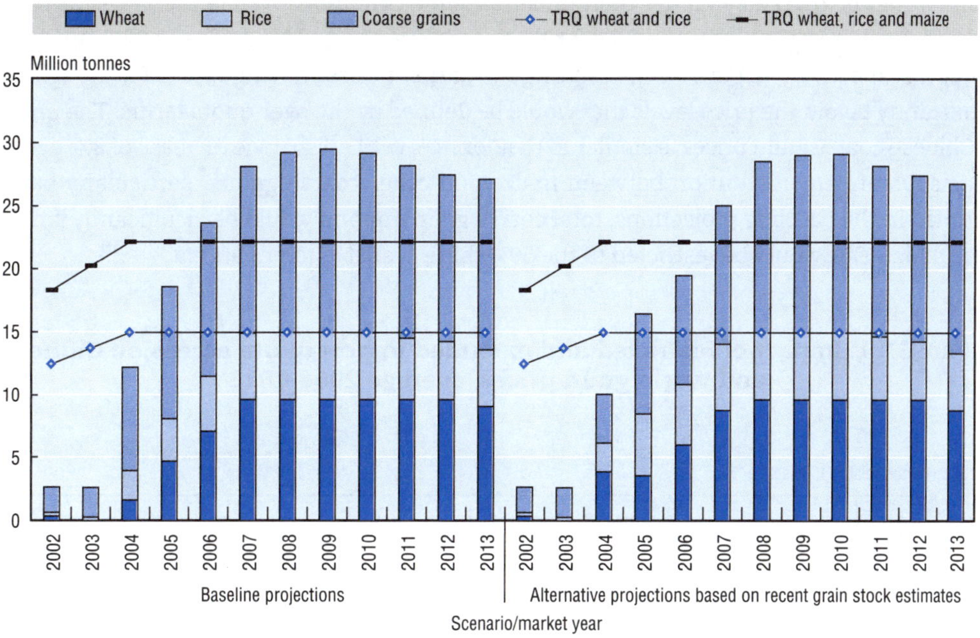

[Chart showing Million tonnes, Baseline projections and Alternative projections based on recent grain stock estimates, 2002-2013, with series: Wheat, Rice, Coarse grains, TRQ wheat and rice, TRQ wheat, rice and maize]

Scenario/market year

Source: AGLINK simulation results, OECD (2004c).

Obviously, there is no reason to assume that the share of TRQs that is left under STE control would remain entirely unfilled even if domestic market conditions would ask for the respective imports. In fact, given the possibility of non-STE traders to import within the unfilled STE quota shares after the third quarter of any given year, it seems very unlikely that the STE shares would remain completely unfilled in a Chinese deficit situation. Nor does this analysis aim to suggest that the non-STE shares would become completely filled under all circumstances. For illustrative purposes, however, the first scenario discussed in this section assumes effective TRQs for wheat, maize and rice to be reduced to the non-STE shares as given in the last column of Table 3.9. In contrast, a second scenario assumes an expansion of import quotas to a degree that would make the TRQs unbinding.

Table 3.9. **Development of Chinese grain tariff rate quotas after WTO accession**

Commodity	Year	Total import quota 1 000 t	Share reserved to STEs %	Remaining non-STE quota 1 000 t
Wheat	2002	8 468	90	847
	2003	9 052	90	905
	From 2004	9 636	90	964
Maize	2002	5 850	68	1 872
	2003	6 525	64	2 349
	From 2004	7 200	60	2 880
Rice	2002	3 990	50	1 995
	2003	4 655	50	2 328
	From 2004	5 320	50	2 660

Source: Schedule of the People's Republic of China.

3. POLICY IMPACTS

Grain imports restricted to the shares reserved for private traders

With effective TRQs and hence imports significantly reduced, domestic market prices for grains would be markedly higher than under baseline conditions. While the TRQs for wheat and rice would remain binding, prices for these food grains would increase by 20% and 9% on average, respectively (Figure 3.10), resulting in market prices moving closer to, though still significantly below, the price levels that would be defined by the over-quota tariffs. The impact on Chinese coarse grain prices is similar to that for rice, with an average increase of 8%: despite some substantial substitution between maize and other coarse grains, particularly barley, assumed in the baseline projections, total coarse grain imports would be significantly smaller if the maize TRQ would be restricted to the 50% share available to private traders.[10]

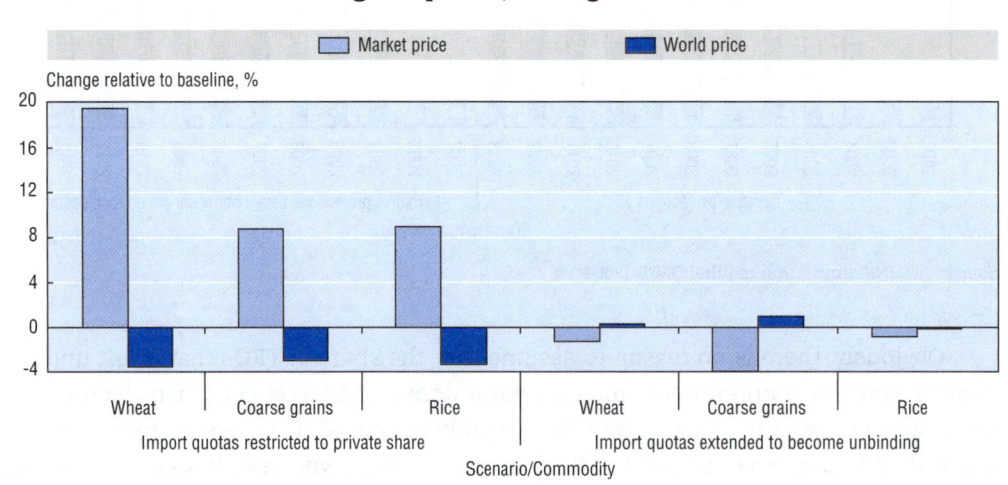

Figure 3.10. **Impact of restricted and extended import quota access on Chinese and world grain prices, average 2004-2013**

Source: AGLINK simulation results, OECD.

Internationally, the decline in Chinese grain imports reduces world cereal prices by 3% to 4% on average for the simulation period. The impact on international prices is particularly strong for wheat – due to the large change in Chinese imports – and rice – due to the smaller world market for this commodity. Impacts on world coarse grain prices would be less pronounced given the smaller impact on Chinese import levels relative to wheat, and the much larger world markets relative to rice. These price effects are strongest during the 2007-2010 when TRQs would be binding in the baseline projections. It should be noted that due to the more restrictive trade regime for wheat, maize and rice, these lower world market prices would be of little relevance to both domestic producers and consumers. With the exception of coarse grains where some substitution between maize and other feed grains could be expected, domestic market prices would be determined internally and largely de-linked from international price developments.

These significantly higher prices create larger incentives for domestic producers. Chinese cereal production would on average be about 2% higher than in the baseline, with stronger effects on wheat output than for coarse grains and rice following the different price effects on the domestic markets. However, there is a strong negative effect on domestic use of grains, as shown in Figure 3.11. In particular, with reduced livestock production (particularly beef, pork and poultry), feed use of cereals would be reduced by

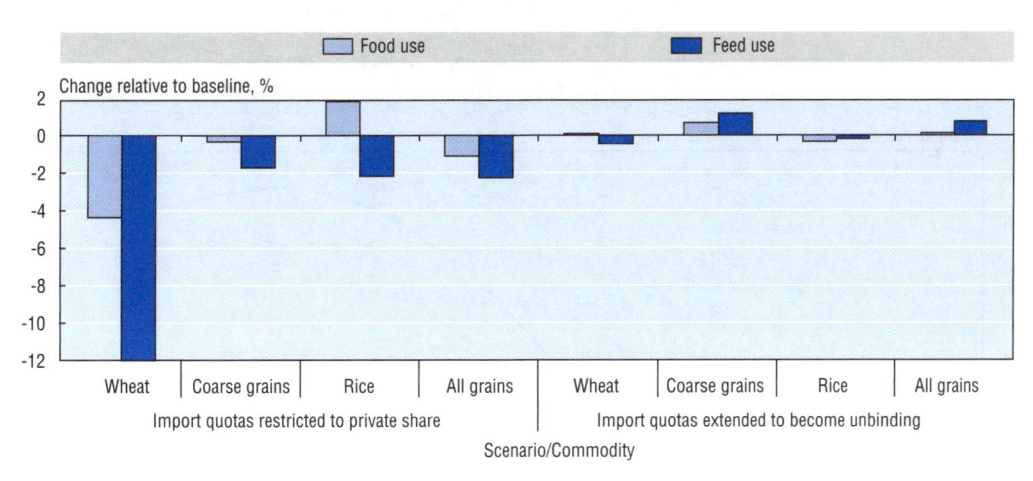

Figure 3.11. **Impact of restricted and extended import quota access on Chinese grain consumption, average 2004-2013**

[Chart: Change relative to baseline, %. Food use and Feed use bars for Wheat, Coarse grains, Rice, All grains under two scenarios: "Import quotas restricted to private share" and "Import quotas extended to become unbinding".]

Source: AGLINK simulation results, OECD.

more than 2% on average. While the strong decline of wheat feed use is of lesser importance given the small share of wheat in the Chinese feed mix, feed use of rice and particularly of coarse grains would fall by about 2%. Food consumption would decline as well, though less strongly. As a consequence of the different price impacts on the domestic market, wheat consumption would be affected most significantly and would average more than 4% below baseline levels. In contrast, rice consumption would be higher than under baseline conditions as relative prices trigger a shift from wheat to rice as the staple food.

The lower meat production and corresponding higher meat prices would also result in a – though smaller – reduction of domestic meat consumption. At the same time, pork exports would decline more quickly, and poultry imports would expand faster. On average, total net meat imports would be some 35% higher than under baseline conditions, even though again these changes would occur on relatively low levels of trade.

Grain imports at in-quota tariffs possible beyond TRQ levels

In contrast to restricted access to TRQ imports, a second policy scenario considers the case of unlimited imports at in-quota tariffs. Allowing for demand-driven imports at in-quota tariffs would obviously result in imports larger than current TRQ commitments. Wheat imports could be up to 11% higher than with binding TRQs, whereas the increase in rice imports could reach 5%. Coarse grain imports, which are only partly determined by the TRQ for maize, would increase by up to 38% if the in-quota tariff were applied to all maize imports (Figure 3.12). Note that in the long run the imports of wheat and rice would be lower than under baseline assumptions (Box 3.1) following cross-commodity effects with coarse grains (see below).

With larger imports for wheat and rice, and particularly for coarse grains, domestic market prices would be lower than under baseline assumptions. For wheat and rice, domestic prices would become determined by world market price levels adjusted by the in-quota tariffs and would decline by about 1% on average relative to the baseline projections. The impact on coarse grain prices is larger at 4% on average (Figure 3.10). Consequently,

Figure 3.12. **Impact of unlimited import quota extension on Chinese grain imports, 2002-2013**

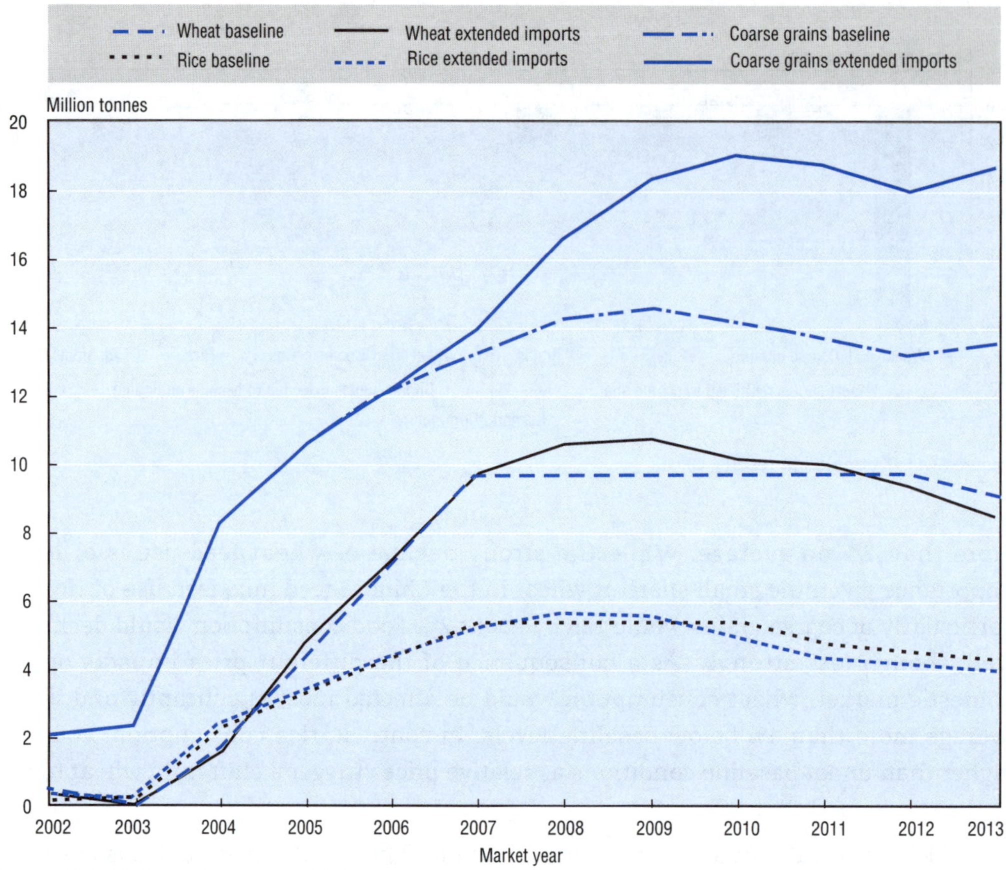

Source: AGLINK simulation results, OECD.

domestic cereal use would be somewhat higher, but the impact is small. Most notably, following the meat production increased by 0.3% on average, feed use of coarse grains would be about 1% higher on average compared to the baseline, more than offsetting the slightly lower feed use of wheat and rice following the higher relative prices for those commodities. In contrast, food grain consumption as well as domestic grain production on average would be barely affected by these changes.

It is interesting to note that in the longer run cross-price effects offset the impact of lower wheat and rice prices on domestic consumption. This is particularly evident for feed use, which in the cases of wheat and rice end 3% and 1% below baseline levels by 2013, but food use would also be reduced in the longer run due to the much lower coarse grain prices. Consequently, given the small response of domestic grain production in general, the extension of in-quota imports would result in lower imports of both rice and wheat as of 2010 and 2012, respectively.

Implications for world market prices are small particularly for wheat and rice. Given the small changes in Chinese imports, world wheat and rice prices would change little. Somewhat stronger impacts are found for world maize prices which on average would be 1% higher than projected in the baseline, with a stronger increase from 2008 when Chinese coarse grain imports would be significantly above baseline import levels. Nonetheless, domestic markets would benefit from cheaper imports as the TRQs particularly for maize would no longer be binding. Meat trade, though again remaining at low levels relative to domestic markets, is affected in the opposite direction: while net pork exports would be slightly higher (+1.3% on average), net poultry imports would be somewhat lower (–1.6% on average) compared to the baseline projections.

Implications for Chinese self-sufficiency rates

China's food security is sometimes defined as 95% self-sufficient in grain production. While it is broadly agreed that such self-sufficiency rates cannot be considered a good indicator of food security (in fact, a high self-sufficiency rate is neither a necessary nor a sufficient condition for all people having reliable access to sufficient and affordable food), they do provide a meaningful aggregate indicator of the overall supply-demand balance.

The Chinese self-sufficiency rate for grains, defined as the total production of wheat, coarse grains and rice divided by the total domestic consumption of these crops, was quite variable in the past. Given a relatively stable growth in domestic use (Figure 3.7), these variations essentially mirror production conditions. During the last three decades of the 20th century, self-sufficiency rates fluctuated between 91% and 109%, with an average just over 100%. More recently, as described above, the Chinese grain markets were characterised by substantially lower rates which were largely offset by stock sales. The lowest self-sufficiency rate for decades was observed in 2003/04 when it dropped to less than 88%.

The baseline projections discussed above (Box 3.1) suggest that the Chinese self-sufficiency rate for grains would remain at levels between 93% in 2004/05 and 94% in 2014/15 (Figure 3.13). This result is only slightly affected by recently changed estimates of Chinese

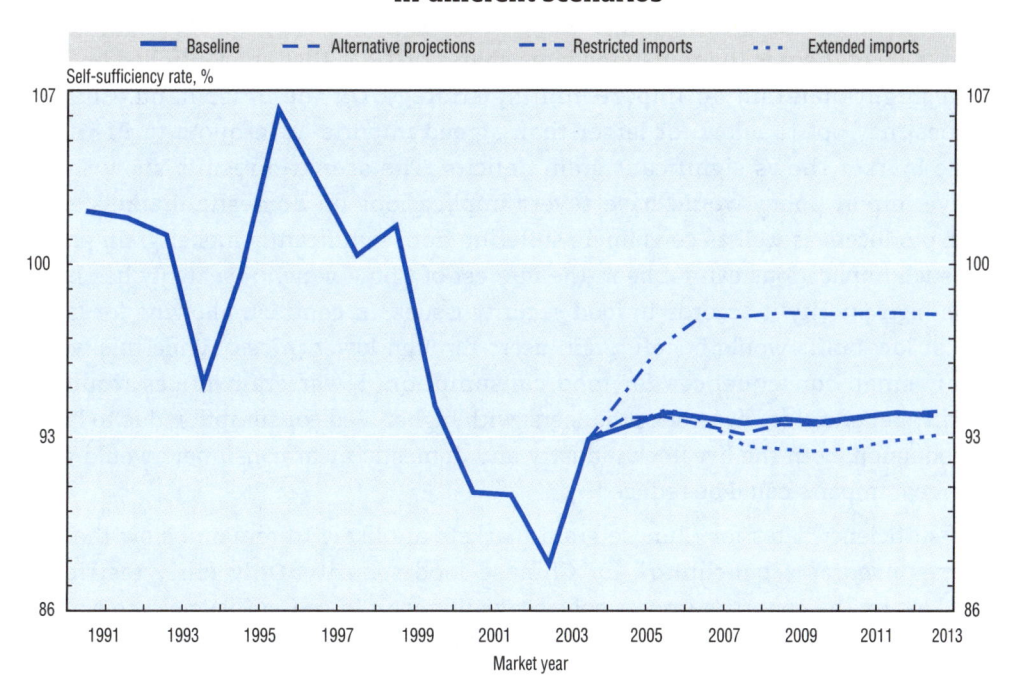

Figure 3.13. **Self-sufficiency rates of Chinese grain markets, 1991-2013, in different scenarios**

Source: Calculated from AGLINK database and AGLINK simulation results, OECD.

grain stocks. Higher stock levels and hence larger supplies from stock sales would slightly reduce average self-sufficiency rates following lower prices, less production and higher domestic consumption, but these differences are small. In other words, the self-sufficiency rate for Chinese grain markets is quite likely to be established below the 95% benchmark. While domestic production will represent almost 98% of domestic consumption in the case of rice, self-sufficiency rates for coarse grains and wheat are likely to remain substantially lower at about 92% and 90%, respectively.

Alternative trade policies could have a major impact on the self-sufficiency rates, as shown in Figure 3.13. If imports were restricted to the TRQ shares reserved for non-STE traders rather than to the full TRQs agreed by China, aggregate self-sufficiency rates would be significantly higher at some 98%, particularly after 2007. These higher self-sufficiency rates are the direct consequence of substantially smaller imports. In contrast, a policy that would allow imports at in-quota tariffs beyond the quota levels would result in somewhat lower rates of about 93%.

Conclusions

With baseline projections influenced by the marked increase of Chinese cereal imports after available supplies from grain stocks run out, the focus of this section is on two main factors that are likely to impact these trade quantities. Alternative projections that take into account recent revisions of Chinese grain stock estimates show that these revisions in general have only a minor impact on the market outlook and that the basic trends outlined in the baseline projections remain valid. The main difference is found in the speed of grain imports filling up the import quotas for wheat, maize and rice agreed for the Chinese accession to WTO. Projections for Chinese grain production, consumption and prices as well as for world market prices remain broadly unchanged.

The two scenarios have focussed on the levels of available cereal import quotas. For several reasons, actual imports could be more limited than suggested by the TRQ quantities even if domestic markets would ask for larger supplies from exporting countries. One of them is the remaining large shares of TRQs that are controlled by STEs and that might maintain an import-limiting strategy. On the other hand, Chinese authorities may opt to allow for larger than agreed imports at in-quota tariffs if the domestic market shows significant grain deficits. The scenario results show that a restrictive import policy would have severe implications on domestic markets, with livestock producers as well as consumers suffering from significantly higher grain prices. Clearly, such implications cannot be in the interest of China which repeatedly has made clear the high priority it accords to food security issues. In contrast, allowing for larger imports at low tariffs would benefit grain users through lower prices. While this would have only small consequences for food consumption, lower grain prices would be particularly beneficial for livestock producers; with higher feed consumption due to larger meat production, both the livestock industry and domestic meat consumers would gain, and net meat imports could be reduced.

Self-sufficiency rates for Chinese grain markets are likely to remain below the 95% sometimes used as a benchmark for Chinese food security. Only fairly restrictive assumptions on the country's import policies would result in self-sufficiency rates above that level. A more open import policy as assumed in the last scenario would result in an aggregate self-sufficiency rate of about 93%. Much of the imports would be of coarse grains for which the international markets are large, and imports of which therefore should remain readily available. At the same time, more liberalised markets would benefit from a more efficient allocation of resources; in particular, farmers close to urban areas and export ports could play off their comparative advantage in more labour intensive products, such as livestock and vegetables, rather than cereal commodities.

Notes

1. In constructing these averages bilateral trade flows were used to weight tariffs actually applied. Moreover, special procedures were used in calculating tariff averages for commodities covered by tariff rate quotas in order to better account for distinctions between out-of-quota and in-quota tariff rates. The combination of trade-weighting and the special procedures for TRQ's means that the estimates obtained may differ from results obtained using other averaging procedures such as, e.g., used for constructing the estimates of China's agricultural tariffs reported in Chapter 2.

2. Even if China were to levy higher tariffs than it faces it might gain from global liberalisation, though its farmers might lose.

3. The survey team was led by Loren Brandt of the University of Toronto, Scott Rozelle of the University of California at Davis, and Linxiu Zhang of the Center for Chinese Agricultural Policy, Chinese Academy of Sciences. The OECD Secretariat thanks them for making the survey data available. Chang (2004) provides information on the survey frame and procedure.

4. The data did not distinguish between agricultural employment and non-agricultural employment.

5. Note that this excludes income from employment in farms, which comes under "wages" and was not separated out in the raw data.

6. The difference is due to Ravallion and Chen's (2004) larger and different data set, application of a rural rather than national poverty line and regional cost of living corrections derived from their own data.

7. All of the products represented in AGLINK are included in the analysis. However, the discussion below focuses only on the crop markets because the interactions between world and Chinese livestock markets are rather limited. This is due partly to China's relatively small trade in some of these markets, partly due to the linkages between China and world markets as represented in AGLINK and partly due to limited representation of Chinese policies in AGLINK.

8. See OECD (2001) p. 21 and OECD (2002a) p. 115, for discussions on possible implications of the introduction of TRQs in China.

9. Note that protection of Chinese markets against grain imports is created not only by TRQs and other tariff measures. For example, differences in the value-added tax (VAT) charged for imported, exported and domestically produced and consumed products, as well as export subsidies, are reported for several commodities. While the tax difference is considered important by some authors (e.g., Huang, Rozelle and Chang, 2004), others note that the actual differences of about 3 percentage points for grains may be considered small (e.g., van Tongeren and Huang, 2004). In Chapter 2 the difference is estimated at 9% (see Box 2.4). In any case, the present analysis does not take into account these tax differences nor export subsidies.

10. Note that the AGLINK model does not represent maize markets explicitly. Instead, the effective import tariff is represented as a function of the in- and over-quota tariffs for maize and the non-TRQ tariff for other coarse grains, as well as the total coarse grains imports relative to the maize TRQ. Results obtained for the coarse grains market therefore need to be interpreted with greater care than those for wheat and rice.

ANNEX A

Labour Mobility and Rural Poverty in China[1]

As discussed in Chapter 1, a precondition for closing a large rural-urban income gap in China is a large outflow of labour from agriculture to other sectors of the economy. Moreover, mobile labour is an important factor contributing to China's welfare gains from multilateral liberalisation. This annex discusses the impact of labour market reforms on rural incomes and poverty.

Rural poverty

Some 99% of China's poor live in rural areas, mostly in the western provinces (Chapter 1 and Table A.1) which are generally characterised by poor infrastructure, underdeveloped social services, fragile natural environments, and a heavy reliance on agriculture. Such areas are the least likely to benefit from either domestic economic growth or increased international trade.

Based on China's official reference measure (USD 0.2 per day at current exchange rate, but USD 0.6-0.7 per day at PPP), rural poverty declined significantly from 250 million to 32 million persons over the 1978-2000 period. However, many rural households have incomes just marginally above the poverty line such that any income reduction from market conditions, natural disasters or illness would push them into poverty again. Using the World Bank measure of poverty (USD 1 per day at PPP), the number of rural poor was estimated to be much higher, at about 88 million at the beginning of the 2000s.

Not surprisingly, studies have shown that poor households have more dependents, a greater reliance on agriculture (especially crop production), less off-farm employment, and lower levels of education. These households not only have limited capacity to improve current incomes, but are also unable to invest sufficiently in their children's development

Table A.1. **Rural poverty rates by region, China, 2002**

Poverty rate	Regions
< 1%	Shanghai, Beijing, Tianjin, Zhejiang, Jiangsu, Shandong, Guangdong, Fujian
1%-5%	Hebei, Liaoling, Jilin, Hubei, Heilongjiang, Hunan, Anhui, Jiangxi, Henan, Guangxi, Hainan, Chongqing, Sichuan
5%-10%	Shanxi, Shaanxi, Gansu, Ningxia, Yunnan, Xinjiang
> 10%	Inner Mongolia, Tibet, Guizhou, Qinghai

Source: *China Rural Poverty Monitoring Report*, 2001; www.sannong.com, 2003.

so that the next generation can participate in China's economic growth, and thereby break the cycle of poverty.

In the early 1980s, the Chinese government introduced a number of agriculture policies that helped alleviate rural poverty. These included the household production responsibility system, raising state procurement prices, extension of advanced technologies, and provision of manufactured farm inputs. Initially, this approach proved effective but less so by the early 1990s with the slowing of growth in domestic demand for farm produce.

As rural poverty became increasingly a regional phenomenon in disadvantaged areas, the government revised its poverty alleviation strategy to focus more on capacity building, funding many regional development projects designed for improving rural infrastructure, production and marketing facilities, technical services and education, subsidised loans for rural households and work for food programmes. For some of the most environmentally fragile areas, there were limited attempts at promoting out-migration. Although the poverty reduction programmes have not proved to be very effective, rural poverty alleviation remains a government priority.

The role of agriculture

Although the importance of off-farm work as a source of income for rural households has increased substantially during the last two decades, agriculture continues to represent a major component of total income of rural households (Chapter 1). As Table A.2 shows, farm income in 2000 represented 76% of the total income of poor rural households. For non-poor farmers the share drops to 56%, while the share imputable to wages doubles from 13% for poor farmers to 24% for non-poor farmers. This suggests that agricultural labour productivity growth is important to both increasing the average labour productivity in rural areas and the labour productivity of farm households.[2] Yet, a broad-based increase in agricultural labour productivity cannot take place without a major out-migration from agriculture towards non-farm jobs.

Since agriculture is the main source of income for poor rural households, poverty rates are closely linked to the performance of the sector. In regions where agriculture growth lagged behind other sectors, poverty reduction was slow. Where there was diversification and rapid growth, farm employment and wage rates increased and rates of rural poverty declined (World Bank, 2000).

Table A.2. **Comparison of income structure in rural areas, 2000**

Income	Poor households		Other farm households	
	Yuan per person	%	Yuan per person	%
Net income	707	100.0	2 536	100.0
Cash income	359	50.8	1 886	74.4
Total gross income	1 203	100.0	3 485	100.0
Wage income	152	12.7	841	24.1
Income from farming	917	76.2	1 939	55.6
Income from non-farm business	91	7.6	491	14.1
Property income	7	0.6	50	1.4
Transfer income	34	2.9	163	4.7

Source: China Rural Poverty Monitoring Report, 2001.

A number of studies (Zhu and Jiang, 1995; Park and Wang, 2001; Kang, 1998; Rozelle, Zhang and Huang, 2000; Fan, Zhang and Zhang, 2002) examined the various poverty reduction schemes and found that government expenditures on rural poverty alleviation had little or no effect. Such programmes often exclude significant groups of the poor population, such as the rural poor who live outside the target area. Instead, these studies conclude that economic growth, especially in the agricultural sector, played a central role in rural poverty reduction, especially in the 1980s when China faced a general shortage of agricultural products.

Quantitative analysis by Tian, Wang and Ke (2003) provides additional insight. Two econometric models regress rural poverty rates on the real growth rates of three economic sectors (primary, secondary, tertiary) with time series observations and on regional agricultural productivity along with other variables using pooled time series and cross-section data. Agricultural GDP growth and agricultural productivity were significant factors in poverty reduction, in close conformity with Rozelle, Zhang and Huang (2000).

While the growth of non-agricultural sectors had little impact, the analysis did reveal that the shift of rural labour to non-agricultural sectors makes a significant contribution to poverty alleviation. This suggests that growth in the non-agricultural sector did not always create additional employment opportunities for rural labour. The modelling results also show that natural disasters lead to a wide occurrence of rural poverty, underlining the fact that rural households are still vulnerable to adverse natural conditions.

These results have some clear policy implications for poverty alleviation. Given the fact that the rate of agricultural growth is expected to decline over time due to growing marginal costs and stagnated demand, further reduction of rural poverty will become increasingly difficult if promotion of agricultural production is used as the only approach. This is especially true considering that the majority of rural poor are located in areas isolated from external markets due to remoteness and/or poor infrastructure.

A better approach to rural poverty reduction would appear to be to facilitate the transfer of rural labour to off-farm jobs through improving basic education, skill training and labour market developments. In some rural areas a complementary option could be the production of high-value "organic" food products, taking advantage of the unpolluted environment.

The role of rural non-farm development

Up to the mid-1990s, a major feature of the transfer of rural labour to off-farm jobs was the rapid development of Township and Village Enterprises (TVEs). Prior to the mid-1980s, all TVEs were owned by rural collectives and received subsidies through preferential financial and taxation arrangements, while TVE workers were primarily members of farm collectives. Since then, TVEs have developed into a diversified system of dominantly private enterprises. Between 1985 and 1995, total employment by TVEs rose rapidly (Box A.1).

In view of their local nature, TVEs allowed many rural workers to find employment out of agriculture without migrating to other areas. The expansion of the TVEs has therefore contributed to relieving partly the low productivity of labour in agriculture. Overall, these changes allowed increasing rural employment opportunities, thereby raising rural incomes while maintaining stable food supplies and food prices in the urban sector.

Box A.1. **The role of rural industries**

Township and village enterprises (TVEs) are rural non-agricultural enterprises, and in addition to collective enterprises include both single owner and other private firms. They are small and medium-size enterprises that specialise in labour intensive products and, along with foreign funded enterprises, produce most of China's exports. Exemptions from central planning restrictions, backing from local governments, business relations with state-owned enterprises (SOEs), greater exposure to market discipline compared to SOEs, and access to cheap rural labour led TVEs to flourish from the mid-1980s. They were the largest contributor to growth in aggregate GDP and employment from the mid-1980s through the mid-1990s, and by 1996 employed 135 million workers, or 20% of the national workforce (Table A.3). The development of TVEs in turn has transformed rural income generation, with almost 50% of rural incomes now coming from non-agricultural activities.

Table A.3. **TVEs in China's economy, 1990-2002**

Year	Number million	Employment million	Share of national employment %	Share of national GDP %	Share of national exports %
1990	18.5	93.0	14.0	13.5	12.0
1995	22.0	129.0	19.0	25.0	28.0
1996	23.4	135.0	20.0	26.0	34.0
1997	20.2	131.0	19.0	27.9	36.0
1998	20.0	125.0	18.0	28.3	35.0
1999	20.7	127.0	18.0	30.3	38.0
2000	20.8	128.0	18.0	30.4	34.0
2001	21.2	131.0	18.0	30.2	36.0
2002	21.3	133.0	18.0	30.9	34.0

Source: *China Statistical Yearbook*, 2003.

The momentum to aggregate growth from TVEs diminished in the second half of 1990s. Between 1996 and 1998, the performance of TVEs deteriorated sharply, and employment fell by about 10 million. This could be partly explained by the 1997 Asian crisis, but mostly by fundamental structural problems. China's TVEs suffered from financial troubles and operating inefficiencies nearly as severe as those afflicting the SOE sector. The exemption from central planning restrictions and sponsorship by local government, which gave TVEs an advantage in the past, became less important as constraints on SOEs were partly relaxed. The disadvantages of TVEs, in terms of distance from infrastructure and other facilities that benefit business in urban areas and which limit the scale of operations of TVEs can achieve, have become more prominent. The degree to which these disadvantages are offset by access to lower cost, but also lower skilled labour, is unclear. In recent years, the performance of TVEs has improved but yearly job creation is about 2 million, compared to 7 million in the first half of the 1990s.

Source: OECD (2002b).

However, TVEs are also plagued by low labour productivity and low competitiveness *vis-à-vis* their urban-based competitors and are unlikely to generate significant increases in rural household income in the future. With stricter pollution controls and the increasing demand for quality by urban consumers, it became harder for TVEs to compete with urban-based firms and enterprises. In the 1990s, it became evident that raising the competitiveness of TVEs required improvements that limited their traditional function of absorbing rural surplus labourers. Rural labourers are increasingly employed by firms established with non-agricultural capital. Many of them are FDI-driven, labour intensive, processing-type and export-oriented in the coastal regions who employ cheap labourers from inland provinces.

As employment opportunities through TVEs declined, an increasing number of rural labourers began searching for off-farm jobs away from their hometown. However, the limitations imposed on the registration of migrants in urban areas perpetuated the labour productivity gap between urban and rural areas.

Restricting migration – the *hukou* system

China's recent experience in migration policy has been one of uncoordinated efforts and of imposition of substantial restrictions. Changes to the present institutional mechanisms regulating migration and the functioning of land markets might facilitate the adjustment of China's agricultural sector.

The main issue is legal restrictions that prevent labour and land markets from adjusting in ways that would raise agricultural labour productivity. Migration and land policies have served as complements in China's strategy of supporting industrial growth. As the share of agriculture as a source of employment and GDP has now declined substantially, it is natural to ask in which direction institutional reforms affecting labour and land circulation will move in the near future.

The enactment of a household registration system (*hukou*) dates back to the mid-1950s when the strategic priority was to transfer resources from the agricultural to the industrial sector. With limited land resources and with scarce capital to be invested in land improvements, the governments organised labour into collectives as means to intensify crop production and raise land productivity. Nearly all surplus farm products were acquired by the government at the low prices set by the government. Thus, increases in agricultural output were obtained at the cost of maintaining labour productivity in agriculture artificially low. The pace of urbanisation was therefore very low when compared to the experience of other developing countries (Tian and Zhang, 2003). The introduction of the household production responsibility system in the 1980s led to increasing use of commercial inputs in farm production and a rapid increase in labour productivity. Since then, the government has gradually relaxed restrictions on the transfer of rural labourers to non-agricultural industries, while the urbanisation process has accelerated notably.

Although China's urbanisation process has accelerated, barriers continue to exist for rural people wishing to enter cities and become urban residents. Under current policies, rural people can change their residency status through a limited number of ways, such as finding employment after graduation from universities or after leaving the army, obtaining urban residency due to confiscation of cultivated land, and purchasing houses in cities on the commercial market.

While rural labourers can go to urban areas for jobs, they are required to pay city governments various fees, such as a temporary residency fee, a family planning fee and an urban size expansion fee (Bai and Song, 2002). Without urban residency, this migrant population is not entitled access to most of the social services provided by city governments or by employers, such as child education, socialised health services and housing. Such sharp differences between rural migrants and urban residents became major causes for conflicts in urban areas.

In 1997, the Chinese government began experimenting with different types of policy reforms of the urban residency system. These experiments concerned 382 counties that were allowed to use small cities and towns to accommodate rural migrants (MOA, *China Agricultural Development Report*, 2002). Under the scheme, rural people who have a residence and a stable income-earning job in such cities or towns can apply for registered permanent residence while still allowed to either retain their contracted lands for their own operation or lease it so as to avoid the land becoming idle.

According to the MOA (2002), by 2001 only 1.3 million rural residents had changed their residency status. In the meantime, some metropolitan areas began to relax restrictions on the granting of residency although the system still favours those migrants endowed with human capital and financial resources. Thus, most prospective rural migrants still face major obstacles in obtaining urban residency.

There are signs that the Chinese government will change the development strategy toward a co-ordinated development of urban and rural sectors. Recently, it has declared building "a well-off society in an all-round way by 2020" as a strategic goal (Jiang, 2002; Zhu, R., 2003).

With regard to the transfer of rural labourers, two basic policies are proposed. The first is to encourage the rural labour force to find work locally, including the development of profitable and labour intensive agriculture alongside non-agricultural industries in rural areas. The second is to guide the rural labour force to find employment in other areas by strengthening information networks, starting employment agency services, offering pre-transfer training to rural workers and organising an orderly flow of the rural labour force. Establishment of a unified and standardised labour market and the household registration system are also considered. As a result of this policy reorientation, it is expected that China's urbanisation process will be accelerated in the years to come.

Due to regional variations in socio-economic conditions, different regions often take different approaches towards rural labour migration. In those economically developed regions where a market-oriented economic system is usually well established, enterprises, especially private firms, tend to employ workers as cheaply as possible. Therefore, most unskilled positions are filled by rural migrant labourers, especially those coming from underdeveloped inland regions.

In such provinces (*e.g.* Guangdong), immigration of rural labourers from other regions is allowed to take place rather freely. However, like other regions, few of the migrants are granted permanent local residency. Consequently, migrant labourers move regularly between those areas that receive migrant workers and their hometowns, giving rise to what is known as a "floating population." Many city governments face pressure to reduce local unemployment and to maintain adequate incomes of urban residents. Ensuring the sustainability of these schemes often generates among local authorities a negative perception of immigration.

Other constraints to migration – land policy and tenure security

A second important institutional factor influencing the incentive to migrate on a more or less permanent basis is the ability of households to exit agriculture without facing a substantial loss in terms of their asset endowments and/or entitlements. As land is the most important asset of farm households, a key question is the extent to which land use rights can be transferred to other farmers without incurring the risk of an entitlement loss. As recently argued by Deininger and Jin (2002), there is evidence of a large demand for land for agricultural production that remains unattended through the land rental market. This reflects a divergence at the margin between the willingness to pay and the willingness to accept a given payment for an additional unit of land rented. Such a divergence might be attributed to the existence of insecurity of tenure which would act as a transaction cost.

There are at least three reasons for which the risk of losing an entitlement to land might deter a migrant from moving permanently to urban areas. On the one hand, land is a key income generating asset. Although farm income has declined substantially as a source of total income for the average rural household, it is still an important source of livelihood. Second, as shown recently by Burgess (2004), in a setting characterised by imperfect factor markets, access to land reduces the shadow price of food and contributes to maintain a higher consumption of calories. Thus, the value of an entitlement to use land exceeds its market rental value whenever imperfect property rights might exacerbate the risk of losing that entitlement as a result of a decision to migrate.

Finally, an entitlement to the use of land might act as the best insurance policy that a farm household can pursue when some of its members migrate to urban areas. The inability to access the social programmes otherwise available to urban residents as a means to weathering shocks to the urban labour market, turn the option of migrating back to the farm into a valuable insurance policy. The latter point can be further reinforced as land policy has traditionally been regarded as complementary to the *hukou*. Limiting access to social welfare programmes by rural migrants and leaving the latter exposed to labour market fluctuations has implicitly reinforced the insurance role of agricultural land. Thus, restrictions concerning the right to transfer land have been regarded as justified by the principle of maintaining social stability in rural areas.

Land policy in China has traditionally imposed severe restrictions on land transactions due to social security and stability concerns. Uprisings by farmers who had lost their land and found no opportunities of being re-employed in other sectors have been recurrent throughout China's history. These concerns have eased somewhat in recent decades as the rural economy underwent important changes. Limitations on land rights transfers have been reduced substantially with the reform in 2002 of the Law for Rural Land Contracts which provides the legal basis for protecting the interests of land users and for voluntary transfer of land use rights. As land cannot be individually owned and can only be leased, protection of rights emerging from leasing arrangements might not be enough to affect the perception by those who are entitled to the original right of use from losing that entitlement should they relinquish possession over an extended period of time. Uncertainties related to fluctuations in urban labour markets and off-farm incomes might further increase the perception of risk attached to a complete exiting from farming (Carter and Yao, 2003).

These considerations suggest that reforming the *hukou* system, in the sense of extending the right to benefit from existing social welfare programmes to migrants in urban areas, might have an important second order effect in terms of further stimulating land rental markets in rural areas. As Deininger and Jin (2002) have shown, present land transactions already serve the purpose of moving land use from farmers with low agricultural ability to more productive farmers. As a result, and in view of the thread that connects land and migration policies, it might be expected that a further opening of the *hukou* system would reinforce pressures to further liberalise land markets.

Migration and the rural/urban income gap

The rural-urban income gap represents a serious structural imbalance affecting China's social development. As described in Chapter 1, a key factor behind the rural/urban gap is the extremely low productivity of labour employed in agriculture, reinforced by the differential treatment of urban and rural areas in the provision of social services (*e.g.* education and health) and in the structure of fiscal incentives.

Migrants from rural areas tend to be younger and better educated than those left back (Taylor *et al.*, 2003). One would expect that better education would reap higher benefits from migration into urban areas and that this would in turn be reflected in more remittances sent to the household. Yet, while out-migration from rural areas has been substantial during the past decades, its impact on the rural/urban income gap has been limited for a number of reasons.

Rural migrants usually lack bargaining power when entering work contracts with employers due to acute competition for jobs among rural job-seekers. Therefore, employers tend to depress wages to the lowest level possible. The *hukou* system reinforces this state of affairs. Inferior working and living conditions, default in the provision of labour insurance as required by relevant government regulations or even in wage payment, and the lack of legal protection of their rights as workers substantially weaken the bargaining power of rural migrants in the urban labour market. This often results in undue dismissal of employees and inhuman treatment of workers.

In addition, migrant labourers are usually unable to protect their own interests through lawful means due to their lack of legal knowledge and ability to enter lawsuits with employers. Thus, to the extent that part of the income earned in urban areas is remitted to the household of origin by the migrant, the *hukou* system will limit the gain that rural households may achieve from migration. Taylor *et al.* (2003) have shown that remittances alleviate the credit constraints that rural households face, thereby allowing them to purchase commercial inputs and raise farm productivity.

In addition, Zhang *et al.* (2003) provide evidence (Tables A.4 and A.5) that the amount of remittances increases with the duration of the migration spell, a fact only partly explained by the increase in income associated with the duration of the spell. Thus, insofar as the *hukou* system increases the incidence of circular versus permanent migration; it limits the long-term benefits rural households may achieve through migration.

Access to land also plays a key role in this respect. As mentioned above, even those households whose major family labourers work away from hometowns still hold their contracted lands given the great uncertainty in off-farm jobs and incomes. In this sense, the contracted land is used as a means for rural households to ensure family food security (Burgess, 2004) as well as to manage the unemployment risk faced by household members in rural areas (Carter and Yao, 1999). However, this leads to an inefficient utilisation of China's scarce land resource and contributes to maintaining low labour productivity in agriculture as a result of scarce incentives to invest in land improvement and machinery.

Table A.4. **Migrants and remittances**

	Migrants who do not remit	Migrants who remit
Number of observations	218	267
Age (years)	23.6	22.8
Trained (%)	30.7	31.5
Male (%)	53.9	54.9
Years of schooling	8.4	8.3
Labour earnings by off-farm job (CNY)	4 236.5	5 906.1
Duration of migration (month)	25.6	32.1
Remittances to household (CNY)	0.0	2 253.5
Remittances to migrants (CNY)	177.4	138.7
Net remittances (CNY)	–177.4	2 099.4

Source: Reproduced from Zhang *et al.* (2003).

Table A.5. **Duration of migration and remittances**

	Number of observations	% of observations	Remittances (yuan/year)	Disremittances (yuan/year)	Net remittances (yuan/year)	Earnings by off-farm employment (yuan/year)
< 1 year	138	28.5	620.4	226.2	394.2	3 171.0
1-2 years	70	14.4	845.7	65.3	780.4	5 626.4
2-3 years	109	22.5	1 592.8	113.3	1 479.5	5 719.1
3-4 years	45	9.3	1 698.7	160.7	1 538.0	5 592.2
> 4 years	123	25.4	1 681.3	165.4	1 482.6	6 455.3

Source: Reproduced from Zhang *et al.* (2003).

In addition to the institutional constraints mentioned above, other factors prevent the income gap from closing further as a result of rural out-migration. For instance, since rural migrants are in general better educated than those who stay home, migration means an export of human capital from rural to urban sector, or from poorer regions to richer regions. On the other hand, although migrant labourers are better educated than those who stay home to farm, their knowledge and ability is often inadequate to find and hold skilled jobs in towns and cities. In fact, most migrant labourers are engaged in low-paid jobs.

Finally, China lacks an efficient job information system. Therefore, migrants move sometimes blindly, leading to substantial loss of time and money. What is more problematic is that some "blind-movers" may resort to illegal means to get money in order to maintain their life while away from home.[3]

The *hukou* system imposes high transaction costs to permanent migration to urban areas. Whether a substantial reform of the *hukou* will help reduce the income gap between urban and rural areas via labour market adjustment depends to a great extent on the pattern of out-migration from rural areas. A major overhaul of the *hukou* is expected to contribute through three main channels. First, by relocating surplus labour from farming and off-farm jobs in rural areas where the marginal productivity of labour is low, the average productivity of labour would be expected to increase (Tian *et al.*, 2003).

Second, migrants are likely to send substantial remittances to their families of origin. There is now enough evidence (*e.g.* Zhang *et al.*, 2003) showing that remittances are an important source of income for those rural households that have experienced the migration of some of their members. Even in those instances in which migration is not permanent, returning migrants contribute substantially to total household income with the savings accumulated during their absence. Thus, expanding the opportunities for migrants from rural areas in urban labour markets will improve incomes of households in source communities.

Third, an increase in the outflow of labour from rural areas is likely to generate dynamic gains that would materialise through higher labour productivity and increased investments in farming activities and TVEs. For example, if households that have a comparative advantage in off-farm employment relative to farming were allowed to relocate to urban areas and transfer their contracted land to households with better farming skills, average labour productivity in agriculture would increase. In turn, farmers expanding their operations would find an incentive, provided their newly acquired land rights were adequately protected, to invest in land improvements, raising the productivity of land as well.

This effect might be reinforced by the inflow of remittances received by households having some of their members working in urban areas. A similar process might also affect the productivity of labour employed in TVEs and, by making the latter more competitive, the incentives to invest and expand their scale of operations. Inflow of remittances might ultimately prove to be a factor contributing to an increase in the number of TVEs.

Rural out-migration and the supply response of agriculture

A final issue concerns whether the removal of the *hukou* will affect the ability of the agricultural sector to meet the growing market demand for food as a result of the urbanisation process, further accelerated by the removal of the *hukou* itself. The main issue in this respect is not how to increase the overall volume of food production, but rather how to meet growing market demand for food that to date has been satisfied largely by own production for on-farm consumption. Some evidence suggests that China's agriculture could in fact meet the challenge. In their recent paper, Taylor *et al.* (2003) closely examine the issue of how remittances influence the composition of farm household income and crop yields using data from two provinces near Beijing.

Their hypothesis is that imperfections in credit, land, and labour markets constrain farm households from allocating their labour resources across various activities (*i.e.* farming, self-employment, and wage employment) in an optimal way. Migration from farm households is therefore an optimal response to such a difficult institutional environment as it allows farm households to obtain cash with which they can purchase commercial inputs or assets for crop production or for production in household-based small scale enterprises. The difficulty is that migration subtracts labour resources and is therefore expected to have a partial negative impact on household income.

The total net effect is a matter of empirical measurement. Taylor *et al.* (2003) find that while migration reduces crop income it provides a positive impact through remittances. The net effect is one of reducing crop income but at the same time increasing per capita household income and crop yields. Farmers substitute migrated labour with commercial inputs purchased with the remittances received. Cash expenses on commercial inputs decrease net income from crops, while the use of the same inputs allows increasing yields. Total per capita household income increases.

De Brauw et al. (2000) examine the impact of market liberalisation on China's agricultural sector using the concepts of flexibility and responsiveness. Flexibility concerns the speed with which quasi-fixed factors of production (labour and land) adjust to changes in the exogenous constraints affecting the sector, such as irrigation, legislation, and prices. Responsiveness refers to the ability by farmers to adjust their mix of inputs and outputs as relative prices change. Flexibility is typically measured by the amount of time that farmers would need to fully adjust to external changes, responsiveness by own and cross price elasticities.

The results in Taylor et al. (2003) and De Brauw et al. (2000) jointly suggest that China's agricultural economy has displayed increasing flexibility during the period that runs from the late seventies to the mid-nineties, a period during which TVEs offered substantial scope for labour to reallocate permanently to off-farm activities. As a reference, the estimated degree of flexibility compares very well with similar computations for the United States.

During the same period, producers became more sensitive to wage and fertiliser price changes, thereby manifesting more responsiveness to changes in relative prices. As the potential of TVEs to absorb more labour is very limited, further improvements in the flexibility of farm labour depend very much on the conditions regulating rural to urban migration. These results suggest that should the rate of urbanisation increase substantially and real wages also increase, China's agricultural sector would be able to adjust by raising land productivity and real output.

Clearly, the final impact on yields and agricultural output will be reinforced to the extent that more productive farmers will be able to adjust not only by substituting labour with commercial inputs but also by expanding the size of their operations through the land markets. As argued above, while there is evidence of an active land market from the demand side, the supply side of the market is still relatively weak leaving, on balance, the land market rather thin.

While more research is needed in order to extend the analysis to other areas of rural China, the initial evidence suggests that a significant reform of the *hukou* system coupled with further reforms in land policy would provide scope for reducing the rural-urban income gap without jeopardising China's food security.

Policy conclusions

Although China achieved remarkable success in alleviating rural poverty over the past two decades, further progress is both difficult and costly since many of the remaining rural poor are located in environmentally fragile areas with limited access to external markets and new technologies. Furthermore, they lack the appropriate physical and human capital to adapt to competitive market conditions. While further growth of agricultural production may improve household food security of the poor, it cannot reproduce the same effect on rural poverty reduction as that in the 1980s.

Restrictions on labour and land markets have resulted in a significant misallocation of resources and contributed to the large disparity between urban and rural incomes. This suggests that the administrative barriers to rural/urban migration, permanent residence, land markets, and tenure security should be removed or at least relaxed.

The issue is not whether to allow rural people to migrate to cities, but how to manage the process properly so as to avoid undesirable consequences. Urban areas would benefit from increases in GDP as would government fiscal revenues, although urban workers would be exposed to acute competition in labour markets with downward pressure on wage rates. The transfer of rural labour to non-agricultural employment would create favourable conditions for the structural adjustment of agriculture, such as farm land consolidation and intensification of farm production.

In general, the gradual transfer of rural people to towns and cities is a necessary step in achieving co-ordinated rural/urban development, which in turn increases China's long-term social stability. This requires fundamental reforms of the policy making process, laws and social institutions as well as the empowerment of rural people through improvement of rural education and vocational training.

The transfer of rural labour depends on both regional socioeconomic conditions and household attributes. The real poor in remote regions are less likely to migrate because of high costs and risks and they are also less likely to succeed in the competitive labour market if they do migrate. Given the current state of development, migration may not significantly alleviate rates of rural poverty, although it would raise average rural income. Thus, complementary measures, such as the establishment of income safety nets, should be considered along with the transfer of rural labour.

At present, the Chinese government places a high priority on improvement of rural income and the alleviation of rural poverty to strengthen social stability. While these objectives are regarded as positive externalities of agricultural growth, the appropriate way to achieve such goals needs to be considered carefully. The FAO Roles of Agriculture study for China argues that promotion of agricultural production, especially increases in conventional grain crops, may not be the most desirable approach. Future growth of agricultural production will most likely be achieved by improving technical efficiency, rationalising output structure and facilitating marketing. This requires substantial innovation in research and extension systems combined with deregulation that allow markets to operate more freely.

Notes

1. This Annex was prepared by the FAO based on its study of China conducted as part of the Roles of Agriculture Project (ROA) carried out by the FAO in 2001-2003. It draws heavily from the "*National Report*" on China by Tian, W. as well as two background reports: "*Social Viability Roles of the Agricultural Sector in China*" by Tian, W., Liu, X., and Kang, X., and "*The Poverty Alleviation Role of Agriculture in China*" by Tian W., Wang, X. and Ke, F.
2. This is especially important in those inland areas and for poorer rural households.
3. More critically in the longer term, rural people, including those migrant labourers, are unable to participate in the political process and express their own interests.

ANNEX B

Agricultural Policies and Support for Individual Commodities

This annex consists of eight sections, one for each major commodity grouping discussed in Section 2.4 of Chapter 2. Each section comprises an overview of domestic support policies for the commodity group, border measures, and trends in producer support estimates. For tea and tobacco, the overview is limited to domestic support policies and border measures as these commodities are not covered by producer support estimates.

Cereals

Domestic policy

As a result of good harvests in 1989 and 1990, China's grain supply was relatively abundant and farmers found it difficult to sell grains profitably. To protect farmers' interests, the special grain reserves scheme, which included a price support policy component, was introduced in 1990 (State Council, 1990a). The government promised to purchase all grains offered for sale by farmers – after the fulfilment of procurement quota obligations – at rates no lower than the announced support prices. While the national government set the reference prices for different types of cereals, sub-national governments were allowed to offer higher prices if they wished to do so (State Council, 1990b). Although this new arrangement was intended to help farmers maintain their incomes, the State Grain Enterprises (SGEs) often did not make timely payments for grains delivered under this policy initiative. At the same time, the government was faced with problems generated by the policy, namely the overstocking of low quality grains and the huge financial burden this imposed.

To cope with the situation, the government decided in 1991 to carry out an ambitious programme of reform of the grain marketing system aimed at gradually liberalising the market. In April 1991, the national government decided to raise the retail sales price of grains for the first time since the establishment of the centrally planned grain distribution system in the mid-1950s (State Council, 1991).[1] Retail prices were raised again in April 1992.[2] Many provincial governments implemented their own reform programmes to liberalise grain marketing either partially or completely during 1992-1993.

State procurement adjustments were also made at this time. In late 1992, the government initiated price premiums for mid- and late-season indica rice and excluded early-season rice from purchase under the special reserve scheme, due to its generally poorer quality. This adjustment was intended to encourage farmers to adjust their output to increase the yield, quality and profitability of grain production (State Council, 1992).

The national government decided in October 1993 that from 1994 the state would only procure a fixed amount of grains and that grain prices would be determined by market forces. In order to stabilise grain markets, grain risk funds were established at both the national and provincial levels. The grain deficit provinces were required to hold grain stocks sufficient for six months consumption.

However, the domestic grain market became volatile in late 1993. Market prices rose quickly and panic buying and hoarding occurred. Under these circumstances, the proposed measures were not put into practice in 1994. Instead, the government turned its attention to market stabilisation, insisting that 70-80% of marketed grains should be controlled by the SGEs. In order to induce farmers to produce and sell more grains, the government raised state procurement prices sharply (Table B.1) by 43% in 1994, by 23% in 1995, and by 22% in 1996. In urban markets, the national government imposed price ceilings for food grains in 1994 and rationing of grains was restored in many cities as an emergency measure.

Commencing in 1994, the national government began to construct commercial grain (and cotton) production bases in 650 counties where conditions were most suitable. A total of CNY 6.5 billion of preferential loans (loans with lower than normal interest rates) were provided for the programme over five years. This programme was extended in 1996 with the decision to establish twenty large-scale commercial grain production bases at the prefecture level, with additional investment of CNY 2.13 billion from both national and sub-national governments (*People's Daily*, November 1996). As discussed previously in Chapter 2, a range of policy measures were instigated to assist the dissemination of functional technology, control conversion of cultivated land to non-agricultural uses, support farm input industries, and improve agricultural infrastructure.

The "governor's grain-bag responsibility system" (GGBRS) was formally instituted in early 1995 (Chapters 1 and 2), imposing a responsibility on provincial governments to

Table B.1. **Changes in grain prices (CNY/kg in current price)**

Year	Rice		Wheat		Corn	
	State-set/ Protective	Negotiated purchase	State-set/ Protective	Negotiated purchase	State-set/ Protective	Negotiated purchase
1990	0.51	0.82	0.51	0.85	0.38	0.63
1991	0.51	0.73	0.51	0.77	0.38	0.55
1992	0.55	0.65	0.59	0.73	0.42	0.55
1993	0.62	0.74	0.66	0.75	0.46	0.64
1994	0.89	1.14	0.89	1.04	0.69	0.90
1995	1.09	1.72	1.08	1.53	0.86	1.38
1996	1.33	1.71	1.31	1.65	1.06	1.39
1997	1.48	1.45	1.46	1.43	1.23	1.10
1998	1.46	1.34	1.44	1.30	1.23	1.17
1999	1.33	1.23	1.31	1.22	1.14	1.05
2000	1.13	..	1.14	..	0.96	..
2001	1.11	..	1.09	..	0.94	..

Note: State-set prices were applied until 1997 and protective prices between 1998 and 2001. For definitions of state-set, protective and negotiated prices, see Section 2.2 in Chapter 2.
Source: MOA, *China Agricultural Development Report*, various issues.

ensure adequate local grain supply and market stability, especially for grain-deficit regions.[3] One of the immediate consequences of this system was that inter-provincial grain trade declined; as provincial governors erected trade barriers as a means of ensuring local grain availability was not diminished.

These measures collectively helped to boost China's grain production. Cereal output reached a new record of over 450 million tonnes in 1996 and remained high for the following three years (Chapters 1 and 3). In 1997, however, Chinese domestic grain prices started to decline under the influence of both abundant domestic supply and falling world market prices (Figure 2.2 in Chapter 2).

In response to falling domestic prices, and in the belief that the decline in world prices would be short-lived, the national government reinstalled guaranteed state purchase of grains at state-set floor prices in late 1997 with the aim of raising farmers' incomes and maintaining grain production capacity. SGEs were ordered to buy whatever quantities farmers wanted to sell. The government committed to subsidise SGEs storage costs for overstocked grains and also agreed to subsidise interest payments made by the SGEs on loans required to fund the additional grain purchases.[4] These subsidies were paid from the grain risk funds established in the early 1990s, of which 60% came from sub-national governments' contributions.

These measures were developed into a new grain policy package issued in June 1998 (State Council, 1998a). The new package consisted of three policy measures:

- In place of the state procurement quota, SGEs were given a monopoly to purchase all grains that producers wished to sell at the state-set or protective price floor.
- SGEs are to sell purchased grains at prices that cover all operating costs.
- All SGE financial transactions must go through appropriate accounts at the Agricultural Development Bank of China (ADBC).

According to the ADBC's "method for managing grain procurements and sales", the ADBC provides loans only if an SGE actually makes procurements, and the amount of the loans is linked to quantities and prices paid by the SGE. The ADBC also keeps records of grain stocks at all storehouses. When an SGE sells grains, it should report to the ADBC information on buyers, volume of grains sold and prices, and the storehouses from which grains were sold. The loan officers of the ADBC are expected conduct regular inspection of SGEs and the storehouses for which they are responsible.

The new policy package also proposed to deepen reforms of SGEs by separating their policy activities from their commercial operations. In essence, this measure simply forces SGEs to maintain separate accounting systems for policy and commercial operations. In addition, SGEs were given VAT exemption for policy-related operations, commencing in August 1999.

However, the implementation of the new grains policy package met similar problems to those that occurred in the early 1990s. Farmers delivered large amounts of low quality grains to SGEs. With slower growth in demand in the domestic market and declining world prices, it became increasingly difficult for SGEs to sell the grains in stock without incurring losses. Overstocking of grains imposed a heavy burden on the budgets of all levels of government, particularly in the grain surplus regions. In order to avoid financial losses, SGEs employed various excuses to depress procurement prices, obviously a deviation from the aim of the policy they were implementing. As a result of the SGEs being swamped with low quality grains, they had reduced capacity to undertake commercial grain operations. Contrary to the expectations of policy makers, the SGEs could not become commercially viable businesses. It is reported that the SGEs incurred huge financial losses.

As the low grain prices persisted, the national government began to alter policies in order to cope with the problems in SGEs due to overstocking, poor grain quality of government stocks, and the continuing threat to farmers' incomes. Protective prices were reduced gradually (Table B.1) and coverage of the scheme was narrowed. For instance, spring wheat, wheat produced in southern provinces and indica rice were phased out, starting from the 2000 harvest, and corn produced in southern China was phased out in February 2002. Also, some non-SGE grain users (*e.g.* grain processors, feed manufacturers etc.) were allowed to purchase grain directly from producers.

The government actively encouraged farmers to increase plantings of superior quality grain types. The national standards for grains were revised to better take into account quality attributes. According to official statistics, during 1998-2002 the area planted to high quality varieties rose from 22.7 million hectares to 38.7 million hectares for early indica rice, from 16.0 million hectares to 72.7 million hectares for wheat, and from 19.3 million hectares to 59.3 million hectares for corn (MOA, *China Agricultural Development Report*, 2003). The successful dissemination of improved quality wheat varieties resulted in China being able to substitute domestically produced wheat for some imported wheat, contributing to China becoming a net exporter of wheat in 2002 and 2003.

With an abundant supply of grains in the latter part of the 1990s, the government took measures to reduce pressure on the environment. Starting from 1998, the government began to implement a "grain for green" programme under which some marginal lands were retired from grain production and converted to pasture or forest. While this programme led to a reduction of grain output in the short term, it is envisaged that it will contribute to maintaining long-run production capacity (see Chapter 2).

China became a WTO member in late 2001 and deregulation of grain marketing was once again on the agenda. In 2001, eight grain-deficit provinces (Beijing, Tianjin, Shanghai, Jiangsu, Zhejiang, Fujian, Guangdong, and Hainan) were allowed to liberalise their regional grain markets. This involved the abolition of compulsory state purchasing (farmers no longer obliged to sell to the State) and the end of state "protective" price setting for grains – government purchases in these provinces are made on the open market at market prices. In 2002, some other provincial governments decided to liberalise their own markets either as a province-wide initiative, or in major grain deficit areas within the province. For instance, Shandong liberalised grain marketing in Qingdao, Yantai and Weihai in April 2002. Further reforms involving more provinces were conducted in 2003; however, no clear national liberalisation policy was enunciated during 2003.

In 2004, following a number of years of declining grain production and a corresponding decline in grain stocks (Chapter 2), the national government initiated new measures intended to restore grain production levels. These were to provide direct payments to grain producers, to subsidise the purchase of higher quality grain seeds, and to subsidise the purchase of farm machinery. In addition, minimum prices for japonica rice and early indica rice were announced in early 2004 in an attempt to assure farmers that their income levels would be maintained. Funding of CNY 10 billion was appropriated from the state grain risk fund to fund these measures to support farmers in the 13 main grain producing provinces and autonomous regions. The Ministry of Finance allocates the funds to special provincial accounts at the Agricultural Development Bank of China and the provincial governments then disperse the money to county-level accounts based on production levels and land use rates, determined at the provincial level. While each province can determine the method of allocation of the direct producer subsidies, most provinces appear to be making payments of CNY 10 per *mu* for area planted to grain (Gale *et al.*, 2005). In an attempt to ensure that farmers benefit from government subsidies, sub-national governments are required to publicise all details about the use of the grain risk fund monies and penalties are in place to dissuade inappropriate use of those funds. The Ministry of Finance reports (Jing, 2004) that during the first five months of 2004, twenty-nine provinces and autonomous regions released an estimated CNY 11.3 billion for direct subsidies for grain growers, and CNY 1.24 billion to subsidise the purchase of improved seeds of rice, wheat, and corn (and soybeans).

In June 2004, the national government issued a new regulation on grain marketing (Chapter 2). The new regulation has, in principle, liberalised grain marketing by allowing qualified non-state firms to buy and sell grains on the open market. Private firms which meet certain criteria are also permitted to engage in grain processing and storing activities. If this new regulation is fully implemented, the monopolistic position of the SGEs will be effectively ended. Coupled with the earlier reforms which appear to be putting an end to the state procurement system, the Chinese government seems to be withdrawing most of its direct interventions in the grain markets. However, the governors' grain-bag responsibility system is, in principle, retained under the new regulations. Although how that responsibility is to be discharged in the newly liberalised market is not yet public.

To some extent, the package of new measures is a response to the perceived insecurity of grain supply. However, it also strengthens a new policy direction to support farmers' incomes. It results in a fundamental change in the approach to supporting the grain sector. In the past, the Chinese government's grain subsidies were paid mainly to the procurement and distributing sectors, and grain growers appear to have received only a reduced benefit from government aid programmes.

While the new, direct subsidies are symbolically important as noted above, examination of the subsidies' role in increasing farm incomes shows only a minor impact. Gale *et al.* (2005) note that the subsidies equate to 1-2% of the average 2003 rural household income; and, of a 6.8% rise in rural incomes in 2004, direct subsidies accounted for only around 5% of the increase – 49% of the increase was attributed to increased product sales and 43% was attributed to non-farm income sources.

Border measures

Throughout the period 1990-2004, trading of grains was subject to state controls. At the beginning of this period the central government planned volumes of imports and exports of grains according to the expected domestic supply situation. Import quotas were allocated according to planned requirements. Grains were imported by state entities in years when a domestic shortage was expected, and exported by the state in years when a domestic surplus was forecast. Imports and exports occurred according to policy directives rather than in accordance with market price signals. In the event of an expected surplus in the Chinese harvest, for example, the relevant state trading company would be directed to purchase a particular quantity of grains, at the Chinese support or negotiation procurement price, in order to export at the going international price. In situations of the sale price being lower than the purchase price, trading losses were borne by the government. Similarly for imports, if a grain deficit province required imports through the state trading system, which cost more than the regulated domestic price of grains, the difference was subsidised by the central government.

In 1993, the arrangement whereby the central government subsidised the provincial government for trading losses was abandoned. The State Planning Commission's central allocation of import and export quotas according to planned requirements was replaced by the agent system for many agricultural commodities. Under this system, sub-national governments became responsible for their own trading results. Trade was controlled centrally according to plans developed in conjunction with sub-national levels of government.

Provincial governors, the central government and relevant Ministries remained responsible for determining distribution plans for grains. Based on these plans, the Ministry of Foreign Trade and Economic Co-operation (MOFTEC) would direct the China National Cereals, Oils and Foodstuffs Import and Export Corporation (COFCO) to undertake import and export activities. For example, if a province was unable to obtain adequate supplies of grain domestically, the provincial governor would send a request to obtain imported grain to the State Administration of Grain Reserves (SAGR) which would either approve or deny the application. If approved, MOFTEC would direct COFCO to arrange imports to be delivered to nominated ports for delivery to the relevant provincial grain bureau. The provincial grain bureau would obtain bank loans and contact COFCO to negotiate the purchase contract (ERS, 1999).

The arrangement for exports was similar. When the annual distribution plan called for exports, MOFTEC would arrange for provincial grain bureaus to purchase grain for export at the Government Fixed Exported Grain Transfer Price. Upon State Council approval, the SAGR allocates export quotas to provincial grain bureaus, and MOFTEC directs COFCO to negotiate purchase contracts with foreign buyers. The provincial grain bureaus then deliver grains to ports for shipment (ERS, 1999).

Thus, China's foreign trade in grains was tightly controlled, with annually determined quotas and state trading companies maintaining a monopoly position in executing trade plans.

Despite the strict government control of and heavy government involvement in grain imports, in 1996 China established import tariff rate quotas (TRQs) for grains (Table 2.10 in Chapter 2). The major grains (wheat, rice and corn) had in-quota tariff rates of 1-3% and above-quota tariff rates of 91-114%.

In December 2001, China acceded to the WTO under an agreement with existing members of the WTO which stipulated particular rules for China's trade activity. Under the Protocol of Accession, China must abide by the general WTO rules of not discriminating against particular trading partners and not supporting exporters with export subsidies. Particular rules were also negotiated to govern China's trade in particular commodities.

Under the terms of its WTO accession, China was permitted to maintain the state trading regime for grains. Under this regime, COFCO remains the nominated STE for grain imports (mainly wheat) and COFCO and the Jilin Grain Import and Export Company are the nominated STEs for exports of corn and rice. Under the WTO modified version of China's state trading regime agreed proportions of the in-quota volume are reserved for the STE(s) for that commodity. If part of the STE allocation for the year remains unused at

15 September, that part of the allocation must be returned to the National Development and Reform Commission (NDRC) for reallocation to private firms.

The WTO accession agreements also stipulated TRQs. These are detailed in Table B.2. The clear trend is for maintenance of strong control over the commodity most likely to be imported (wheat) and lessening of state control over those grains more likely to be exported. High above-quota tariffs are in place to discourage excessive imports – although the quota fill-rates in 2003 were 5% for wheat, 6% for rice, and there were no imports of corn. The quotas also rose in volume during the WTO transition period (2002-2004 inclusive). The Foreign Agricultural Service of the United States Department of Agriculture reports that quota fill-rates for 2004 are significantly higher for wheat (75% of quota), and rice (14% of aggregate quota, 28% of long grain quota), but the fill-rate for corn remained at 0%.

WTO accession also meant an end to export subsidies for China. As noted in Chapter 2, China subsidised corn and rice exports from the late 1990s until the end of 2001.

Table B.2. **China's WTO grain trade commitments**

		2002	2003	2004	2005
Wheat	In-quota tariff[1]	1%	1%	1%	1%
	Over-quota tariff	71%	68%	65%	65%
	STE share of quota	90%	90%	90%	90%
	Quota (million tonnes)	8.5	9.1	9.6	9.6
Corn	In-quota tariff[2]	1%	1%	1%	1%
	Over-quota tariff	71%	68%	65%	65%
	STE share of quota	68%	64%	60%	60%
	Quota (million tonnes)	5.9	6.5	7.2	7.2
Rice	In-quota tariff[3]	1%	1%	1%	1%
	Over-quota tariff	71%	68%	65%	65%
	STE share of quota	50%	50%	50%	50%
	Quota (million tonnes)	4.0	4.6	5.3	5.3

Note: The tariffs noted in the table are those of the tariff lines most likely to be imported for food use.
1. Simple average in-quota tariff for all wheat tariff lines is 4.7%.
2. Simple average in-quota tariff for all corn tariff lines is 6%.
3. The simple average in-quota tariff for all rice tariff lines is 3.3%.
Sources: WTO; NDRC (cited by USDA).

Producer support trends

Basic trends in the level of cereal producers' support were presented in Section 2.4 of Chapter 2 and can be summarised as follows:

- As the level of budgetary support for cereal producers is low, the level of support is driven by the price gap between producer prices and reference prices adjusted to the farm gate level.
- Trends in producer and reference prices were for most of the analysed period aligned: sharply growing between 1993 and 1996; sharply falling in 1997 and 1998; stabilising between 1999 and 2002; but then diverging in 2003 with domestic prices growing more than reference prices, in particular for maize and rice (Figures B.1, B.2 and B.3).
- Delays in adjustments in domestic prices resulted in some fluctuations in the level of support, in particular until 1997.

ANNEX B

- High level of border protection in the 1990s was not transmitted to domestic prices and the impact of tariffs was superseded by other more direct regulations such as state trading, grain quota system and state-fixed prices for grains.
- Since 1998, the level of support for grains has tended to increase with the striking difference that wheat producers are implicitly taxed, and rice and maize producers are implicitly supported (Figures B.1, B.2 and B.3).
- Most likely, the differences in the level of support across different types of grains in the 2000s are due to export-import decisions still made by the government and driven by the level of strategic stocks and expected production trends of various grains rather than by prospects of profits based on price differentials.

Figure B.1. **Percentage PSEs, producer and reference prices for wheat, 1993-2003**

Source: OECD Secretariat.

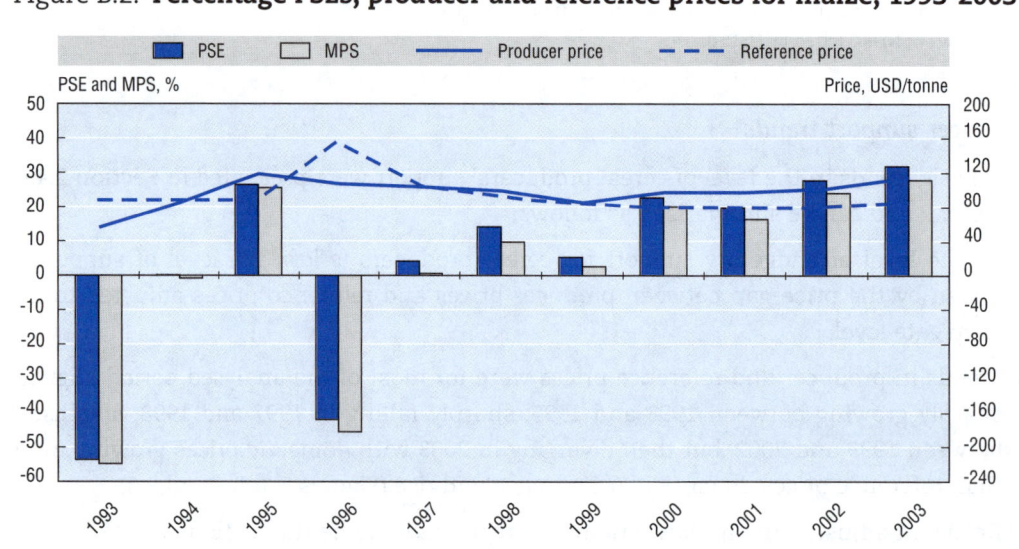

Figure B.2. **Percentage PSEs, producer and reference prices for maize, 1993-2003**

Source: OECD Secretariat.

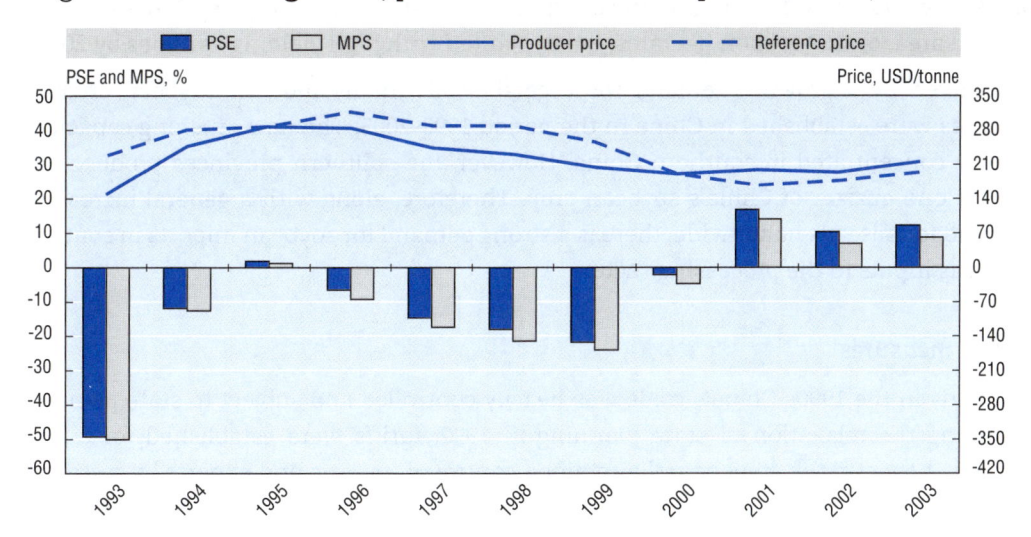

Figure B.3. **Percentage PSEs, producer and reference prices for rice, 1993-2003**

Source: OECD Secretariat.

Oilseeds

Domestic policy

China is one of the largest producers of oilseed crops in the world. Soybean, peanut and rapeseed are the major oilseed crops in China. Similar to the situation with grains, oilseeds were subject to state procurement and state-set pricing until the early 1990s.

As a result of reforms of the grain and oilseed marketing system in the early 1990s, oilseeds and vegetable oils were generally removed from the national quota procurement and pricing regime. The exception was soybean production and marketing. In the main soybean production areas, such as in Heilongjiang province, soybeans remained subject to state controls similar to those applied to the production and marketing of cereal grains. At the beginning of the 1990s, the "three links scheme" (Chapter 2) also applied to soybean production to ensure availability of product for state procurement.

In the mid-1990s, China switched from being a net exporter to a net importer of soybeans, and the volume of soybean imports rose rapidly, growing from 52 000 tonnes in 1994 to 2.9 million tonnes in 1997. In response to this situation, in 1998, the central government launched a programme to improve soybean productivity. Measures included the dissemination of improved seed varieties and production technologies in major producing regions, especially in Heilongjiang. Also in 1998, the state procurement price for soybeans became a protective price (Figure 2.2 in Chapter 2) under the changes detailed in China's new grain (and soybean) policy package (State Council, 1998a). While the national protective price was issued until 1999, some provincial governments issued "guidance" prices for local major oil crops and vegetable oil products until 2000-2001.

Since WTO accession, the national government has changed support for soybean development. The Soybean Revitalisation Plan was launched in 2002; a project intended to raise both soybean yield and oil content, through large-scale demonstration projects in the northeast provinces and Inner Mongolia. Under the project, the government provides seed subsidies to farmers adopting high oil content varieties and assists the extension of advanced technologies to soybean producers, harvesters and processors. Funding for the project was initially set at CNY 100 million. The programme also encourages contract production practices. The programme is expected to result in around 5% of China's soybean planted area, or 0.67 million hectares, being planted with high yielding varieties by 2007.

Driven by a growing demand for vegetable oils, many modern soybean crushing factories were established in China in the period 1990-2003. The new crushing capacity is heavily concentrated in southern China. However, the southern provinces produce only around one quarter of China's soybean crop. Therefore, along with a general increase in demand for oilseeds nationwide, there is a strong demand for soybean imports in southern China as inputs to the processing sector.

Border measures

Early in the 1990s China's trade was heavily controlled and subject to state planning. Following the relaxation of state planning in 1992, tariffs were established for oilseed imports; however, the state actually retained control of imports and exports by restrictive state trading and annual quota arrangements, similar to the trade arrangements for grains.

In 1995, to encourage the development of the domestic intensive livestock industries, China exempted soy meal imports from the 13% VAT. This led to an influx of soy meal which depressed the price at which domestic soy crushers could sell Chinese soy meal. Owing to the fact that soy oil and soy meal are joint products derived from soybeans, the declining price of soy meal impacted negatively on the domestic crushing industry, which cut-back production. As a consequence, soy oil, which was not freely importable, became relatively scarce and the price rose dramatically. In 1999, the soy meal VAT exemption was removed.

China established TRQs for soybeans and rapeseed in 1996, replacing the tariffs which had previously applied. Upon accession to the WTO, China abandoned the TRQ arrangement for soybeans and rapeseed, maintaining only a 3% import tariff for soybeans and a 9% tariff for rapeseed. Tariff arrangements for cottonseed and groundnuts were not changed, with the tariff for both commodities remaining at 15%.

Additionally, China's accession protocol does not allow for specific state trading arrangements for imports of raw seeds. While state trading companies may engage in imports of oilseeds, they have no particular regulated advantages over private firms. Conversely, exports of soybeans are subject to state trading.

Although imports of raw oilseeds are not discouraged by trade barriers, imports of vegetable oils are subject to TRQs and state trading arrangements. These arrangements provide protection to China's domestic oilseed crushing industry.

Producer support trends

As for grains, the direction of changes of China's domestic prices for oilseeds followed changes in border reference prices. However, while for importable soybeans and rapeseeds domestic prices remained above the border prices for most of the period, those for exportable peanuts were almost constantly below border prices and then fully aligned with them in 2002 and 2003.[5] In addition, while in recent years the price gap for rapeseed is close to zero, the gap for soybeans remained large and in 2003 even increased as domestic prices increased much more than border prices (Figures B.4, B.5 and B.6). As a result, the level of support for rapeseeds and peanuts is low in the 2000s, but remains high for soybeans. In fact, the average level of support for soybeans at 17% on average between 2000 and 2003, is one of the highest among all product covered by the PSE calculations for China (Figure 2.17 in Chapter 2).

The large price gap for soybeans might be surprising taking into account China's huge imports of soybeans (almost 21 million tonnes in 2003) and a low official tariff applied on soybean imports (3%, see above). One explanation might be that as soybeans were one of the "strategic" commodities[6] covered by strict government control, it is likely that not all previous state trading arrangements for soybeans have effectively been dismantled to allow domestic prices align with those on world markets. It is also possible that the price gap is due to a differentiated market existing. If imported beans are used for oil extraction and meal, and domestic beans are largely used as a food product (not for oil), it is possible that a price premium exists for domestic beans.

Figure B.4. **Percentage PSEs, producer and reference prices for soybean, 1993-2003**

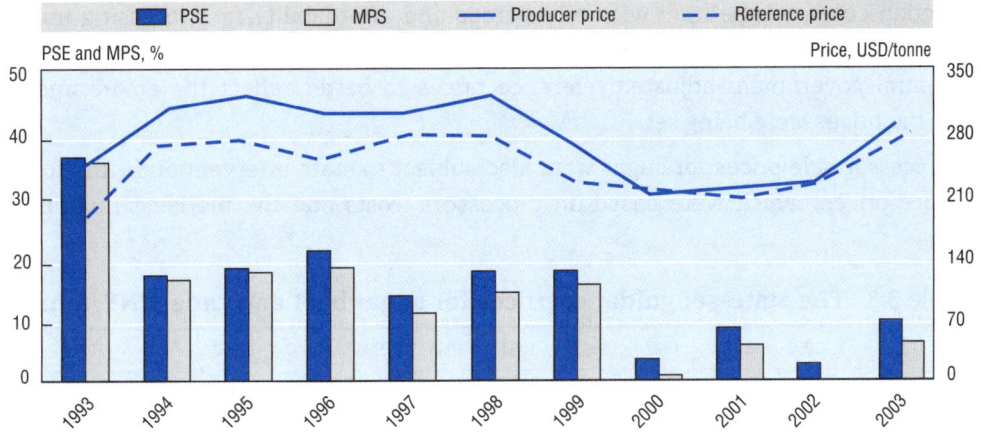

Source: OECD Secretariat.

Figure B.5. **Percentage PSEs, producer and reference prices for rapeseed, 1993-2003**

Source: OECD Secretariat.

Figure B.6. **Percentage PSEs, producer and reference prices for peanuts, 1993-2003**

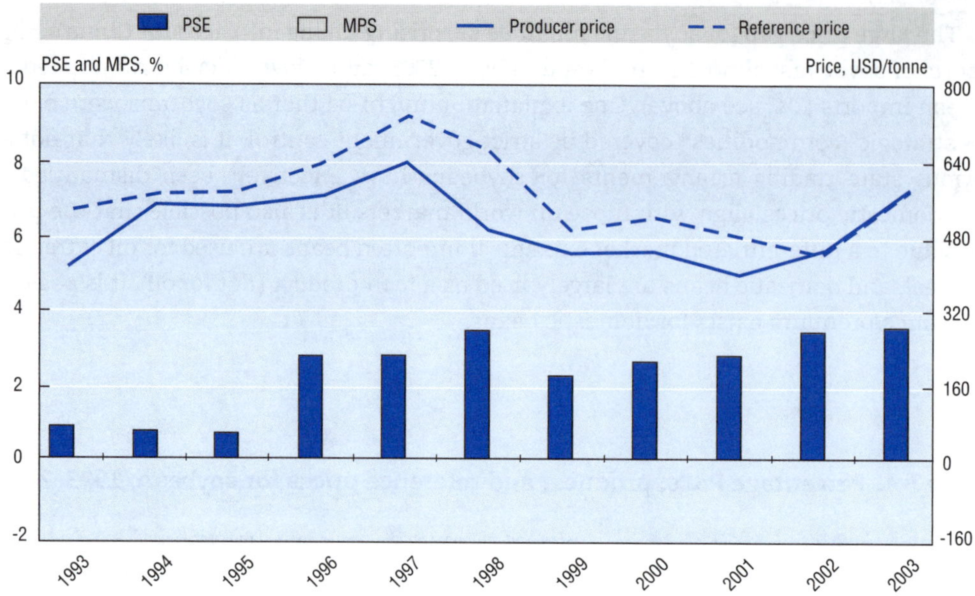

Sugar

Domestic policy

China produces sugar beet in northern China (mainly Inner Mongolia, Heilongjiang and Xinjiang) and sugar cane in southern China (mainly Guangxi, Guangdong and Yunnan), with the latter being the major source for sugar production. State control over production of sugar crops was relaxed in the late 1980s with reforms of the marketing system. However, while production controls were relaxed, state-owned sugar mills dominated the sugar industry and marketing of sugar was subject to planning control.

Entering the 1990s, the national government continued to issue guidance purchase prices for sugar beet and sugar cane and the provincial governments in production regions decided local purchase prices (reference prices) within the allowed range (guidance price ±10% in the mid-1990s). In principle, the reference prices were set taking into account production costs, competition with other crops and affordability for processing mills. In practice, however, this was not necessarily the case. On several occasions during the 1990s, the central government adjusted reference prices to better reflect the environment in which the prices were being set.

Processor sale prices for sugar were also subject to state intervention in the form of guidance prices, which were based on processors' costs and the market situation. The

Table B.3. **The state-set guidance prices for sugar beet and cane (CNY/tonne)**

Year	1990	1991	1992	1993	1994	1995	1996	1997	1998	1999	2000	2001
Sugar beet	155	155	155	155	190	280	310	304	280	255	220	250
Sugar cane	140	140	140	140	170	230	229	220	210	160	180	200

Sources: NDRC (to 1995); CAAS survey (from 1996).

government also established a system of state sugar reserves at the national and provincial levels. This initiative was intended to stabilise the sugar market.

Throughout the period 1990-2004, the government provided assistance intended to improve sugar crop production in major producing regions (Guangxi, Yunnan, Guangdong, Xinjiang) under a comprehensive agricultural development project. The measures included extension of advanced production technologies, facilitating producer access to credit, and improving input supply. In addition, as part of the strategy to develop western China, sugar production bases are being established in Yunnan and Xinjiang.

In June 2002, the central government issued a regulation which encourages vertical integration in the sugar industry by means of contract production. It also stipulates that processors are obliged to make advanced payments and provide technical services to growers in the areas supplying their mills. The regulation encourages the adoption of a two-stage payment system, under which farmers receive two payments for their cane or beets – the initial payment is a base price payment, and the second payment is determined according to the sale price of sugar achieved by the processor.

Sub-national governments are responsible for supervising contracts between the growers and processors. Provincial governments can set either procurement prices or guidance prices through an appropriate public consultation process. The central government does not determine prices received by cane and beet producers.

Border measures

Trade policy in the sugar sector relates to raw or refined sugar, rather than to sugar cane or sugar beets. China perceives sugar as an important food and controls trade in that commodity.

Sugar imports are usually in the form of raw sugar and processed in China's sugar refineries, and then either exported or sold on the domestic market. A significant proportion of China's raw sugar imports are from Cuba, under the terms of a long-standing trade agreement.

Until 2001, China controlled sugar trade with a combination of state-trading and a quota-licence mechanism similar to that used for soybean and grain trade. Quotas were determined according to the calculated domestic supply and sugar reserves situation.

Under the terms of accession to the WTO, China was permitted to maintain the state-trading regime for sugar imports, but agreed to eliminate controls on exports.

The permitted controls on sugar imports are summarised in Table B.4. The quota fill rates for 2002 and 2003 were 67% and 42% respectively.

Table B.4. **China's WTO sugar trade commitments**

		2002	2003	2004	2005
	In-quota tariff	20%	20%	15%	15%
Sugar	Over-quota tariff	65.9%	58%	50%	50%
	STE share of quota	70%	70%	70%	70%
	Quota (mil. tonnes)	1.8	1.9	1.9	1.9

Sources: WTO; MOFCOM.

Producer support trends

The Chinese sugar sector benefits from considerable support with the %PSE average at almost 40% between 2000 and 2003, the highest among all commodities covered by the PSE calculations (Figure 2.17 in Chapter 2 and Figure B.7). In 2000, the level of support was at 68%, when reference prices decreased strongly and this fall was not transferred to domestic prices. Since then, the level of support strongly fluctuated but, in general, tended to fall and in 2003 was slightly above 20% (Figure B.7). The high level of support is provided mostly through border protection combining tariffs, TRQ and a high share of STE within quota (Table B.4).

Figure B.7. **Percentage PSEs, producer and reference prices for sugar, 1993-2003**

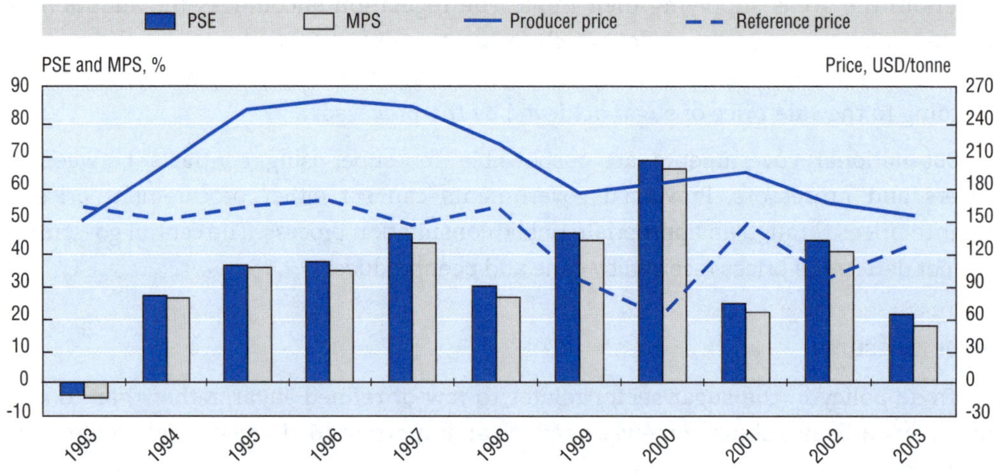

Source: OECD Secretariat.

Cotton

Domestic policy

Cotton is regarded as a "strategic" commodity in China. As a result, its production, marketing and consumption have been subject to state planning. Until 1998, the government issued state procurement prices and the Supply and Marketing Co-operatives (SMCs) were designated as the only buyers of cotton, making purchases on behalf of the State. In a control and planning regime similar to that which was in place for grains, the national government assigned production, purchasing and interregional allocation plans for both cotton producing regions and cotton textile producing regions. Prices at each marketing level were set annually, with a procurement price stipulated for SMC purchases from farmers and a set allocation price at which SMCs could sell to textile mills.

Under the cotton production system, there was no upper limit on the quantity of cotton which the State would purchase from producers, provided farmers also fulfilled their grain delivery obligations. From 1985 to 1998, cotton farmers where obligated under a production "contract" to supply the SMC with a certain quantity of cotton at the planned procurement price; they could also sell any above-quota quantity of cotton to the SMC.

Similar to the situation with grains, there was a surplus of cotton during the early 1990s. In response to the surplus, in 1992 the national government proposed a reform programme that was intended to significantly reduce state intervention in the cotton market by 1995. Trials were carried out in Shandong, Henan and Jiangsu provinces in 1993, and it was planned to extend such trials to other provinces in subsequent years (SPC, 1992). The proposal tested consisted of a number of elements: both the raw cotton purchase prices and processed cotton sale prices of the designated processing enterprises would be determined by market conditions, rather than stipulated by the State; raw cotton purchases would be based on contracts between cotton producers and purchasing/processing enterprises; enterprises other than the SMCs would be allowed to purchase cotton directly from producers; cotton trading centres would be established to facilitate transactions between regions; and, to ensure market stability, cotton reserves at both the national and provincial levels would be established.

However, cotton production levels in 1993 were very low due to an outbreak of cotton bollworm, the major insect pest associated with cotton production in northern China. As a result, the textile industry experienced a shortage of cotton. The de-regulation of the cotton processing sector increased competition for raw cotton and SMCs were unable to secure adequate supplies. The cotton market became chaotic, with reports of panic buying and industrial sabotage. The government abandoned the reform programme and re-instituted stringent administrative control in 1994 in an attempt to restore order to the cotton market. SMCs retained their monopolistic position in the primary cotton market until 1998.

A scheme for paying higher prices for the above-quota deliveries to the State was initiated in 1994. Under this scheme, farmers could be paid above the state-set procurement price, as a means of inducing an increased supply of cotton. Cotton was also covered by the provincial governor responsibility system, implemented in 1995.

The "three links scheme" was applied to cotton, as well as grains and oilseeds, in the early 1990s. Under this scheme, at the time of planting, farmers received an advance payment of 20% of the expected value of their cotton crop. They also received supplies of fertilisers and diesel oil. This scheme was intended to ensure that farmers had the means to produce the desired crop. During the 1990s, the government also established cotton production bases in selected counties.

The government also implemented measures to improve the efficiency of the cotton market, and to guide upgrading of the textile industry. Beginning in 1996, SMCs were allowed to trade cotton directly with local textile mills (partially replacing the state planned allocation of cotton to textile mills) and the national government organised the first national cotton trade fair in Zhengzhou (in central China's Henan province). At the fair, cotton users (textile processors in importing regions who were authorised to import) and sellers (cotton companies of the SMCs in regions where planned out-shipment quotas were assigned) could bid for raw cotton. While the buyers and sellers could choose with whom they would trade, the tradable quantities were still bound by assigned quotas for importing and exporting regions. The two parties to a trade were allowed to decide the terms of trade within a specified price range (state-set cotton allocation price ±4%). These new arrangements allowed the market mechanism to function to some extent.

During the 1990s, Xinjiang became China's largest cotton producing region. Xinjiang is, however, the most western of China's provinces and is far removed from the major cotton processing centres. As a consequence, transportation costs are a limiting factor for cotton production and sale. In response to this geographic difficulty, in 1997 the national government provided a special subsidy for transporting cotton from Xinjiang. The subsidy, of CNY 1.5 per kg of shipped cotton, was only available for the 1997 harvest.

In April 1998, the government changed the cotton purchase price from a government-set price to a guidance price for all newly harvested cotton. This action was intended to overcome the problem of surplus in the domestic cotton market. The guidance price was set at CNY 12.35 per kg (± 5%) and the government committed to purchasing all cotton that farmers wished to sell at that price (Table B.5).

In December 1998, a fundamental reform of cotton marketing system was undertaken (State Council, 1998b). Beginning in September 1999, cotton prices were determined by market forces, while the government issued a reference price late in each year to guide production in the next year. While cotton prices are generally liberalised, the guidance price is also a floor price for cotton for Xinjiang. Xinjiang continues to produce a large proportion of China's cotton crop, producing one third of China's cotton crop in 2003. Thus, government price intervention remains a potentially significant factor in China's cotton industry.

The SMCs' exclusive rights to purchase cotton from producers ceased in 1999. Instead, state farms and qualified textile mills were allowed to engage directly in cotton purchasing, processing and marketing. The government intended to use state reserves as a tool to stabilise the cotton market in conjunction with a state controlled international trading arrangement for cotton. Development and maintenance of national reserves is funded by special loans from the Agricultural Development Bank of China, with costs being borne exclusively by the central government. Provincial governments were permitted to decide whether to establish their own regional reserves or not. Further reforms of SMCs were also included in the decision – it is intended to transform the SMCs from state institutions to producer co-operatives. The reforms of the cotton marketing system introduced in late 1998 remain in place in China.

Implementation of the new policies led to a significant fall in regional variation in cotton prices within China. For instance, the purchase prices of one kilo of new raw cotton in September 2002 was CNY 8.6 in the Yangtze river area, CNY 9.0 in central China and CNY 7.4 in Xinjiang, a price diversity that better reflects regional variations in demand and supply.

Table B.5. **The state-set prices of cotton**
Unit: CNY/kg of standard grade

Marketing year	1990-91	1991-92	1992-93	1993-94	1994-95	1995-96	1996-97	1997-98	1998-99	1999-00	2000-01	2001-02
Planned	6.00	6.00	6.00	7.28	10.00	14.00	14.00	14.00				
Guidance									12.35	10.00	8.00	9.00
Actual price										7.62	10.34	7.20

Source: NDRC.

Border measures

Until 2001, China's trade in cotton was subject to tariffs and state-trading. Under the state-trading arrangement for cotton, initially only the China National Textiles Import and Export Company (Chinatex) had the right to engage in international cotton trade on behalf of China. At the beginning of the 2000s, the monopoly was broken with a number of other STEs being given the right to trade in cotton, but trade was still controlled by the State in terms of quantities imported and exported.

Despite the existence of state-trading, China established a tariff of 3% for cotton imports in the 1990s. In early 2001, a TRQ was implemented with an in-quota tariff of 3% and an over-quota tariff of 90%.

China agreed to reduce the in-quota and over-quota tariff rates upon accession to the WTO. China also agreed to reduce the over-quota protection and increase the quota volume in its WTO transition period and agreed that two-thirds of import activity could be undertaken by non-STEs (Table B.6).

Table B.6. **China's WTO cotton trade commitments**

		2002	2003	2004	2005
Cotton	In-quota tariff	1%	1%	1%	1%
	Over-quota tariff	54%	47%	40%	40%
	STE share of quota	33%	33%	33%	33%
	Quota (mil. tonnes)	0.82	0.86	0.89	0.89

Sources: WTO; NDRC (cited by USDA).

The importance of cotton as an input to China's industrial sector is highlighted by the TRQ fill rates. In 2002, cotton imports were only 22% of the available quota, while in 2003 the fill rate was 102% – a quantity greater than the quota was imported at the in-quota tariff rate. This appears to be a response to a domestic shortage of cotton in 2003 (Butterworth and Wu, 2004), and signals China's intention to allow market forces to maintain an equilibrium in the domestic cotton market.

In order to promote cotton exports, the government in 2002 decided to exempt these from VAT. Under this arrangement, VAT reimbursement can be claimed at the point of export. Cotton exports are also subject to state-trading. Exports of cotton are not expected to be significantly large, however, in the near future as a result of strong domestic demand and the limited capacity to increase domestic production (Butterworth and Wu, 2004).

Producer support trends

In the early period of implementation of the 1998 reforms, producer prices fell significantly, even below border reference prices. As a result, cotton producers were implicitly taxed in 1999 and again in 2001. Recently, the expanding apparel industry has stimulated demand for cotton and cotton imports increased quickly, reaching over the TRQ level in 2003. As cotton domestic prices are generally liberalised, they adjusted to the new market situation by doubling between 2001 and 2003 (Figure B.8). As a result, the %PSE increased strongly from –13% in 2001 to 53% in 2003.

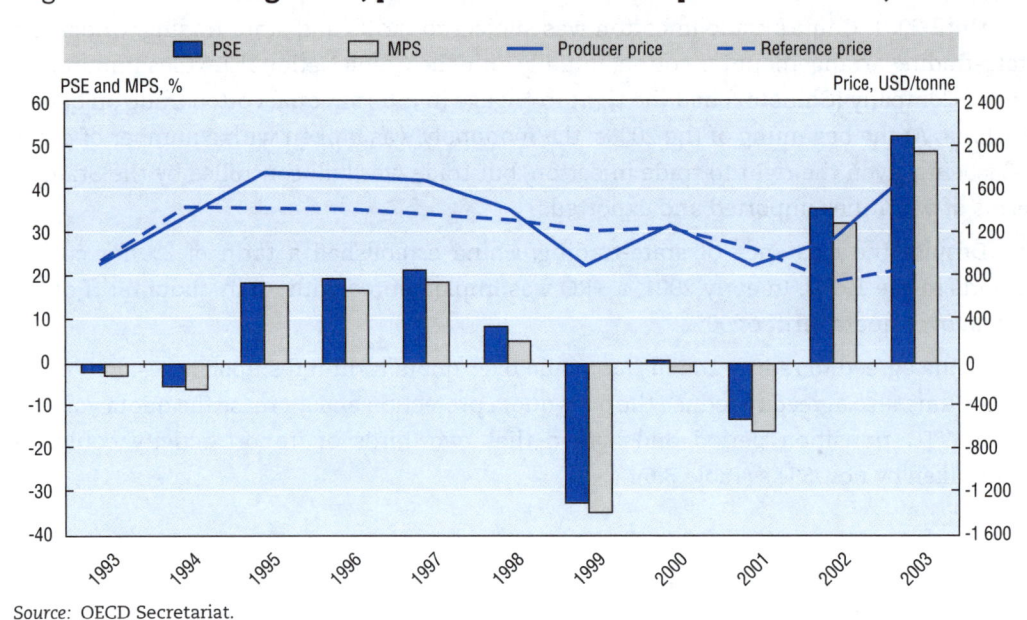

Figure B.8. **Percentage PSEs, producer and reference prices for cotton, 1993-2003**

Source: OECD Secretariat.

Tea

Domestic policy

China is one of the world's major tea producers. While tea is an important commodity for both Chinese domestic consumption and exports, there are few specific domestic policies regarding tea. During the period 1990-2004, most government support was in the form of assisting research on production technologies. Distribution and sale of tea to some remote regions occurred under a state-pricing scheme until 2001.

In recent years, safety of tea products has become a public concern and government involvement with tea production has been focused on improving safety inspections of the commodity. In major producing areas, local governments have encouraged farmers to establish large-scaled production bases and to produce safer products. Organic tea products are promoted and there is encouragement for farmers to adopt production techniques that require reduced inputs of hazardous chemicals.

The main government intervention in this sector has been in the form of taxation. When the tax on special agricultural products was introduced in 1983, tea was one of the farm products subject to this tax, the rate being determined by provincial governments. In 1994, the national government revised the Special Agricultural Tax regulations and set a 16% tax rate for tea. In 1997, the rate was reduced to 12%. Since 2004, this no longer applies to tea.

Border measures

Tea is an important part of the Chinese diet and trade in tea has been controlled to ensure that an adequate domestic supply is maintained. Throughout the period 1990-2004, China exported significantly greater quantities of tea than it imported. According to FAO data, the ratio of volume of exports to volume of imports for 1990-2002 averaged around 17:1, and the ratio of value of exports to imports averaged approximately 25:1.

During this period, tea imports were subject to a tariff and to state-trading arrangements. The latter constituted a monopoly on tea trading, with trading rights given to the China National Native Products and Animal By-Products Import and Export Company.

The tariff on tea imports declined from 80% in 1992 to 27% immediately prior to China's WTO accession. Accession to the WTO led to China further reducing the tariff on tea imports. In 2002, the tariff was reduced to 21%, dropping to 18% in 2003, and 15% in 2004 (the final bound tariff).

Tea exports have remained subject to state-trading. According to MOFCOM's 2003 Catalogue of Commodities Subject to Export Licensing Controls, unfermented tea (green and oolong) trading is a state-trading monopoly. The WTO accession protocols list the China National Native Products and Animal By-Products Import and Export Company as the designated STE for tea.

While trade in unfermented tea is tightly controlled, black tea is not subject to state-trading. Most black tea is produced for export.

Tobacco

Domestic policy

The tobacco industry in China operates under a state monopoly, with rigid state control over tobacco processing, cigarette manufacturing, and sale of tobacco products.

During most of the 1990s, tobacco was subject to state pricing. The State Planning Commission and the National Tobacco Bureau jointly set the state purchase prices for medium grade tobacco and the National Tobacco Bureau then determined the purchase prices for different varieties and all grades.[7] Provincial governments were not allowed to adjust prices. However, in practice, they did provide production subsidies that effectively increased the state set prices. In 1999, the State Council issued an instruction requiring all provinces to remove local subsidies, while the state purchase price was raised from CNY 4.84 per kg in 1998 to CNY 7.0 per kg in 1999. The government continued to set purchase prices until 2001.

The Law of Monopolized Marketing of Tobacco was enacted in June 1991 and covered all tobacco-based products. It laid the basis for state management of production, marketing and trade of tobacco-based products.

In order to generate budgetary revenue as well as to dampen consumption, the government imposed high taxes on tobacco products. Tobacco was subject to a 31% tax rate under the Tax on Special Agricultural Products, the highest rate among all taxable agricultural products. This rate was reduced to 20% in 1999. From 2004, tobacco is the only product covered by this tax and the rate remained at 20% (Table 2.6 in Chapter 2). Moreover, production and marketing of tobacco-based products is subject to value-added tax at a rate of 17%. In addition, cigars and other tobacco products are subject to a consumption tax. Before 1999, the rate of consumption tax was 40% for all types of products. In 1999, a multi-rate consumption tax for tobacco products was introduced with rates ranging from 25% to 50%.

Border measures

Tobacco trade policy during 1990-2004 consisted of reducing tariff rates and maintaining the state monopoly on tobacco and tobacco products.

The State Tobacco Monopoly Administration has the monopoly right to market tobacco in China and controls imports and exports according to domestic marketing requirements. In addition to the domestic monopoly on tobacco, China applies a tariff to imports. The unmanufactured tobacco tariff fell from 50% in 1992 to 34% immediately prior to WTO accession. In 2002, the tariff was further reduced to 22%, declining to 16% in 2003, and 10% in 2004. Under the terms of China's WTO accession protocol, tobacco is agreed to be subject to state-trading arrangements.

As a result of the monopolistic nature of the tobacco processing and marketing sector in China, exports of tobacco are state-controlled.

Fruits and vegetables

Domestic policy

The market for vegetables and fruits was liberalised in the mid-1980s. Since then, the government has made very few interventions in the market. The "vegetable basket project", launched in 1988, included a component to promote vegetable production and thus improve supply of vegetables to urban markets. Under this project, the government invested in marketing facilities in both producing regions and urban areas, developed information networks, and established quality inspection stations. Shipment of vegetables has been given priority in the transport system to facilitate the distribution of these perishable foodstuffs.

During the late 1980s and early 1990s, relatively high returns to fruit production raised concerns that land previously sown to grain, may be withdrawn from grain production to be used for fruit growing. The introduction of the Tax on Special Agricultural Products in 1985 was intended to prevent such a consequence. The rate of this tax for oranges and apples was 15% in 1989 and 12% after 1994. From 2004, the Tax on Special Agricultural Products no longer applies to fruits and vegetables.

As a result of continuing concerns regarding the transfer of crop lands to other uses the State Council issued the Regulation on the Protection of Basic Farmland in 1998 and reinforced the terms in 2004. The Regulation forbids the transfer of land from grain production unless special approval from the Ministry of Land and Resources is obtained. This Regulation creates a barrier to the development of new orchards, ponds for aquaculture, commercial forests, and prohibits the use of arable land for development of intensive livestock enterprises.

In response to the growing awareness of food safety issues, the government has tightened regulations on the use of pesticides in fruit and vegetable production, and promotes and certifies "organic" products and "green food" products (produced using minimal inputs of hazardous chemicals).

The state has also been active in disseminating advanced production and product handling techniques.

Border measures

Trade policy for fruits and vegetables has varied only slightly during the period 1990-2004. Throughout the period, horticultural products have not been subject to state-trading or to product specific licensing arrangements, but exports have been subject to general trade business licensing (Chapter 2), and imports have been subject to tariffs.

Throughout the period 1990-2004, imports were subject only to an *ad valorem* tariff. Different tariffs applied to different products, but the trend has been for the level of tariffs to decline. In 1992, for example, fruits typically had a tariff of 80%, which declined to 30-40% in 2001, 13-20% in 2002, and 10-13% in 2004. Vegetables have followed the same trend, with the tariffs being 30-50% in 2001, 13-29% in 2002, and 10-15% in 2004. Clearly, WTO accession in 2001 coincided with an accelerated decline in the tariff rate.

Producer support trends

Apples are the only commodity among all fruits and vegetables for which a complete set of data has been collected to calculate the level of support (Figure B.9). This is an export-oriented product in which China has a comparative advantage. Reflecting lack of any particular market price support policies for apples, the price gap has been set at zero (Box 2.7 in Chapter 2). The measured level of support, with %PSE ranging between 1% and 4%, does not reflect any specific support for apple producers, but rather more general measures such as input subsidies provided to crop producers.

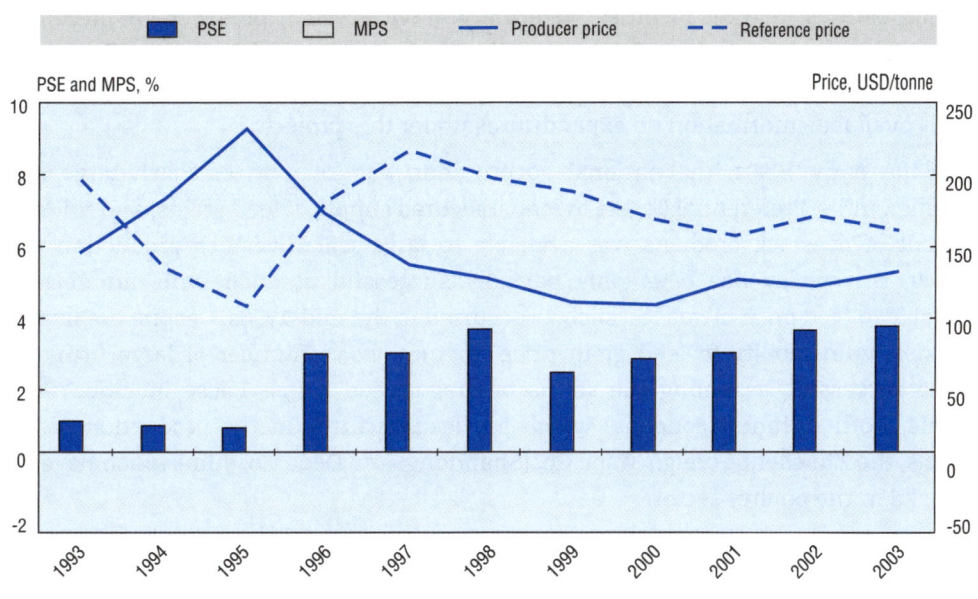

Figure B.9. **Percentage PSEs, producer and reference prices for apples, 1993-2003**

Source: OECD Secretariat.

Livestock products

Domestic policy

The livestock products markets were largely liberalised in 1985 and producers have since made decisions on production and sale of livestock products based on market conditions. During the late 1980s and early 1990s, the government still exercised some intervention on prices of pigs in rural markets, and on prices of pork, eggs and milk in urban markets, however, the effects of these interventions tended to weaken over time. There has been no formal policy announcement to withdraw these price interventions, but they have been redundant since 1997 as a result of the decline in food prices.

China has no tradition of designing policy programmes for specific livestock products. However, sub-national governments have tended to adopt measures to support development of certain livestock products, based on regional comparative advantages. Examples include the development of the dairy industry in Inner Mongolia and Heilongjiang, development of the beef cattle industry in Henan and Anhui, and development of the poultry sector in Shandong and Jilin. Assistance provided to these industries includes preferential access to loan funds, a government-secured supply of feed grains, and subsidies for purchasing equipment and breeding stock.

During the period 1990-2003, the government collected an Animal Slaughtering Tax when animals were sold. Sub-national governments determined the rate of this tax. In principle, this tax was to be paid by the buyers but deviations from that principle were widespread. It is reported that in some jurisdictions, Animal Slaughtering Tax obligations were calculated according to the land area contracted by a household, while in other jurisdictions, a government revenue target under this tax was simply allocated to all households, whether they raised livestock or not. As noted in Chapter 2, the Animal Slaughtering Tax ceased to be collected by 2004.

Under the "vegetable basket project", launched in 1988, the government assisted development of the livestock sector through direct investment in supporting facilities (breeding centres, market facilities, veterinary services, etc.), provision of preferential loans to producers, extension of new technologies, and preferential tax rates for the feed industry. This project is still in place, currently funded at the provincial level. There is no publicly available information on expenditures under this project.

In the early 1990s, the national government was keen to establish large-scaled production units. Preferential access to loans, secured supply of feed grains, and priority in shipment of livestock products was available to large-scale livestock production units. However, this policy has been only partially successful as such units are relatively vulnerable to feed price increases, such as occurred in the mid-1990s. Despite the livestock industries' vulnerability to feed grain price fluctuations, a number of large firms have become successful, assuming the status of dragonhead firms. These include Yili and Mengniu (both in Inner Mongolia) which are leading firms in the production of dairy products, the Zhucheng Foreign Trade Co. (Shandong) and Deda Co. (Jilin) which have been successful in the poultry sector.

Faced with feed grain supply constraints, the government adopted a suggestion in 1992 to develop a straw-fed cattle production system and extend successful techniques across the grain producing regions. As a result, a treatment process which improves the digestibility of various types of straw and other non-grain feed materials has been made available, meaning grain field stubble can now be used for feed rather than burnt or ploughed under. This development has been concentrated in the crop regions of central China. To some extent, the extension of this technique has helped the formation of the "beef-belt" in Henan, Hebei, Shandong and Anhui.

In 2001, the Ministry of Agriculture made a proposal to accelerate development of the livestock sector. It highlights ideas on possible adjustments of production structures and regional distributions of livestock production, and makes suggestions on how best to increase government support to the livestock sector (MOA, 2001). The document is a statement of intent to develop the livestock industries, and no firm actions or detailed plans have been published.

In order to meet the food safety requirements of importing countries, set on the basis of the WTO SPS and TBT Agreements, as well as to improve food safety in the domestic market, the national government has recently begun to improve the regulation of livestock production practices, particularly the use of feed additives. In December 1997, the State Council issued a regulation on live pig slaughtering, requiring that all animals be slaughtered at designated slaughterhouses, to ensure quarantine inspection, and appropriate tax collection.

Beginning in 2001, the Ministry of Agriculture began to collaborate with provincial governments to establish disease-free areas in selected regions. Shandong, Liaoning, Sichuan, Jilin and Hainan provinces and Chongqing municipality (formerly part of Sichuan), were initially selected for demonstration projects. Measures being tested include controls on stock movement, livestock vaccinations, and the creation of isolation areas and buffer zones to prevent the transmission of diseases.

It is difficult to assess the impact of China's recent efforts to control animal diseases, with diseases of poultry in particular continuing to manifest in 2003 and 2004. China continues to rely on vaccinations to manage a wide range of animal diseases.

Border measures

Trade policy for livestock products has varied only slightly during the period 1990-2004. Throughout this period, livestock products were not subject to state-trading but imports were subject to tariffs. In addition to tariffs, imports were also subject to a range of sanitary measures designed to minimise the risk of livestock disease incursions.

Imports were generally subject to *ad valorem* tariffs, although some poultry products have been subject to a specific tariff since 2000-2001. Different tariffs applied to different products, but the trend has been for the level of tariffs to decline. In 1992, for example, meats typically had a tariff of 50%, which in 2001 declined to 40% for beef, 22% for sheepmeat and 20% for pork and poultrymeat. In 2004, the tariffs ranged from 15% for sheepmeat to 20% for beef, pork and poultrymeat. Live animal imports have followed the same trend, with the tariffs being 20-40% in 1992, and 10% in 2002-2004. Dairy products and eggs have followed the same trend.

There have been no livestock specific export-related policies in place during the period 1990-2004; although exporters have been required to meet the minimum requirements stipulated for businesses engaged in foreign trade (Chapter 2).

Producer support trends

There is an important difference in support trends for exportable and importable livestock products. For exportables such as beef and veal, pigmeat, poultry and, to a smaller extent, eggs producer prices fluctuate around border prices reflecting time-lags in adjustments, but in a longer term both trends and levels of the two price series align (Figures B.10, B.11, B.13 and B.15). As for these products there are no identifiable market price policies and export subsidies are not applied, the price gap has been set at zero (Box 2.7 in Chapter 2). A taxing effect of feed adjustment, mainly due to higher producer prices for maize than border prices, results in a small negative %PSE for beef and veal and pigmeat, not compensated by limited budgetary support provided to livestock producers. %PSE for poultrymeat and eggs is marginally positive.

Figure B.10. **Percentage PSEs, producer and reference prices for beef and veal, 1993-2003**

Source: OECD Secretariat.

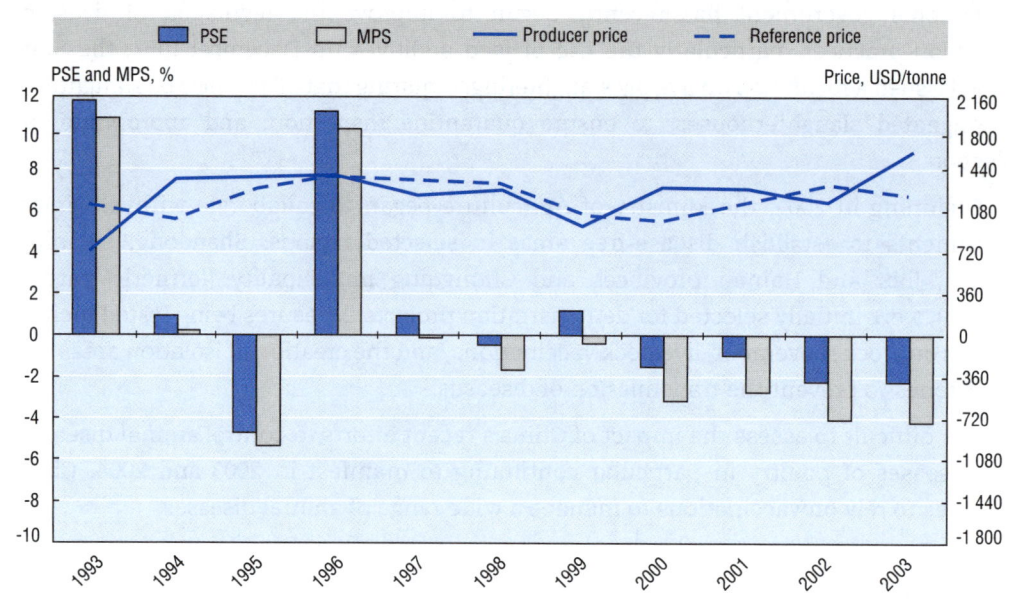

Figure B.11. **Percentage PSEs, producer and reference prices for pigmeat, 1993-2003**

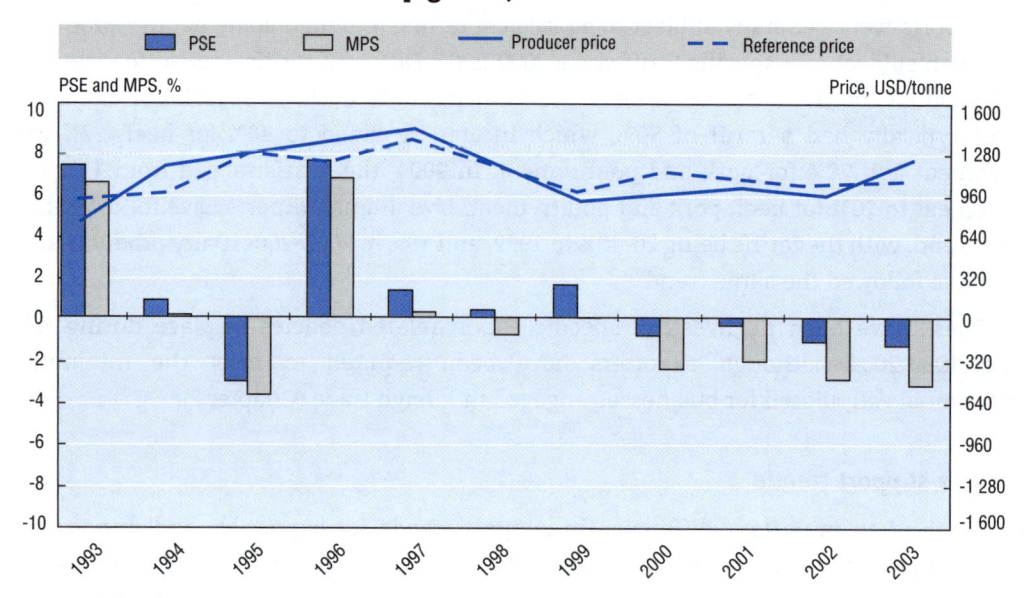

Source: OECD Secretariat.

Figure B.12. **Percentage PSEs, producer and reference prices for sheepmeat, 1993-2003**

Source: OECD Secretariat.

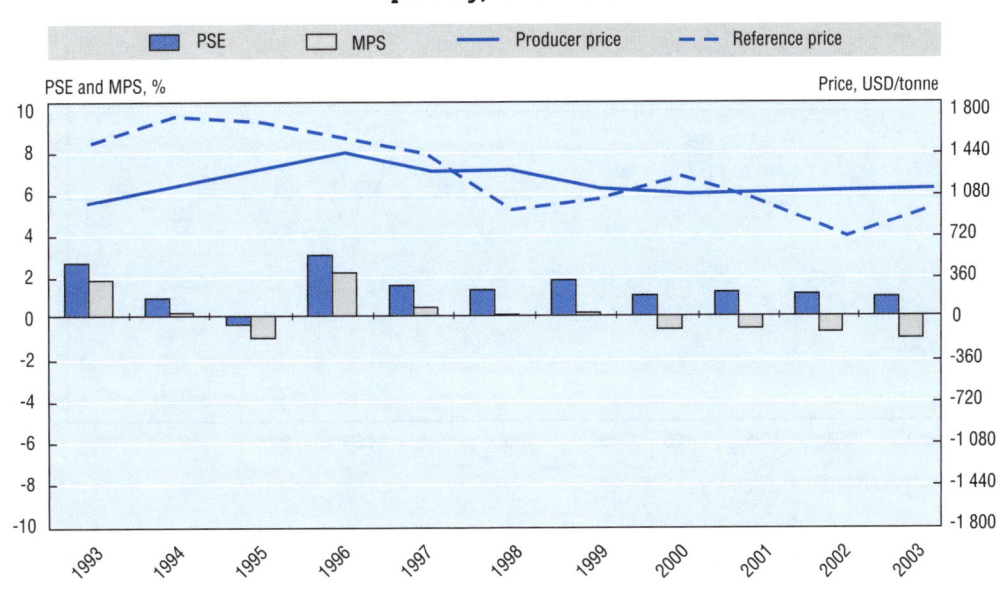

Figure B.13. **Percentage PSEs, producer and reference prices for poultry, 1993-2003**

Source: OECD Secretariat.

Figure B.14. **Percentage PSEs, producer and reference prices for milk, 1993-2003**

Source: OECD Secretariat.

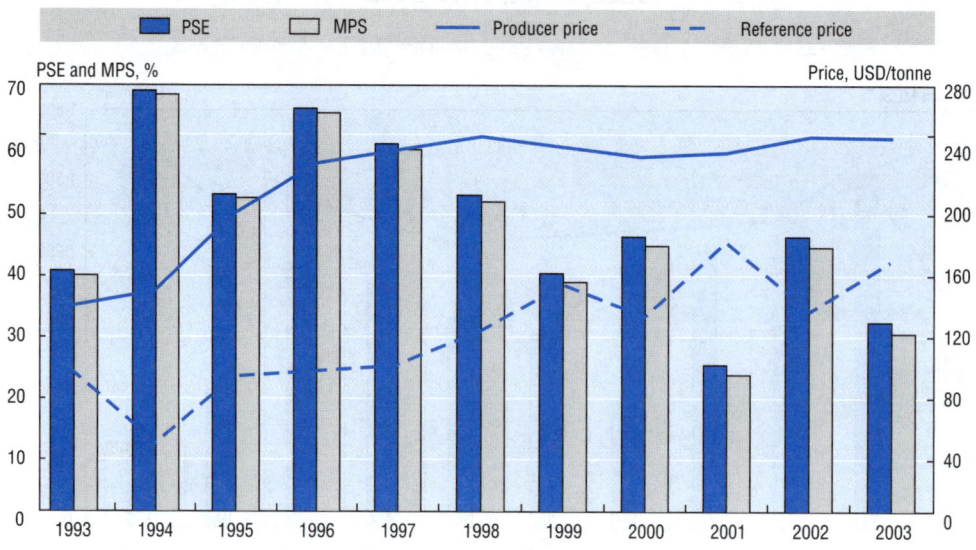

Figure B.15. **Percentage PSEs, producer and reference prices for eggs, 1993-2003**

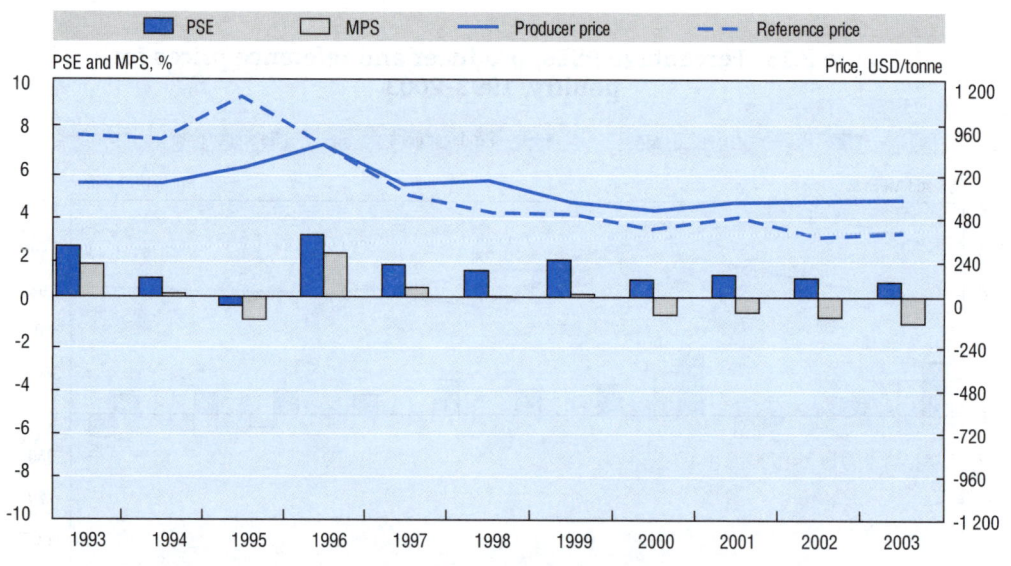

Source: OECD Secretariat.

Importables such as sheepmeat and milk enjoy a relatively high level of support, almost exclusively through border protection measures. The average %PSE for milk was high at 36% and for sheepmeat at 31% between 2000 and 2003 (Figures B.12 and B.14).

Notes

1. China's urban consumers were entitled to purchase a ration of grains at state-set retail prices. Quantities in excess of the ration could be purchased at free market prices. The government also decided to provide urban consumers with a lump-sum subsidy as partial compensation for the price rise.

2. The average price of wheat flour, rice and corn of medium grade was raised by CNY 0.2 per kg in 1991 and CNY 0.22 per kg in 1992. The prices of vegetable oils were also raised by an average CNY 2.7 per kg in 1991.
3. Those regions which did not produce adequate quantities of grain were prone to significant supply fluctuations.
4. The rate of operating cost subsidy for overstocked grains was set at CNY 120 per tonne in 1997.
5. While Figure B.7 reflects farm gate prices and border FOB prices for peanuts adjusted to the farm gate level, the price gap for peanuts has been set at zero as no export subsidies and no other market price support policy has been identified (see Box 2.7 in Chapter 2).
6. As explained in Chapter 2, in China, "strategic" agricultural commodities are: cereals, oilseeds and vegetable oils, sugar, cotton, tobacco, tea and silk.
7. Tobacco prices are set for 5 regions and over 40 grades.

ANNEX C

China's Approach to Food Safety*

Introduction

China has traditionally been preoccupied with producing the maximum quantity of food possible to feed its large population. For example, high yield varieties of crops were favoured over high quality ones. Today, China has achieved basic self-sufficiency in food and the country has changed the emphasis from quantity to safety of food. This emphasis on safer food has been given a boost by China's membership in the World Trade Organisation (WTO). The changes required at the field level to produce safer food for export also benefit the domestic market. For example, skills acquired to allow a more targeted use of pesticides and animal medications to avoid unsafe residues in export foods can also be used to reduce residues in domestic food.

The domestic food safety system

Overview

Food standards exist at the national, local and enterprise levels. They are managed at the government level by a national commission on the management of standards, with the co-operation of the Ministries of Health, Agriculture and the National Administration for the Quality Supervision, Inspection and Quarantine (AQSIQ). Quite a few standards in China are below international levels. For example, maximum residue limits for pesticides are not always compatible with international standards. Further challenges arise from the fact that some important standards have not yet been formulated and some old standards need to be revised. China is making a significant effort to reform and upgrade its standard management systems and that its standards are based on good science. Special emphasis is being placed on developing tolerance levels for pesticides, animal medications, important organic contaminants, food additives, feed additives, and certain biological agents. However, there is still much to be done to reach compatibility with the international standards relevant to international trade. By the end of the Tenth Five Year Plan (2005), China hopes to establish a new standards system for the food industry designed to meet the requirement of imports and exports, and in conformity with the international food standards system. Specifically, it is planned that for agricultural

---

* This annex draws on "Study on China's National Food Safety Strategy", a Project Group of the Development Research Centre of the State Council, PRC, distributed at the Global Food Safety Forum, Beijing, 18-19 November 2004.

products 50% of Chinese standards will conform with international standards and for processed food products the rate will be 55% by the end of 2005 (DRC, 2004).

The mandate for food safety in the domestic food chain remains divided between ministries, which include, the Ministry of Agriculture for food production, the Ministry of Health for food safety, and the Ministry of Commerce for food distribution. This mandate is further split between these central government ministries and similar units at the provincial level. It should be noted that provincial staff are under the control of the provincial governments as they are hired and paid for by provincial authorities. This division of responsibilities has important implications for food safety. For example, suppose a food poisoning event occurs domestically at an event where a catering company has provided the food consumed and persons have been poisoned following the consumption of that food. Since people have suffered food poisoning, the Ministry of Health may be charged with the subsequent investigation. However, the causal factor may be contamination at the food production level of the food chain. To correct such a problem at the food production level the mandate falls within the responsibilities of the Ministry of Agriculture. Indeed, it may well be the provincial agriculture staff that has to take corrective action. Thus, corrective action could require co-ordination between the Ministry of Health, the Ministry of Agriculture and provincial agencies. Corrective action under such circumstances is arguably more difficult than if a single food inspection agency had both central and provincial control of food safety for both the domestic and international market.

A significant strength that China has as it works to improve its systems for ensuring food safety and quality is that all parties seem to accept the concept of "farm-to-table" food system. This concept suggests that ensuring food safety requires dealing with food production, transportation, storage, processing, and distribution as a single integrated system. For example, all parties seem to accept that a contaminant at the farm level can affect food safety at the consumption level. All seem to accept that safety cannot be "inspected" in the final stage of food production; safety has to be built in at each step of the food chain. This means that while the mandate for food inspection is divided, there is a general consensus that ministries and provinces must work in a co-ordinated way if food safety is to be ensured. In order to facilitate co-ordination, China has recently established a body to co-ordinate the actions of all government agencies involved in food safety. This new body is known as the State Food and Drug Administration (SFDA, see below).

Legislation

China has legislation affecting each major phase of the food chain from farm to table. Of key importance to food safety are the Food and Health Law, the Agriculture Law and the Standardisation Law which set out general food safety requirements. These laws were drafted at a time when the focus was on producing food in greater quantity and need to be amended if they are to meet current consumers' food safety needs. Problems also exist due to gaps and duplication in the legislation. The laws contain somewhat different requirements for food safety and health. As a result, those enforcing the laws have difficulty understanding exactly which requirements they should follow when implementing legislation. In addition, the penalty provisions for violation of the laws are, arguably, too low. These problems can be compounded by inconsistent application between the various government agencies.

The current body of legislation covers the following domains:

- *Food and Health Law*: Under the authority of the Food and Health Law are found central government regulations, local government regulations, food and health standards, and inspection procedures. Following China's entry into the WTO, the Ministry of Health undertook a revision of the food and health standards governed by this Act. This revision covered food safety, food additives, food containers, food packaging, and provisions affecting food-catering premises. It also clarified some mandate issues in relation to food safety inspection.

- *Agriculture Law*: The Agriculture Law has as its purpose the regulation and economic development of the agricultural sector, and the protection of farm workers.

- *Standardisation Law*: The Standardisation Law sets out standards for many sectors including food. It provides guidance on the formulation and implementation of standards.

- *Supporting Legislation*: The above pieces of legislation are supported by the Inspection of Imported and Exported Goods Law, the Prevention of Animal Epidemics Law, the Quarantine of Animals Crossing the Border Law, the Border Health and Quarantine Law, the Environmental Protection Law, and the Protection of the Consumers Rights and Interest Law.

The existence of this portfolio of legislation, and its associated problems, has led to discussion in China suggesting that the country would benefit from having a single Food Act. This could be designed to eliminate legal gaps and overlaps, as well as to provide some clarification of legal mandates between Ministries.

Certification

In order to meet the demand for safer food, China has instituted control programmes starting at the farm level. These include the programmes described hereunder:

- *Hazard-Free Agricultural Products Certification*: This system is sometimes known as "Non Public Hazard Food" and is basically of Chinese origin. It is designed to help ensure that foods are produced, for example, at the farm level in a manner that is not likely to present a hazard to consumers. In 2003, the Ministry of Agriculture listed the first set of products that had gained certification as "Hazard-Free Agricultural Products". This status was awarded to about 200 products from about 150 local production areas. These products are identified with a special logo authorised by the Ministry of Agriculture's product safety certification centre.

- *Green Food Certification*: While similar programmes exist in other countries, this system is, again, basically of Chinese design. The Green Food certification system was initiated in 1990. Its focus is on producing safe food in an environmentally friendly manner. The Chinese Green Food Development Center has responsibility for this programme and it operates through the delegation of programme implementation to institutions that it accredits. Several thousand products and food enterprises have been accredited to date.

- *Organic Food Certification*: This system, created in 1994, is based on international practice and its purpose is to ensure that organic food production techniques are used. Overall control of this programme is in the hands of the Ministry of Agriculture, which delegates certification authority to third party certification institutions. At present, a relatively small number of products and food enterprises have received certification.

- *Hazard Analysis, Critical Control Point (HACCP) Certification*: HACCP is internationally recognised. Increasingly, it is becoming required practice in national food safety systems. As its name suggests, it involves identifying the "hazard" to be controlled, and then focusing on ensuring safety at the "control points" that are "critical" to ensuring product safety. China's State Commission on Supervision of Certification issued regulations on HACCP certification for food manufactures in 2002. This provided China with a base from which to continue building its HACCP system. The Ministry of Agriculture has placed emphasis on making this system available to the aquaculture industry, given this sector's importance in China. Nevertheless, implementation of HACCP at the domestic market level has faced difficulties as the HACCP systems in place have, in some cases, not been tailored to Chinese conditions. Certification by some parties has been of questionable quality. In addition, in relation to the size of China's food system, few enterprises have implemented HACCP.

Other internationally recognised systems used to help improve product safety include Good Manufacturing Practice (GMP), Good Laboratory Practice (GLP), Total Quality Management (TQM), International Standards Organisation (ISO) 9 000 series, and ISO 14 000 series. These are being used to varying degrees in China. While these certification programmes have the capacity to improve food safety in China, they do face difficulties. There is a need for institutions that are designed to provide guidance and training to those wishing to obtain certification. Without good advice, it is difficult for food enterprises to gain certification, especially internationally recognised certification. Even when certification is received, its value can be diminished by lack of mutual recognition between institutions. Some food buyers will accept certain certificates, but not others. From a domestic point of view, these certification systems have mixed levels of public recognition and acceptance.

Institutions

In the central government, the management of food safety is jointly in the hands of the SFDA, the Ministry of Health, the Ministry of Agriculture, the AQSIQ, and the Ministry of Commerce. These departments are independent of each other. Mirror image organisations exist at the provincial, municipal and county levels.

- *State Food and Drug Administration (SFDA)*: Directly affiliated with the State Council, the State Food and Drug Administration is responsible for the oversight of the safety of food, drugs, health foods, and cosmetics. Given that China's food safety mandate is divided between several central government ministries and their provincial counterparts, this new body has been charged with the overall co-ordination of inspection activities within the food chain. In this role, it is responsible for organising the relevant agencies in order to co-ordinate the division-of-work in the food chain.

- Independent Provincial Food Inspection departments have corresponding units at the provincial, municipal and county levels. With the exception of AQSIQ, these departments are generally under the "line" control of the local governments and under the "functional" control of the central government. The local units are, for the most part, hired and financed locally. This can lead to differences in priorities between central and local officials.

- *Ministry of Health (MOH)*: The Ministry of Health is responsible for overall food safety and public health policies. This Ministry is responsible for formulating the overall regulations and standards governing food safety and public health, and for monitoring food processing and food distribution. It is also charged with food contaminant monitoring and for monitoring outbreaks of food-borne diseases.

- *Ministry of Agriculture (MOA)*: The Ministry of Agriculture is responsible for domestic oversight of the quality and safety of agricultural products at the farm level. The Ministry launched a programme in 2001 to promote "Hazard-Free Agricultural Products" designed to improve the safety and quality of agricultural products. It is based on safety and quality standards, and on the implementation of acceptable testing, inspection, and certification practices. The Ministry of Agriculture is responsible for the testing and quarantine of domestic animals and plants. It is also responsible for monitoring agricultural inputs and the environment of agricultural food producing areas.

- *National Administration for Quality Supervision, Inspection and Quarantine (AQSIQ)*: AQSIQ is a ministerial level government agency responsible for the control of food safety in the production and processing of food destined for international trade. It has "line" control over its own staff, at both the central government and provincial levels. It is self-contained in that it has its own inspection staff and laboratories at all levels of the food chain.

- *Ministry of Commerce (MOFCOM)*: With a mandate in the domain of food transportation and distribution, the Ministry of Commerce is responsible for establishing and improving the inspection system for food safety. It is also responsible for monitoring the sanitation, safety and quality of food circulating in the market, and of agricultural products under transport for export. It has launched a campaign to create a green market, and has worked to standardise and improve food transportation and distribution. A market place inspection system, where product sampling is being conducted, is being established. This system will focus on large wholesale markets for agricultural products.

- Two Divisions under the department of WTO Affairs in the MOFCOM have responsibility for managing China's "Inquiry Point". These are the Division of Inquiry and Technical Support and Division of Review and Notification. Because of China's membership to the WTO is only recent, it is still in the "early days" in the development of this system. The usefulness of China's "inquiry points" is, however, facilitated by the existence of a single import/export agency, AQSIQ, at the operational level on which MOFCOM can contact for detailed information. For notification procedures, MOFCOM's Division of Notification has the lead.

Other related institutions include the State Administration for Industry and Commerce which is responsible for ensuring good business practices in commercial transactions. It is responsible for the supervision of business practices throughout the food chain. This organisation conducts an inspection and approval process before a health and safety permit is issued to a food-based business. The Ministry of Science and Technology has duties associated with scientific research. To help improve food safety, the Ministry established a programme known as Key Technologies in Food. Its purpose is to fund research projects that can help improve food safety. The Chinese Green Food Development Center has delegated authority to deliver its programme, at the field level, to a number of operational bodies. The system imposes standards on the food production environment and methods, and on product quality, packaging, and storage. In 2002, the National Center of Prevention and Control of the Diseases was established to help with public health emergencies.

Implementation and compliance

In 2000, China had about 171 800 enterprises producing food. Of these, about 60 085 were under AQSIQ inspection. About 95% were small enterprises with less than 100 workers, and about 80% of these had less than 10 people. The small scale of the industry and its geographically dispersed nature make supervision difficult. In addition, given the lack of an adequate highway network and appropriate infrastructure for cold storage and transportation, ensuring food safety is a major challenge for this industry.

From 2000 to 2002, China's Center for the Prevention and Control of Diseases conducted testing for contamination of raw meat, cooked meat, milk and dairy products, aquatic products and vegetables. The results demonstrated that microbiological contamination was the number one cause of food poisoning, followed by chemicals and toxins. China has been working hard to improve this situation. In this context, it is interesting to note that over a similar time frame, the percentage of vegetables with residue levels above the tolerance level dropped from 37.5% to 15%. AQSIQ recently conducted a survey of food enterprises and found that 82.5% had no capability of in-house food testing. This finding suggests that, for the immediate future at least, there may be limited scope for self-regulation by the average food industry enterprise.

As mentioned previously, different government departments involved in food chain compliance administer their affairs independently and, as such, it is difficult to co-ordinate activities. It is not uncommon to find one subject covered by two standards. Different departments sometimes inspect the same enterprises and the same products according to their own departmental standards. As a result, food enterprises do not always know which standard to comply with. The underlying principles of many standards are, in general, not based on scientific risk assessment principles. The problem is especially important in the area of testing food for residues of harmful substances.

It is difficult to conduct effective animal disease control in China because of the geographically dispersed nature of animal husbandry and the vast area of the country. This is compounded by the fact that China has a land border with many countries, across which animal disease can spread. In China, animal diseases such as tuberculosis, foot and mouth disease, swine fever, and Newcastle disease continue to present a challenge. Until China's trading partners are convinced that diseases such as these are controlled in accordance with international trade standards established by the International Office for Epizootics (OIE), exports of cattle, sheep, hogs, poultry and their fresh products will be severely limited.

A significant proportion of China's animal medication plants have not been registered. Supervision of production and marketing is limited, as is testing food products for residues. Residues in meat of drugs, such as valium, olaquindox, terramycin, aureomycin, clenbuterol and furazolidone, are of particular concern and can result in rejection of a product in the export trade.

AQSIQ has introduced HACCP as a food safety enhancing measure. AQSIQ has also introduced a "market access" system. This requires that special pre-qualification requirements be met before an enterprise is allowed to enter the food market. The Department of Health has introduced a "performance measurement" system that involves awarding enterprises grades to indicate their level of regulatory compliance. Despite these current compliance efforts, bacterial contamination, chemical residues, and food-borne disease still present a threat to China's food system.

The central ministries and provincial officials have been very active in training their food system staff, both in Chinese and in overseas institutions. Training is being given in the functioning of international bodies such as the WTO, Codex Alimentarius, the OIE, and the International Plant Protection Convention (IPPC). Training is also being provided in technical areas such as Risk Assessment, HACCP, laboratory testing, and Good Laboratory Practice (GLP). AQSIQ has been especially active in offering training to their staff. International agencies, such as the World Bank, and individual countries have also been working with China on a bilateral basis. China is also drawing on the experience of developed nations such as the United States, the EU, Japan, and Canada, which are important trading partners for China.

Food safety and trade

Generally speaking, the Chinese food certification systems are not well integrated into the international system. From an export point of view, this has limited the value of certification. To overcome this problem, enterprises wishing to export from China often have to hire foreign institutions to conduct the necessary examination of the enterprise, and provide internationally accepted certification. This can impose a significant extra cost on both Chinese exporters and suppliers to large food supermarket chains operating in China.

China has recently become a member of the WTO and has taken steps to join the international organisations that set sanitary and phytosanitary standards for trade in crops, animals and food. China is also working hard to draw upon the laws, regulations, standards, guidelines and technologies related to food production, processing and distribution developed by such international organisations as the World Health Organization (WHO), Food and Agriculture Organization (FAO), the OIE for animals/animal products, International Plant Protection Convention (IPPC) for crop/crop products and the Codex Alimentarius Commission (CAC) for food. China has established a group to co-ordinate Codex activities and the Ministry of Health draws experts from across China to provide advice and to represent China's interests at Codex conferences. China has been especially active on files related to the use of food additives, soybean products, aflatoxin, and lead residues in fish.

Membership in the WTO has placed a new emphasis on food import and export issues. China's borders must be controlled to ensure that food exports could be certified with confidence for safety, from farm production, through the food chain right to the port of export. The borders must also be controlled to ensure that imports meet international and Chinese standards. China has created a single food inspection agency for imports and exports. As noted above, the single food inspection agency, AQSIQ, has been given direct control of food produced for export on the farm, in food processing plants, and at border crossings. It has direct control over its own staff, at the farm, food-processing and port levels. For the domestic market, the traditional inspection system, involving division of duties between central ministries and the provinces, remains in place. This organisational change allowed international trade to be inspected and certified without a major disruption of the mandates of the central ministries or provinces.

WTO membership created great expectations in certain sectors of China's agriculture and food industry for new export success in areas such as aquaculture, floriculture, and fruit and vegetables. These expectations have met with mixed success due to problems associated with product safety. Unless China can reduce the problem of chemical residues in crops, control animal medication residues in meat, eliminate animal diseases of international trade significance, and control bacteria in foods, growth of the export market will be unattainable.

To help reduce these problems, Chinese government agencies are working to improve the registration system for farm inputs such as pesticides, animal medications, feeds, and fertilisers. These agencies are also working to improve the certification system for foods produced at the farm level. This is being done with the introduction of the systems mentioned earlier, *e.g.* the Hazard-Free Agricultural Products Certification programme. In addition, as of the late 1990s, AQSIQ has, for example, introduced HACCP in enterprises exporting aquatic products, poultrymeat, red meat, and vegetables and fruit juices.

Again, to help reduce the problems mentioned above, and to help meet the safety demands of the export market, China has favoured the use of "key enterprises" just as it has done for some time in the domestic market. This often involves a food processing company working in partnership with a large number of farmers. The "key enterprise" provides market analysis, marketing skills, capital, equipment and food production, handling, transportation and processing technology. Together, the group can also achieve some economies of scale. In some cases, the "key enterprise" is an association of Chinese and foreign interests. The foreign partner often has a ready market for the product in question in its home country. This provides the operation with good knowledge of the safety requirements of the destination market and the technical capacity to meet them. The system of "key enterprises" is being used to provide competitive products for both the export and domestic markets. This concept is used in China's domestic dairy market to produce competitive products where Yili Dairies, for example, has assembled a group of farmers to supply it with milk. Yili provides the farmers with all the technology required to ensure a supply of safe, high quality milk, and also makes available to the farmers a high-tech milking parlour. This allows sharing of equipment and bulk buying of farm inputs, such as feed and fertiliser. Yili handles all aspects of product marketing.

To generate the food safety data necessary to negotiate certification requirements for exports, the Ministry of Health began conducting in the early 1980s systematic surveys to monitor food safety. This same data is helpful in making informed policy decisions on domestic food safety. The results indicate that samples meeting China's domestic standards have risen from about 60% initially to close to 90% at present (DRC, 2004). It will be interesting to follow the results of these surveys as China gradually moves to the adoption of international standards. Two forces will come into play: China's food safety standards will become more demanding and control systems will also improve. Such hard data can be important in relation to exports as it can be used to help potential importers assess the risks associated with Chinese imports.

International food safety standards and negotiations related to food trade disputes increasingly require a scientifically supported "risk assessment" to justify the provisions included in new regulations and is becoming a necessary tool in the successful negotiation of difficult trade issues. A scientifically supported "risk assessment" generally requires, as its base, hard data on both the "hazard" in question and on the "exposure" of the human, animal or crop population at risk. For example, in setting a tolerance for a chemical food residue, data is required on the exposure of the population to this chemical. This requires survey knowledge of the food consumption profile of the population. This type of data is difficult and expensive to generate. At present, China's export efforts are made more difficult by the lack of such data; however, the country is working to train its staff in this area and to generate better data.

To minimise possible damage to its export sector, China needs a system to rapidly identify disease outbreaks and health threats, such as those described above, and to bring these under control as quickly as possible. Thus, sustainable success in the export field requires the establishment of an emergency control system. Such a system is normally designed to include both an early warning system, to alert authorities to a potential crisis and an emergency response system to bring the crisis under control. China's experiences in dealing with the recent outbreak of Severe Acute Respiratory Syndrome (SARS) and the avian influenza in 2004 provide valuable experience from which such a system could be improved. In this context, AQSIQ established in 2002 a committee involving several ministries on the Analysis of Cross-border Animal and Plant Quarantine Risks to China. AQSIQ also produced a report covering Detailed Implementation of Early Warning and Rapid Response Regulations, on Animal and Plant Quarantine Risks.

Conclusions

The current system for ensuring food safety in China is changing rapidly. China's progress in establishing institutional support for food safety is reasonably good. Areas in which further improvement might be considered are as follows:

- Correcting legislative gaps and overlaps, perhaps passing an overall Food Law.
- Consolidating and then improving co-ordination between the domestic food inspection ministries.
- Increasing the penalties for violators of food safety laws.
- Improving the Emergency Response System.
- Improving the system for tracing problems to their point of origin.
- Improving the capacity to use the latest test methods.
- Improving testing capacity at the local level.
- Improving inspection coverage for China's numerous small food enterprises.
- Introducing a degree of self-regulation to industry.
- Harmonising food standards with international standards.
- Improving public recognition and acceptance of internationally recognised certification systems.
- Improving the registration systems for pesticides, fertilisers, animal feeds, and animal medications.
- Increasing the testing for residues of unsafe substances in food.
- Reorganising an "agricultural extension" system and refocusing it from training farmers on how to produce more food to producing safer food in an environmentally acceptable manner.
- Increasing the use of "key enterprises" to give technical assistance to farmers and quality assurance to consumers.
- Improving the generation of the data necessary for risk analysis.
- Improving skills in conducting risk analysis.

- Training food safety staff, both in China and abroad.
- Improving information sharing and co-operation between government, industry, consumers, the media, educational, and research institutions.
- Giving priority to solving problems that act as a bottleneck to exports and threaten domestic public health. These are the problems associated with:
    - Pesticide residues in food products of plant origin.
    - Animal drug residues in food products of animal and aquatic origin.
    - Bacterial contamination in all food products.
    - Control of animal diseases of international trade significance.

Acronyms and Abbreviations

ABC	Agricultural Bank of China
ACFSMC	All-China Federation of Supply and Marketing Co-operative
ADBC	Agricultural Development Bank of China
AGVA	Agricultural Gross Value Added
AQSIQ	National Administration for Quality Supervision, Inspection and Quarantine
ASEAN	Association of South-East Asian Nations
CCCPC	Central Committee of the Communist Party of China
CCTV	China Central TV
CEREOILS	China National Cereals, Oils and Foodstuffs Import & Export Corporation; now COFCO
CIF	Cost, Insurance and Freight
CITES	Convention on International Trade and Endangered Species
CNY	Yuan Renminbi
COFCO	China National Cereals, Oils and Foodstuffs Import & Export Corporation
CPC	Communist Party of China
CSE	Consumer Support Estimate
DRC	Development Research Centre of the State Council
EU	European Union
FAO	Food and Agriculture Organization of the United Nations
FAOSTAT	FAO statistical database
FDI	Foreign Direct Investment
FOB	Free on Board
GAO	Gross Agricultural Output
GDP	Gross Domestic Product
GGBRS	Governor's Grain-Bag Responsibility System
GMO	Genetically Modified Organisms
GSSE	General Services Support Estimate
GVA	Gross Value Added
HACCP	Hazard Assessment Critical Control Point
HPRS	Household Production Responsibility System
ISO	International Standards Organisation
JGIEC	Jilin Grain Group Import and Export Company
MFN	Most Favoured Nation
MLR	Ministry of Land Resources
MOA	Ministry of Agriculture
MOF	Ministry of Finance
MOFCOM	Ministry of Commerce
MOFTEC	Ministry of Foreign Trade and Economic Co-operation; now MOFCOM

MOH	Ministry of Health
MPS	Market Price Support
MWR	Ministry of Water Resources
NBSC	National Bureau of Statistics of China
NDRC	National Development and Reform Commission
NPC	National People's Congress
NTBs	Non-Tariff Barriers
OECD	Organisation for Economic Co-operation and Development
OIE	International Office for Epizootics
PBC	People's Bank of China
PPP	Purchasing Power Parity
PRC	People's Republic of China
PSE	Producer Support Estimate
RCCs	Rural Credit Co-operatives
RCRE	Research Centre of Rural Economy
SARS	Severe Acute Respiratory Syndrome
SAGR	State Administration of Grain Reserves
SASAC	State-owned Assets Supervision Administration Commission
SEPA	State Environmental Protection Administration
SFDA	State Food and Drug Administration
SGA	State Grain Administration
SGEs	State Grain Enterprises
SINOGRAIN	China Grain Reserve Corporation
SMCs	Supply and Marketing Co-operatives
SOEs	State Owned Enterprises
SPS	Sanitary and Phytosanitary (measures)
STEs	State Trading Enterprises
TBT	Technical Barriers to Trade
TFP	Total Factor Productivity
TRQ	Tariff Rate Quota
TSE	Total Support Estimate
TVEs	Township and Village Enterprises
VAT	Value Added Tax
WB	World Bank
WTO	World Trade Organization

ISBN 92-64-01260-5
OECD Review of Agricultural Policies
China
© OECD 2005

Bibliography

AMAD (2004), Agricultural Market Access Database, OECD, Paris.

ASEAN (2002), Framework Agreement on Comprehensive Economic Co-operation between the Association of South East Asian Nations and the People's Republic of China, www.aseansec.org/13196.htm.

Aubert, Claude (2004), "Consommations Alimentaires: l'Inconnue Chinoise", in *Déméter 2005, Economie et Stratégie Agricoles*, Club Déméter, Paris, pp. 19-41.

Aubert, Claude (2005), "China: A Peasant Economy at the Crossroads", mimeo, Beijing, 15 April.

Aubert, Claude and Xiande Li (2002), '"Peasant Burden": Taxes and Levies Imposed on Chinese Farmers', in *Agricultural Policies in China After WTO Accession*, Proceedings of the Workshop on Agricultural Policy Adjustments in China After WTO Accession, Beijing, China, 30-31 May, 2002, OECD, Paris.

Bai, N. and H. Song (2002), *Returning to Villages or Entering Cities: Study on Return Migration of Rural Laborers in China*, China Economic and Financial Press, Beijing.

Bean, Ralph (2003), *People's Republic of China: Retail Food Sector Report 2003*, USDA Foreign Agricultural Service, Global Agricultural Information Network Report CH3825, November 12.

Browne, Andrew (2004), "Peasants Bloom", *Far Eastern Economic Review*, 11 October.

Burgess, R. (2003), "Access to Land and Hunger: Opening the Black Box in China", London School of Economics, mimeo.

Butterworth, Jim and Xinping Wu (2004), *People's Republic of China: Cotton and Products Annual 2004*, USDA GAIN Report, Number CH4013.

Cai, F. and C. Zhang (2000), *Report on China's Population 2000 – Rural Population Issues and Options*, Social Sciences Documentation Publishing House.

Carter, Colin (2001), "China's Trade Integration and Impacts on Factor Markets", *China's Agriculture in the International Trading System*, OECD, Paris.

Carter, Colin A., Jing Chen and Baojin Chu (2003), "Agricultural Productivity Growth in China: Farm Level Versus Aggregate Measurement", *China Economic Review* 14, pp. 53-71.

Carter, M. and Y. Yao (1999), "Specialization without Regret. Transfer Rights, Agricultural Productivity, and Investment in an Industrializing Economy", *Policy Research Working Paper* 2202, World Bank.

Chang, Hongxiao (2004), "The 'Black Hole' in China's Farmland Data", *Caijing*, no. 122, 13 December.

Chang, M. (2004), "Impact of Commodity Trade Liberalization on Rural Food Security in China", paper presented at the European Association of Agricultural Economists, Florence, September 8-11.

Chen, S. and M. Ravallion (2004), "Welfare Impacts of China's Accession to the World Trade Organization", *World Bank Economic Review* 18(1): 29-57.

Chen, Xiwen (2002), "China's WTO Accession and its Agricultural Policy Reform", *Agricultural Policy Adjustments in China After WTO Accession*, OECD, Paris.

Cheng, Guoqiang (2004), "Study on the Strategy of the Agricultural Exports in China", Research Report, Development Research Centre of the State Council.

China Animal Husbandry Yearbook 2003 (2004), Agricultural Press.

China Chemical Industry Newsletter (2004), 21 April.

China Daily, various editions.

China Environment Communiqué (2000; 2003).

Chinese Press, *China Food Industry Yearbook*, various editions.

CSLS (Center for the Study of Living Standards) (2003), "China's Productivity Performance and its Impact on Poverty in the Transition Period", background paper prepared for the 2004 World Employment Report of the International Labor Organization, *CSLS Research Report 2003-07*, CSLS, Ottawa.

Customs General Administration (2003), *China Customs Statistical Yearbook 2002* (and previous issues), Beijing.

Customs General Administration (2004), *Customs Import and Export Tariff of the People's Republic of China 2003* (and previous issues), Economic Management Publishing House, Beijing.

de Brauw, A., J. Huang and S. Rozelle (2000), "Responsiveness, Flexibility, and Market Liberalization in China's Agriculture", *American Journal of Agricultural Economics* 82(5), 1133-39.

Deininger, K. and S. Ju (2002), "Land Rental Markets as an Alternative to Government Reallocation? Equity and Efficiency Considerations in the Chinese Land Tenure System", *Policy Research Working Paper* 2930, World Bank.

DRC (2004), "Study on China's National Food Safety Strategy", Project Group of the Development Research Centre of the State Council, PRC, Distributed at the Global Food Safety Forum, Beijing, November 18-19, 2004.

Du, Qinglin (ed.) (2003), *Strategic Adjustment of Agricultural and Rural Economic Structure in China*, China Agriculture Press.

Du, Ying et al. (eds.) (2002), *Impact of Rural Population Change on the Reform of the Land Tenure System in China*, China Financial and Economic Publishing House, May.

ERS (1999), "State Trading and Management of Grain Marketing in China", in *Agricultural Outlook*, June-July 1999, The Economic Research Service of the United States Department of Agriculture, Washington, DC.

Fan Shenggen, Linxiu Zhang and Xiaobo Zhang (2001), "Growth, Inequality, and Poverty in Rural China: The Role of Public Investment", mimeo, February, 2001.

Fang, Cheng, Francis Tuan and Funing Zhong (2002), "How Might China Protect its Agricultural Sector?", in Gale, F. (ed.) *China's Food and Agriculture; Issues for the 21st Century*, The Economic Research Service of the United States Department of Agriculture, Washington, DC.

FAOSTAT, FAO, Rome.

Felloni, Fabrizio, John Gilbert, Thomas I. Wahl and Philip Wandschneider (2003), "Trade Policy, Biotechnology and Grain Self-Sufficiency in China", *Agricultural Economics*, Vol. 28, pp. 173-186.

Fogel, Robert W. (2004), "High Performing Asian Economics", University of Chicago, mimeo.

Gale, Fred (2002a), "China Statistics: Are they Reliable?", China's Food and Agriculture: Issues for the 21st Century, Economic Research Service/USDA, Washington, DC.

Gale, Fred (2002b), *China's Corn Exports: Business as Usual, Despite WTO Entry*, Economic Research Service/USDA, Washington, DC, www.ers.usda.gov/publications/fds/dec02/fds1202-01/fds1202-01.pdf.

Gale, Fred, Bryan Lohmar and Francis Tuan (2005), *China's New Farm Subsidies*, Economic Research Service/USDA, Washington, DC, www.ers.usda.gov/publications/WRS0501/WRS0501.pdf.

Hertel, T. (1997), *Global Trade Analysis: Modeling and Applications*, Cambridge University Press, UK.

Hertel, T., F. Zhai and Z. Wang (2004), "Implications of WTO Accession for Poverty in China", in Bhattasali, Deepak, Shantong Li and William J. Martin (eds.), *China and the WTO: Accession, Policy Reform, and Poverty Reduction Strategies*, The World Bank and Oxford University Press, Washington, DC.

Hertel, T. and F. Zhai (2005), "Labor Market Distortions, Rural-Urban Inequality, and the Opening of China's Economy", mimeo, World Bank.

Holz, Carsten (2005), "Institutional Arrangements for the Production of Statistics", *China in the Global Economy: Governance in China*, OECD.

Hsu, Hsin-Hui (1998), "China's Livestock Economy: Introduction", *China International Agriculture and Trade Report*, US Department of Agriculture, Economic Research Service, July.

Hu, Biliang (2004), "Active Informal Financing in Rural Areas: A Case Study of Rotating Savings and Credit Associations in a Chinese Village", *Rural Finance and Credit Infrastructure in China*, OECD, Paris.

Hu, Dinghuan, Thomas Reardon, Scott Rozelle, Peter Timmer and Honglin Wang (2004), "The Emergence of Supermarkets with Chinese Characteristics: Challenges and Opportunities for China's Agricultural Development", *Development Policy Review* 22(5).

Huang, Jikun and C. Chen (1999), *Effects of Trade Liberalization on Agriculture in China: Institutional and Structural Aspects*, United Nations ESCAP CGPRT Centre, Bogor, Indonesia.

Huang, Jikun and Scott Rozelle (2002a), "Agricultural Protection and Regional Market Integration in China: The Impact of China's WTO Membership on the Agricultural Sector", *Agricultural Policies in China After WTO Accession*, OECD, Paris.

Huang, Jikun and Scott Rozelle (2002b), "China's Accession to the WTO and Shifts in Agricultural Policy", *Working Paper* No. 02-002, Department of Agricultural and Resource Economics, University of California, Davis.

Huang, Jikun and Scott Rozelle (2004), "Chinese Agricultural Policy", in Tongeren, F.W. Van and Jikun Huang (eds.), *China's Food Economy in the Early 21st Century: Development of China's Food Economy and its Impact on Global Trade and on the EU*, Agricultural Economics Research Institute (LEI), The Hague.

Huang, Jikun, Scott Rozelle and Min Chang (2004), "The Nature of Distortions to Agricultural Incentives in China and Implications of WTO Accession", in Bhattasali, Deepak, Shantong Li and William J. Martin (eds.), *China and the WTO: Accession, Policy Reform, and Poverty Reduction Strategies*, The World Bank and Oxford University Press, Washington, DC.

Ianchovichina, E. and W. Martin (2004), "Economic Impacts of China's Accession to the WTO", *World Bank Economic Review* 18(1).

Jiang, Z. (2002), "Build a Well-off Society in an All-round Way and Create a New Situation in Building Socialism with Chinese Characteristics", Report delivered at the 16th National Congress of the Communist Party of China.

Jing, Jianming and Dongxia Jing (2004), "New Concepts and New Measures for the Ecological Construction and Ecological Protection", *Environment and Protection*, No. 2.

Jing, Renqing (2004), "Speech of the Financial Minister Jing Renqing on Financial Assistance to Agriculture", obtained from www.xinhuanet.com.

Johnson, David Gale (2000), "Reducing the Urban-Rural Income Disparity in China", Office of Agricultural Economics Research, The University of Chicago, Paper: 00-06, mimeo.

Kang, X. (1998), "Review of the Antipoverty Actions in China", *Working Paper*.

Krusekopf, C.C. (2001), "Diversity in Land Tenure Arrangements Under the Household Responsibility System in China, Has China Become a Market Economy?", International Conference on the Chinese Economy, Clermont Ferrand, 17-18 May.

Kwiecinski, Andrzej and Xiande Li (2002), "The Role of Agricultural and Other Policies in Raising Rural Income in China", *Agricultural Policy Adjustments in China After WTO Accession*, OECD, Paris.

Lardy, N.R. (2001), *Integrating China in the Global Economy*, Brookings Institution, Washington, DC.

Li, Xiande (2005), *Le Village de Lijia, Aspects de l'Économie Paysanne en Chine*, China Agricultural Science and Technology Press, Beijing.

Liu, Jing and Lubiao Zhang (2004), "SARS and Agricultural Environmental Safety", *Annual Report on Economic and Technological Development in Agriculture 2003*, China Agriculture Press.

Liu, Zuo and Tieying Liu (2004), *Tax System of the People's Republic of China*, Intellectual Property Press, China.

Lohmar, Bryan, Agapi Somwaru and Keith Weibe (2002), "The Ongoing Reform of Land Tenure Policies in China", *Agricultural Outlook*, USDA, September.

Lohmar, Bryan and David Skully (2003), "China's TRQ Administration One Year on: Crossing the River by Feeling for Stones", paper presented at the International Conference: Agricultural Policy Reform and the WTO: Where are we Heading? Capri (Italy), June 23-26, 2003.

McCulloch, N. (2003), "The Impact of Structural Reforms on Poverty: A Simple Methodology with Extensions", *World Bank Policy Research Working Paper* 3124, Washington, DC, World Bank.

MOA (Ministry of Agriculture People's Republic of China) (2001), "Suggestion on Accelerating Development of Livestock Production", Ministry of Agriculture People's Republic of China, Government Document.

MOA, *China Agricultural Development Report,* various editions, China Agricultural Press, Beijing.

MOA, *China Agricultural Yearbook*, various editions.

MOF (Ministry of Finance People's Republic of China), *China Financial Yearbook*, various editions.

MOF (2005), *Report of the 2004 Budget Implementation and 2005 Budget Plan to the NPC*, Ministry of Finance, Beijing.

MOFTEC (2002), *Compilation of the Legal Instruments on China's Accession to the World Trade Organization*, Law Press, Beijing.

National Office for the Agricultural Census (1998), *Abstract of Materials of China's First Agricultural Census*, China Statistical Press, Beijing.

Naughton, B. (1995), *Growing Out of the Plan: Chinese Economic Reform, 1978-92*, Cambridge University Press, Cambridge, New York and Melbourne.

NBSC (National Bureau of Statistics of China) (2005), www.stats.gov.cn/tjfx/jdfx/t20050125_402225538.htm.

NBSC, *China Development Report 2004*.

NBSC, *China Labour Statistical Yearbook*, various editions.

NBSC, *China Rural Statistical Yearbook*, various editions.

NBSC, *China Statistical Abstract 2004*.

NBSC, *China Statistical Yearbook*, various editions.

NDRC (National Development and Reform Commission), *China Price Yearbook*, various editions.

NDRC (2004), "Report on the Implementation of the 2003 Plan for National Economic and Social Development and on the 2004 Draft Plan for National Economic and Social Development", National Development and Reform Commission, www.10thnpc.org.cn/english/government/90791.htm.

Nyberg, A. and S. Rozelle (1999), *Accelerating China's Rural Transformation*, World Bank, Washington, DC.

OECD (2001), *China's Agriculture in the International Trading System*, OECD, Paris.

OECD (2002a), *Agricultural Policies in China After WTO Accession*, OECD, Paris.

OECD (2002b), *China in the World Economy: The Domestic Policy Challenges*, OECD, Paris.

OECD (2003), *Farm Household Income: Issues and Policy Responses*, OECD, Paris.

OECD (2004a), *Agricultural Policies in OECD Countries: At a Glance*, OECD, Paris.

OECD (2004b), *Income Disparities in China: An OECD Perspective*, OECD, Paris.

OECD (2004c), *OECD Agricultural Outlook 2004-2013*, OECD, Paris.

OECD (2004d), *Rural Finance and Credit Infrastructure in China*, OECD, Paris.

OECD (2005a), "Institutional Framework for Effective Agricultural Policy: Current Issues and Future Challenges", prepared by Andrzej Kwiecinski and Brad Gilmour, *China in the Global Economy: Governance in China*, Paris.

OECD (2005b) "The Market and Welfare Effects of Agriculture and Non-Agriculture Reform", paper presented at the OECD Joint Working Party on Agriculture and Trade, 6-8 April, 2005.

OECD (2005c), *OECD Economic Surveys – China*, OECD, Paris.

Park, Albert and Sangui Wang (2001), "China's Poverty Statistics", *China Economic Review* 12(2001).

Park, Albert, Sangui Wang, and Guobao Wu (2002), "Regional Poverty Targeting in China", *Journal of Public Economics*, Vol. 86, No. 1, pp. 123–153.

Park, Ji-hyun (2002), "Agricultural Trade between Korea, China and Japan: Disputes and Counter Measures", *East Asian Review*, Vol. 14, No. 3, Autumn 2002, pp. 49–66.

Popkin, Barry M. and Shufa Du (2002), "Dynamics of the Nutrition Transition Toward the Animal Foods Sector in China and its Implications: A Worried Perspective", paper presented at the Conference "Animal Source Foods and Nutrition in Developing Countries" held in Washington, DC, June 24-26, 2002.

Popkin, Barry (2004), "The Nutrition Transition in the Developing World", in S. Maxwell and R. Slater (eds.), *Food Policy Old and New*.

Ravallion, M. and W. Chen (2004), "China's (Uneven) Progress Against Poverty", *World Bank Policy Research Working Paper* 3408, September.

Rozelle, Scott, Linxiu Zhang and Jikun Huang (2000), "China's War on Poverty", *Working Paper* No. 60, Center for Economic Research on Economic Development and Policy Reform, Stanford Institute for Economic Policy Research, Stanford University.

Rozelle, Scott (2001), "Trade and Investment Liberalisation and China's Rural Economy: Impacts and Policy Reponses", Report for OECD.

Rural Survey Organization of National Bureau of Statistics of China (2004), *China Yearbook of Rural Household Survey*, China Statistics Press.

Sen, Amartya (1981), "Ingredients of Famine Analysis: Availability and Entitlements", *Quarterly Journal of Economics*, Vol. 96, No. 3. (August, 1981), pp. 433–464.

Shen, Minggao, Scott Rozelle and Linxiu Zhang (2004), "Farmer's Professional Associations: State Dominated or New State-Private Partnership?", Thomas Wahl and Brenda Sternquist (eds.), *China's Agricultural Sector: Markets, Myths and Mavens*, IMPACT Center, Washington State University, Pullman.

Shi Xinzheng (2002), "Empirical Research on Urban-Rural Income Differentials: A Case of China", mimeo, Peking University.

Song, Hongyuan et al. (2002), *China's Agriculture and Rural Economic Policies during the Ninth Five-Year Plan Period* (in Chinese), China Agricultural Press, Beijing.

State Council (1990a), "Decisions on Establishing National Special Grain Reserve System", *Chinese Grain Economy*, No. 6, pp. 1-2.

State Council (1990b), "Decisions on Strengthening Grain Procurement and Distribution", *Chinese Grain Economy*, No. 5, pp. 1-3.

State Council (1991), "Decision on Adjusting Unified Sale Prices of Grains and Vegetable Oils", April 19, Government Document.

State Council (1992), "Decision on Development of High-Yield, High-Quality and High-Return Agriculture", Government Document.

State Council (1996), "The Grain Issue in China", *www.china.org.cn/e-white/grainissue/index.htm*.

State Council (1998a), "Regulation on Grain Procurement", Government Document.

State Council (1998b), "Decision on Deepening Reforms on Cotton Marketing System", Government Document.

State Council (2004a), "The Suggestions of the Central Committee of the Communist Party of China and the State Council on Policies for Boosting Growth in Farmers' Income".

State Council (2004b), Regulations on Grain Circulation Management.

State Forestry Administration (2004), Government Plans to Convert 60 million mu of land to forestry in 2004, *http://news.xinhuanet.com*.

State Grain Bureau (2002), *Study on Construction of Grain Storage Facilities in the Northeast*, China Planning Publishing House, Beijing.

State Planning Commission (SPC) (1992), "Suggestions on Reforming Cotton Marketing System", Government Document.

Statistical Bureau of Zhejiang Province (2004), Survey Results, *www.zj.stats.gov.cn/content/nr.asp?keyvalue0=3304*.

Su, Ming (2004), "Study on Fiscal Support to Rural Basic Education", in Chen, X.W. (ed.), *Study on Reforming Usage and Management System of Agriculture Support Funds by Chinese Government*, Shanxi Publishing House, Taiyuan, China.

Sun, Xinzhang, Sunkui Cheng and Qinwen Ming (2004), "Rural Ecological Engineering: New Ideas for Resolving China's 'Three Nong' Issues", *Research of Agricultural Modernization*, Vol. 25, No. 2.

Tan, Shuhao, Nico Heerink and Qu Futian (2005), "Land Fragmentation and its Driving Forces in China", *Land Use Policy*, forthcoming.

Taylor, Daniel (2005), "People's Republic of China: Food Processing Ingredients Sector", USDA Foreign Agricultural Service, *Global Agricultural Information Network Report* CH5607, June 16.

Taylor, E., S. Rozelle and A. deBrauw (2003), "Migration and Incomes in Source Communities: A New Economics of Migration Perspective from China", *Economic Development and Cultural Change*, 52(1). pp. 75-101.

Tian, Weiming, Zhang Liqin and Zhou Zhangyue (2002), "Experience and Issues in Measuring the Level of Agricultural Support in China", paper presented at the International Workshop on Agricultural Policy Adjustments in China After WTO Accession, 30-31 May 2002, Beijing, China.

Tian, W., X. Liu and and X. Kang (2003), "Social Viability Roles of the Agricultural Sector in China", Roles of Agriculture Project, FAO.

Tian, W., X. Wang and F. Ke (2003), "The Poverty Alleviation Role of Agriculture in China", Roles of Agriculture Project, FAO.

Tian, W. and L. Zhang (2003), "Recent Economic and Agricultural Policy Development – China", Roles of Agriculture Project, FAO.

Thomas, Bella (2003), "What the World's Poor Watch on TV", *World Press Review*, Vol. 50, No. 1.

Tongeren, F.W. Van and Jikun Huang (eds.) (2004), *China's Food Economy in the Early 21st Century: Development of China's Food Economy and its Impact on Global Trade and on the EU*, Agricultural Economics Research Institute (LEI), The Hague.

Tuan, Francis C. and Ke, Bingsheng (1999), "A Review of China's Agricultural Policy: Past and Present Developments", in *Agriculture in China and OECD Countries: Past Policies and Future Challenges*, Proceedings of the Workshop on Agricultural Policies in China: Review, Outlook and Challenges, Paris, France, November 1998, OECD, Paris.

Uchida, Emi, Jintao Xu and Scott Rozelle (2004), *Grain for Green: Cost-Effectiveness and Sustainability of China's Set-Aside Program*, Department of Agricultural and Resource Economics, University of California, Davis.

Vermeer, E.B. (2001), "Readjustment Issues in Rural China in the 21st Century", The International Symposium on the WTO and Chinese Rural Development in the 21st Century, Maoming, China, 26-29 June.

Wade, John, Adam Branson and Xiang Qing (2002), "China's VAT on Domestic and Imported Agricultural Products Unequal", *GAIN Report* CH2013, Foreign Agricultural Service of the United States Department of Agriculture, *www.fas.usda.gov/gainfiles/200205/145683553.pdf*.

Wang, Nong and Yujin Li (2004a), "Some Suggestions to Enhance the Agro-Ecologic Environment in China", *Environment and Protection*, No. 2.

Wang and Li (2004b), "Study on Financial Inputs to the Comprehensive Agricultural Development Projects", in Chen, X.W. (ed), *Study on Reforming Usage and Management System of Agriculture Support Funds by Chinese Government*, Shanxi Publishing House, Taiyuan, China.

World Bank (2000), *China Overcoming Rural Poverty*, Washington, DC.

World Bank (2001), *China: Air, Land and Water*, Washington, DC.

World Bank (2003), *China Country Economic Memorandum: Promoting Growth with Equity*, Washington, DC.

World Bank (2004a), *Commodity Price Data*, *www.worldbank.org*.

World Bank (2004b), *World Bank Development Report 2004*, Washington, DC.

World Bank (2004c), *World Development Indicators*, CD Rom, Washington, DC.

World Bank/OECD (2004), *Achieving Ukraine's Agricultural Potential: Stimulating Agricultural Growth and Improving Rural Life*, World Bank, Washington, DC and OECD, Paris.

WTO (2001), Accession of the People's Republic of China – Decision of 10 November 2001, *http://docsonline.wto.org/DDFDocuments/t/WT/L/432.doc*.

Wu, Guobao (1997), "Policy on Subsidised Poverty Loans in China: A Discussion", paper presented at the Regional Expert Meeting on Capability-Building to Alleviate Rural Poverty, UNESCAP, Beijing, 25-28 March.

Wu, Laping (2002), "Grain Market Integration and Marketing Margin in China", *Working Paper* CEP02-3, Asia Pacific School of Economics and Management, Asia Pacific Press, Canberra.

Xu, Boyuan and Fushang Liu (2003), *The Second Wave of Development of the Wholesale Market for Agricultural Products in Taiwan Straits Under the WTO*, China Price Publishing House, Beijing.

Yang, D. and H. Zhou (1996), "Rural-Urban Disparity and Sectoral Labour Allocation in China", Research Papers in *Asian Pacific Studies*, Duke University, Durham, North Carolina.

Ye, X.J., Z.Q. Wang and Q.S. Li (2002), "The Ecological Agriculture Movement in Modern China", *Agriculture, Ecosystems and Environment* 92.

Zhai, F. and T. Hertel (2005), "Impacts of the Doha Development Agenda on China: The role of Complementary Education Reforms", mimeo, World Bank.

Zhai, F. and S. Li (2000), "Quantitative Analysis and Evaluation of the Impact of Entry to WTO on China's Economy", *China Development Review*, Development Research Centre of the State Council of the People's Republic of China.

Zhang, Amei (2003), "Poverty Alleviation in China: Commitment, Policies and Expenditure", United Nations Development Programme Occasional Paper No. 27, UNDP, New York.

Zhang, Hongyu (1999), "Land Tenure Policy in Rural China: Development and Prospects", *Agriculture in China and OECD Countries: Past Policies and Future Challenges*, OECD, Paris.

Zhang, L., C. Liu and Q. Li (2003), "The Buffer Role of Agriculture in China's Economic Transition", Roles of Agriculture Project, FAO.

Zhang, Weili, Shuxia Wu, Hongjie Ji and H. Kolbe (2004), "Estimation of Agricultural Non-Point Source Pollution in China in Early 21 Century", *Agricultural Science in China*, No. 7.

Zhu, Kongying (2004), "Pollution Status and Management Countermeasures in Livestock Sector", *Arid Environmental Monitoring*, Vol. 18, No. 1, March.

Zhu, Ling and Zhongyi Jiang (1995), "Yigon-Daizhen in China: A New Experience with Labor Intensive Public Works in Poor Areas", *Employment, Poverty Reduction and Food Security*, edited by Joachim von Braun, Washington, DC: International Food Policy Research Institute.

Zhu, Lizhi Julie (2003), *People's Republic of China: Food Processing Ingredients Sector Report 2003*, USDA Foreign Agricultural Service, Global Agriculture Information Network Report CH3608, June 6.

Zhu, R. (2003), "Report on the Work of the Government", delivered at the First Session of the 10th National People's Congress on March 5, 2003.

Zhu, Tiequn (2000), "Prevention and Control of Water Pollution Caused by Agricultural Non-Point Sources in China", *Rural Eco-Environment*, No. 3.

Zhu, Y. (2004), "The Floating Population's Household Strategies and the Role of Migration in China's Regional Development and Integration", *International Journal of Population Geography* 9, 485-502.

OECD PUBLICATIONS, 2, rue André-Pascal, 75775 PARIS CEDEX 16
PRINTED IN FRANCE
(51 2005 10 1 P) ISBN 92-64-01260-5 – No.54239 2005